# A Developmental Approach To Algebraic Concepts

Second Edition

# A Developmental Approach To Algebraic Concepts

Second Edition

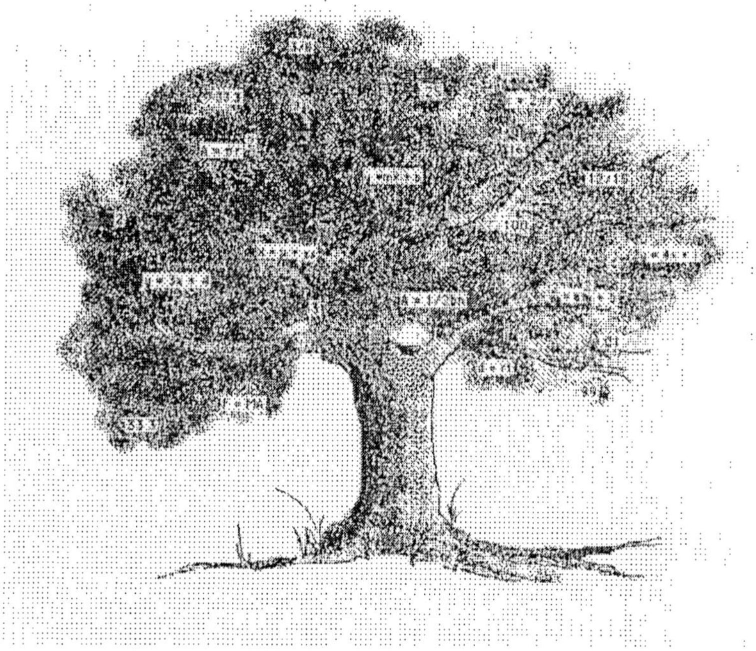

Mauro Cassano
Kathleen McKiernan
Angelo Scordato

Nassau Community College

KENDALL/HUNT PUBLISHING COMPANY
4050 Westmark Drive Dubuque, Iowa 52002

ISBN 0-7872-1353-5



Printed in the United States of America
10 9 8

# Contents

The authors wish to thank . . .

**PROFESSOR ALICE BERRIDGE**
**PROFESSOR DOUGLAS BROWN**
**PROFESSOR CARMINE DE SANTO**
**PROFESSOR ELLIS VON ESCHEN**

. . . for their advice and encouragement

**MEMBERS OF THE MATHEMATICS DEPARTMENT**
**NASSAU COMMUNITY COLLEGE**

. . . for classroom testing the material

**LOUISE CASSANO**

. . . for the cover design

**CHRISTINE SCORDATO**

. . . for proofreading and assisting with typing

**SISTER KATHRYN WALSH, O.P.**

. . . for preparing the manuscript for print.

# Preface

**A Developmental Approach to Algebraic Concepts** is an easy to read text designed for the student whose mathematical background and skills are not compatible with career goals, either because of insufficient course work on the secondary level, or because skills were lost with the passage of time since graduation from secondary school.

While the text is thorough in its coverage of essential topics in algebra, geometry and trigonometry, the writing style is designed to encourage rather than discourage the skills deficient student. The approach is unpretentious but informative and, while maintaining mathematical precision, the prose is not overly technical. Where highly technical prose cannot be avoided, several examples are worked in detail, enabling the student to fully grasp the concept before advancing to the next.

The student's greatest difficulties often stem from psychological barriers which generate self-defeating attitudes, so at the start of each chapter, the authors provide a **vocabulary preview** by which students are encouraged to call upon past experience and "intuition" to define words that will appear in the chapter text. The vocabulary preview is not meant to be threatening; on the contrary, it is meant to point out that students have already seen and used many of the terms, often in a non-mathematical context.

The use of calculators is encouraged, and Chapter One provides explanation and practice exercises to be used as an aid to developing skill in their use. Subsequent chapter exercises presuppose calculator use.

Chapters Two through Five deal with operations involving real numbers, where the fundamental properties of real numbers are applied to integers, decimal forms and fraction forms.

Basic concepts of algebra are covered in Chapters Six through Eight. In Chapter Eight, the student is encouraged to READ the problem and WRITE the linear equations BEFORE being shown how to solve them. Graphing linear equations (Chapter Nine) then, logically follows Chapter Eight.

Algebraic Fractions (Chapter Eleven) and Quadratic Equations (Chapter Twelve) follow Chapter Ten (Factoring) as application.

Chapters Thirteen and Fourteen provide for application of algebraic skills to geometry and trigonometry.

Chapters Fifteen through Eighteen expand the content of previous chapters to include material of a more difficult nature. Chapter Fifteen (Systems of Linear Equations) is an extension of Chapters Eight and Nine. Chapter Sixteen extends content to include linear inequalities. Chapter Seventeen is an extension of Chapters Ten and Eleven (Factoring and Algebraic Fractions).

Every chapter section is followed by a group of exercises, and each chapter is followed by chapter exercises in sufficient number to reinforce the concepts learned in that section of the text.

A chapter test Form A follows each chapter. Form B tests are provided as an appendix to the book, and Form C tests can be found in the Instructor's Manual.

To help the student maintain skills acquired throughout the course, **cumulative exercises** and **cumulative tests** are provided after each of chapters four through eighteen. Additional Cumulative Tests can be found in the Instructor's Manual.

# To the Student

This text is written in such a way that you are sure to be successful, if you follow the text page by page and attempt all of the exercises. The prose is easy to read, and the exercises are meant to reinforce what you have learned in each section of a chapter.

You are encouraged to use a calculator. Learn yours well, and use it whenever possible.

READ the chapter text section by section. Carefully follow each example, reading one step at a time. Do not go on to another example until you understand the one before. Try the section exercises only after you fully understand the sample examples. If there are exercises which you do not understand, or which you cannot work, be sure to ask your instructor for help.

Review each section of the chapter before attempting the chapter exercises. Notice that the chapter exercises are divided in such a way that you can easily identify any content which requires better understanding and more practice.

When you are convinced that you know the material from the chapter, try the chapter test. If you find some problems more difficult than others, spend additional time and energy on them. Rework the test, or try the Form B chapter test from the back of the book.

By the time you have worked and reworked a chapter, you will be ready for the next.

Success comes ONE STEP AT A TIME. Each chapter builds on skills acquired in the chapter before, so if you learn the material in earlier chapters well, you will retain your skills and move easily through the later chapters.

Be sure to do the cumulative exercises and cumulative tests. They are designed to help you maintain the skills you have already acquired as you move through the text.

You can and will succeed if you are willing to READ, WORK and STUDY the text.

We wish you well in your endeavors.

THE AUTHORS

CHAPTER

# 1 Basic Operations Using Calculators

## 1.0 *Vocabulary Preview*

Define these words without using a dictionary. Formulate their meanings from past knowledge or experience:

| | | |
|---|---|---|
| **addend** | **function** | **product** |
| **calculator** | **imaginary** | **quotient** |
| **decimal** | **integer** | **reciprocal** |
| **denominator** | **minuend** | **rounding** |
| **difference** | **negative** | **root** |
| **equivalent** | **non-integer** | **square** |
| **estimate** | **numerator** | **subtrahend** |
| **factor** | **percent** | **sum** |
| **fraction** | **power** | **symbol** |

## 1.1 *Using the Calculator to Add or Subtract*

To **add** numbers, simply enter the numbers (called **addends**), with a "+" between them. When all the numbers have been entered, strike the = key, and their **SUM** will be displayed.

*Example:* 4 + 6.7 + 5.7 + 32.7 + 136.89 + 0.098 = **186.088**

*Example:* 13.4 + 67.898 + 643.9 + 89.07 + 432 = **1246.268**

To **subtract** numbers, simply enter the numbers with a "−" between each. When all the numbers have been entered, strike the = key, and their **DIFFERENCE** will be displayed.

*Example:* 876.90 − 10.87 − 67.89 − 175.87 − 34.55 = **587.72**

*Example:* 78.55 − 150.87 = **−72.32**

Notice that the answer is negative. This is because you subtracted a larger number from a smaller one.

**Exercise Set 1.1**

Add or subtract as indicated: If the answer is negative, explain why.

1. $25.12 + 10.09$
2. $34 + 98.56$
3. $45.5 - 8 + 45.2$
4. $68.90 + 34.75$
5. $155.98 + 27.3$
6. $67.78 - 87.34$
7. $99.9 + 34.2$
8. $12.35 + 56.3$
9. $0.09 + 3.09$
10. $138.5 - 130.5$
11. $51.15 + 64.78$
12. $87.64 - 51$
13. $18.54 - 24.50$
14. $73.21 - 85.6$
15. $115 - 176$
16. $33.12 + 45.32$
17. $104 + 25 - 150$
18. $59.63 - 79.76$
19. $223.5 - 100 + 23.56$
20. $10.09 + 34.21$
21. $22.13 - 76.54$
22. $66.4 + 77.9 - 108$

## 1.2 *Using the Calculator to Multiply or Divide*

To **multiply or divide** numbers, enter them with the appropriate operation sign between them. Press = when all numbers have been entered, and the **PRODUCT** or **QUOTIENT** will be displayed.

*Example:* $8 \times 5 \times 9 \times 7 \times 2 = \mathbf{5040}$

*Example:* $80 \div 5 \div 4 \div 2 = \mathbf{2}$

*Example:* $5 \div 4 = \mathbf{1.25}$

*Note:* the answer is not an **integer** (whole number) because 5 does not divide evenly by 4.

*Example:* Calculate $5 \div \frac{1}{4}$.

First change $\frac{1}{4}$ to a decimal by entering $1 \div 4$ into the calculator. It displays 0.25. Now enter $5 \div 0.25 =$. The calculator shows the answer to be 20.

*Example:* $256 \times 1.5 =$ **384**

*Example:* $256 \div 1.5 =$ **170.666 . . .**

Since the digit 6 keeps repeating, **round off** the answer to 170.7

*NOTE:* Sometimes a bar is written over a number to show that it repeats; as $170.\overline{6}$.

**Exercise Set 1.2**

Multiply or divide as indicated. If the quotient is a repeating decimal, put a bar over the repeating part.

1. $32 \times 85.2$
2. $167.83 \div 3$
3. $56.34 \times 0.2$
4. $132 \div 34$
5. $65.4 \times 3.2$
6. $7.98 \times 2.1$
7. $204 \times 0.5$
8. $0.78 \times 1.23$
9. $22.2 \div 2.2$
10. $15 \times 7$
11. $32 \div 85.2$
12. $167.83 \times 3$
13. $56.34 \div 0.2$
14. $132 \times 34$
15. $65.4 \div 3.2$
16. $7.98 \div 2.1$
17. $204 \div 0.05$
18. $0.78 \div 1.23$
19. $22.2 \times 2.2$
20. $15 \div 7$
21. $1.0 \times 99.9$
22. $1.0 \div 99.9$

## 1.3 *Squares and Other Powers*

When a number is **squared,** it is multiplied by itself; that is, it is used as a factor twice.

*BE CAREFUL* that you don't mistake squaring for multiplying by 2.

*SQUARING,* which is multiplying a number by itself, is symbolized by putting a small 2 above and to the right of the number:

$$8^2 = 8 \times 8 = 64$$
$$12^2 = 12 \times 12 = 144$$

Numbers used as factors three times are said to be "cubed". 8 cubed is symbolized $8^3$ and means $8 \times 8 \times 8$ which is 512

$8^3$ is sometimes referred to as "8 to the third power".

$5^6$ is "5 to the sixth power".

*Example:* Evaluate each: $3^2$, $7^3$, $4^5$

$3^2 = 3 \times 3 = 9$

$7^3 = 7 \times 7 \times 7 = 343$

$4^5 = 4 \times 4 \times 4 \times 4 \times 4 = 1024$

*NOTE:* There might be a special key, $y^x$, on your calculator which will enable you to find the answers without the repetition of multiplication.

*Example:* Use your calculator to evaluate $(7)^3$.

Enter 7.
Strike the $y^x$ key and enter 3.
Strike the = key.
The answer 343 appears.

**Exercise Set 1.3**

*Evaluate the following:*

1. $4^2$
2. $4^3$
3. $5^3$
4. $2^5$
5. $7^2$
6. $2^3$
7. $(0.5)^2$
8. $(1.2)^2$
8. $10^4$
10. $(1.1)^2$
11. $(10)^6$
12. $12^2$
13. $(6.1)^2$
14. $(0)^{10}$
15. $10^2$
16. $1^5$
17. $0^7$
18. $1^{20}$

## 1.4 *Finding Square Roots*

If you do not have a square root key, $\sqrt{\phantom{x}}$ on your calculator, you must use a chart, or you can estimate the **root.**

Finding square root and "squaring" are opposite operations.

*Examples:*

a. $5 \times 5 = 25$; $\sqrt{25} = 5$

b. $10 \times 10 = 100$; $\sqrt{100} = 10$

c. $72 \times 72 = 5184$; $\sqrt{5184} = 72$

d. $0.5 \times 0.5 = 0.25$; $\sqrt{0.25} = 0.5$

e. $23 \times 23 = 529$; $\sqrt{529} = 23$.

Look at the list of squares below and continue on your own:

| | | |
|---|---|---|
| 1 squared is 1 | 2 squared is 4 | 3 squared is 9 |
| 4 squared is 16 | $5^2 = 25$ | $6^2 = 36$ |
| $7^2 = 49$ | $8^2 = 64$ | $9^2 = 81$ |
| $10^2 = 100$, | $11^2 = 121$ etc. | |

Similarly, a list of square roots follows:

| | | | | |
|---|---|---|---|---|
| $\sqrt{1} = 1$ | $\sqrt{4} = 2$ | $\sqrt{9} = 3$ | $\sqrt{16} = 4$ | $\sqrt{25} = 5$ |
| $\sqrt{36} = 6$ | $\sqrt{49} = 7$ | $\sqrt{64} = 8$ | $\sqrt{81} = 9$ | $\sqrt{100} = 10$ |
| etc. | | | | |

TO ESTIMATE A SQUARE ROOT find the squares which are just above and just below the number in question.

To find $\sqrt{136}$, we look at the square just above 136; that is 144 (or 12 squared).

Then look at the square just below 136; that is 121 (or 11 squared).

So, $\sqrt{136}$ is greater than $\sqrt{121}$ but less than $\sqrt{144}$. That is: $\sqrt{136}$ is between 11 and 12.

TO USE YOUR CALCULATOR to find $\sqrt{136}$: first enter the number 136, then press the $\sqrt{\ }$ key.

**Exercise Set 1.4 A**

Find the square roots:

| | | |
|---|---|---|
| 1. $\sqrt{36}$ | 2. $\sqrt{289}$ | 3. $\sqrt{100}$ |
| 4. $\sqrt{144}$ | 5. $\sqrt{109}$ | 6. $\sqrt{400}$ |
| 7. $\sqrt{361}$ | 8. $\sqrt{625}$ | 9. $\sqrt{13}$ |
| 10. $\sqrt{6.25}$ | 11. $\sqrt{64}$ | 12. $\sqrt{121}$ |

**Exercise Set 1.4 B**

Perform the indicated operations:

| | | |
|---|---|---|
| 1. $(-5)^2$ | 2. $\sqrt{25} + \sqrt{49}$ | 3. $(\sqrt{16})^2$ |
| 4. $10^2 - 2^2$ | 5. $\sqrt{4} - 2$ | 6. $\sqrt{100} - \sqrt{9}$ |
| 7. $25 - 5^2$ | 8. $\sqrt{100} - \sqrt{121}$ | 9. $2^2 - (\sqrt{4})^2$ |
| 10. $\sqrt{64} + \sqrt{16}$ | 11. $\sqrt{-4}$ | 12. $\sqrt{-100}$ |

NOTE: The last two answers should have registered ERROR on your calculator. WHY?

## 1.5 *Changing Fractions to Decimals*

To change a number from fraction form to decimal form, divide the **numerator** by the **denominator.**

*Example:* Change $\frac{5}{3}$ to a decimal.

Entering $5 \div 3$ into the calculator yields 1.666666. . . which may be "rounded" to 1.67 or written $1.\overline{6}$

*Example:* Change $\frac{3}{5}$ to a decimal.

Enter $3 \div 5$ into the calculator. It shows 0.6; so, $\frac{3}{5} = 0.6$.

**Exercise Set 1.5**

Change the fractions to decimals:

| | | | |
|---|---|---|---|
| 1. $\frac{6}{7}$ | 2. $\frac{7}{6}$ | 3. $\frac{14}{3}$ | 4. $\frac{3}{14}$ |
| 5. $\frac{15}{4}$ | 6. $\frac{4}{15}$ | 7. $\frac{1}{6}$ | 8. $\frac{6}{1}$ |
| 9. $\frac{7}{8}$ | 10. $\frac{8}{7}$ | 11. $\frac{11}{2}$ | 12. $\frac{2}{11}$ |
| 13. $\frac{10}{6}$ | 14. $\frac{6}{10}$ | 15. $\frac{0}{5}$ | 16. $\frac{5}{0}$ |

*NOTE:* Your calculator should have registered ERROR on that last problem. DIVISION BY ZERO IS IMPOSSIBLE.

## 1.6 *Changing Decimals and Percents*

**PER CENT** means "out of 100".

**To change a decimal form to a percent, simply multiply the decimal form by 100; that is, move the decimal point two places to the right.**

**To change a percent to a decimal, simply divide by 100;** that is, move the decimal point two places to the left.

*Example:* Change .019 to a percent.

$.019 \times 100 = 1.9$ so $.019 = 1.9\%$

*Example:* Change 35% to a decimal.

$35 \div 100 = .35$ so $35\% = .35$

**Exercise Set 1.6 A**

Change the decimals to percents, and the percents to decimals.

| | | |
|---|---|---|
| 1. .25 | 2. 2.5 | 3. 12% |
| 4. .12 | 5. 1.3 | 6. 40% |
| 7. .08 | 8. .35 | 9. 87.5% |
| 10. 1.0 | 11. .125 | 12. 75% |
| 13. 2.9 | 14. .29 | 15. 5% |
| 16. 1.10 | 17. 11.0 | 18. 110% |

**Exercise Set 1.6 B**

Change the fractions to decimals, then to percents.

1. $\frac{1}{3}$
2. $\frac{5}{10}$
3. $\frac{4}{5}$
4. $\frac{7}{9}$
5. $\frac{6}{8}$
6. $\frac{3}{4}$
7. $\frac{3}{12}$
8. $\frac{2}{6}$
9. $\frac{1}{12}$
10. $\frac{8}{100}$
11. $\frac{1}{4}$
12. $\frac{4}{2}$
13. $\frac{1}{8}$
14. $\frac{7}{5}$
15. $\frac{9}{3}$

## 1.7 *Order of Operations*

In algebra, the use of parentheses ( ) or brackets [ ] or { } indicates the order in which calculations should be done.

*Example:* Suppose you want to calculate $10 + (4 \times 5) - (8 \div 2)$.

If you ignore the parentheses and enter the numbers and the operations in the order given, the calculator will display 31 as the result.

However, the ( ) indicates that you should **first** find $4 \times 5$ and $8 \div 2$ **then** do the addition and subtraction.

So the result is $10 + 20 - 4 = 26$.

*Example:* Simplify $12 \times (3 + 5) - (3 \times 4)$

First: Add $3 + 5$ and multiply $3 \times 4$.

Then: $12 \times 8 - 12 = 96 - 12 = 84$

*Example:* Simplify $3 \times (5 - 7) + 2 \times (8 - 4)$

First: $(5 - 7) = -2$ and $(8 - 4) = 4$

Second: Find $3 \times (-2)$ and $2 \times 4$

Then: $-6 \quad + \quad 8 = 2$

*Example:* Evaluate the following: Note the difference the ( ) make.

a. $(7 + 5) - (8 \times 10) - (6 \div 2) =$
$12 - 80 - 3 =$
$12 - 83 = -71$

b. $(7 + 5) - [8 \times (10 - 6)] \div 2 =$
$12 - [8 \times 4] \div 2 =$
$12 - 32 \div 2 =$
$12 - 16 = -4$

NOTE: When there are no parentheses or brackets to be simplified, FIRST MULTIPLY or DIVIDE from left to right, THEN ADD or SUBTRACT from left to right.

c. $7 + 5 - 8 \times 10 - 6 \div 2 =$
$7 + 5 - 80 - 3 =$
$12 - 83 = -71$

**Exercise Set 1.7**

Simplify:

1. $14 - 6 \times 2 + 20 \div 4$
2. $14 - (6 \times 2) + (20 \div 4)$
3. $(14 - 6) \times 2 + (20 \div 4)$
4. $(14 - 6) \times (2 + 20) \div 4$
5. $[(14 - 6) \times 2] + [20 \div 4]$

## 1.8 *Solving Word Problems*

Solving word problems is like working a puzzle:

> you must have all the pieces "sorted out", you must assemble the pieces one at a time, and you must adjust pieces which have been incorrectly positioned.

Before attempting to solve word problems, ask yourself the following questions:

1. WHAT INFORMATION HAVE I BEEN GIVEN? (You must have *enough* information, but you must disregard information which isn't necessary for the solution of the problem.)
2. WHAT AM I LOOKING TO FIND FROM THIS INFORMATION?
3. WHAT STEPS MUST I USE TO GET THE RESULT I WANT?

After you have thought through the problem carefully, you are ready to do the necessary calculations.

When the calculations are completed, ask yourself:

1. DID I ANSWER THE QUESTION?
2. DOES MY ANSWER SEEM "REASONABLE"? (Check your answer whenever possible).

*Example:* A student saves \$35.00 a week in order to buy a second hand car for \$800.00. At that rate, how long will it take to save the money?

Given: Cost of car: \$800.00
Weekly savings \$35.00

Find: Number of weeks needed to save \$800

How?: Divide \$800.00 by \$35.00.

The solution is 22.857142. That means it would take about 23 weeks to save the \$800.00.

*Check:* If the student saved \$35.00 a week for 23 weeks, the savings would equal 35 times 23 or \$805.00.

The solution is reasonable!

*Example:* A student received the following grades on math exams: 78, 65, 90, 72, 80.

What was the student's average grade?

Given: 5 exam grades.

Find: average grade

How?: Sum the 5 grades and divide by 5.

The solution is 385 ÷ 5 or 77.

*Check:* 78 is better than 77 by 1 point
80 is better than 77 by 3 points
90 is better than 77 by 13 points.
That totals 17 points.

On the other hand:

65 is 12 points lower than 77 and 72 is 5 points lower than 77. This would also total 17 points.

*Good solution!*

*Example:* A coat listed at $135.00 is to be discounted 25%. How much is the discount? How much is the cost after discount?

Given that 25% of the original cost of the coat is to be deducted:

Multiply $135.00 by 25% to find the discount. Subtract that amount from the original cost to find the price after discount.

First change 25% to a decimal: 0.25

Then: $135 × .25 = $33.75 This is the amount of discount.

$135.00 − 33.75 = $101.25. This is the price of the coat after discount.

*Check:* If I paid $101.25 for the coat after 33.75 was deducted, the original price must have been $101.25 + 33.75 or $135.00.

### Exercise Set 1.8

Solve the following problems using the "strategy" described in this section. Check your answers.

**Answers**

1. Jo invests $2000 in a savings account which yields 6.25% each interest period. How much is her investment worth after the first period? — 1. ________

2. At a rate of 8.5% sales tax, what would a $35.69 purchase cost after tax? — 2. ________

3. Mrs. Smith's weekly grocery bills amounted to: $125.35, $97.29, $102.76, $93.27. What was the average amount she spent on groceries? — 3. ________

4. A lumber yard offered a 10% discount on all items purchased during a one-day sale. Mrs. Foley bought four items priced at $1.39, six priced at $3.59, and one item priced $11.98. How much did she pay after discount? — 4. ________

5. The area of a triangle is equal to $\frac{1}{2}$ the product of its base and height. If a triangle has a base of 12 inches and a height of 4 inches, what is its area? — 5. ________

6. Jim saved $25 per week for 12 weeks, then spent $100 on a ski trip. How much did he have left? — 6. ________

7. You decide to start a small business. During the first business week: you gained $125, lost $35, gained $39, lost $120 and gained $75. What was the final outcome of your efforts? — 7. ________

8. At noon, the temperature in Merryville was 87 degrees. In Twin Lake it was 7 degrees cooler. Doonesville was 15 degrees warmer than Twin Lake. What was the temperature in Twin Lake? What was the temperature in Doonesville? — 8. ________

9. Jack earns three times as much as Bill, but only half as much as Sarah. If Jack earns $45.00, what does Bill earn in the same time period? What about Sarah? — 9. ________

## CHAPTER EXERCISES

Use your calculator to simplify the following:

### Section 1.1

1. $16.7 + 7.32$
2. $23.86 - 8.4$
3. $2.72 + 6.85 - 3.21$
4. $89.5 - 2.53 - 6.7$
5. $.01 + .275 + .875$
6. $15 - 2.65 + 0.03$
7. $28 + 36.1 - 17.2 + 18.6$
8. $2.73 - 8.425 - 0.45$
9. $-20 - 3.46$
10. $-15.1 + 2.25 - 1.5$

### Section 1.2

11. $24 \times 37$
12. $360 \times 25.2$
13. $7.78 \times .631$
14. $2.75 \times .125 \times .875$
15. $330 \div 2.6$
16. $14.2 \div 2.7$
17. $.32 \div 4$
18. $1.32 \div .05$
19. $55.1 \times 6.2 \div 2$
20. $10.78 \div 3.1 \times 11.5$

### Section 1.3

21. $8.1^2$
22. $2.3^4$
23. $(.15)^2$
24. $6^3$
25. $11^3$
26. $2^5$
27. $3^4$
28. $(.3)^3$
29. $10^6$
30. $3^7$

### Section 1.4

31. $\sqrt{25}$
32. $\sqrt{144}$
33. $\sqrt{169}$
34. $\sqrt{1000}$
35. $\sqrt{.49}$
36. $\sqrt{1.36}$
37. $\sqrt{21.7}$
38. $\sqrt{90}$
39. $\sqrt{.025}$
40. $\sqrt{.003}$

**Section 1.5**

Change the following fractions to decimals:

41. $\frac{7}{8}$

42. $\frac{5}{16}$

43. $\frac{3}{4}$

44. $\frac{4}{7}$

45. $\frac{21}{14}$

46. $\frac{81}{3}$

47. $\frac{10}{7}$

48. $\frac{16}{0}$

49. $\frac{2.1}{6}$

50. $\frac{17.3}{9}$

**Section 1.6**

Change the following decimals to percents:

51. .17
52. .152
53. .03
54. .005
55. .625
56. .7
57. .3
58. 2.7
59. 7.1
60. 6

**Section 1.6**

Change the following fractions to percents:

61. $\frac{7}{8}$

62. $\frac{1}{16}$

63. $\frac{2}{5}$

64. $\frac{5}{8}$

65. $\frac{10}{17}$

66. $\frac{21}{3}$

67. $\frac{7.2}{3}$

68. $\frac{12}{.2}$

69. $\frac{1.2}{.3}$

70. $\frac{7.8}{6}$

**Section 1.7**

Simplify:

71. $7 - 5 + 6 + 5$

72. $21.1 - 3.2 - 6.1$

73. $23 - (4 + 11) \div 2$

74. $(61 - 27) \div (8 + 2)$

75. $7 + (2 \times 3) \div 5$

76. $28 \times 3 + (81 \div 3)$

77. $[6 \times (7 - 4) \div 2]$

78. $(18 + 3) \div 7 + 2 \times (8 \div 4)$

79. $2 + [3 \times (21 \div 7)]$

80. $423 \times (25 \div 6) - 1.5$

**Section 1.8**

Solve the following problems. Check your answers when possible.

81. A student earned \$5.25 an hour at a part-time job during the winter break. She worked 3 hours a day, 4 days a week for 4 weeks. How much did she earn?

82. When John Smith was born, his father was 28 and his mother was 26 years old. John is now 21. How old are his parents?

83. The measures of the 3 angles of a triangle must sum to 180 degrees. Two angles measure 60 degrees each. What is the measure of the third angle?

84. Find the average amount spent on a Friday night party if seven students each spent \$12.50, \$10.90, \$9.65, \$7.80, \$12.40, \$11.87 and \$13.56.

85. The cost of two hamburgers and milk shake is \$2.30. A hamburger costs \$.90. How much is the milkshake?

86. A suit is marked "$\frac{1}{3}$ OFF". If the original price was \$89.00, what is the cost of the suit after discount?

87. A 40 ft. length of wire must be cut into 3 equal pieces, to be measured in feet. To the nearest foot, what is the longest piece which can be cut, and how much will be wasted?

88. Frank earned \$5.50 an hour at his part time job. He worked twelve hours a week for three weeks, then worked 20 hours for the fourth week. What did he earn in the month?

89. Use the data from #88 and calculate how many additional hours Frank would have had to work, if he wanted to reach \$500.00?

## CHAPTER TEST *Form 1A*

Use your calculator to answer the following questions: Find the value of:

1. $173.72 - 12.1 + 27$
2. $168.4 \div 31$
3. $21.3 \times .73$
4. $17^2$
5. $\sqrt{225}$
6. $3 \times 4.2 + 7$
7. $.22 - .7 + 2.4 - .005$
8. $\sqrt{51}$
9. $8 \times (7.25 + 5.1)$
10. $32 \div (8.1 - 3)$
11. $2.1^4$
12. $(25 \times 3) \div (5.1 - 2)$
13. $\sqrt{5.25}$
14. $\sqrt{49} + (13 \div 2)$

Change the following to percents.

15. .753
16. $\frac{12}{25}$
17. 1.5

Change the following to decimals:

18. $\frac{84}{27}$
19. 5%
20. 17.6%

Solve the following problems. Check your answers.

**Answers**

21. The sum of the angles of a triangle is 180°. If all three angles of a certain triangle are equal, what is the measure of each angle? 21. ________

22. A fourteen foot length of wood is cut into four equal pieces. How long is each piece? 22. ________

23. Find the average height if the following heights are recorded: 23. ________
    5 ft. 2 in. 6 ft. 1 in. 4 ft. 10 in.
    6 ft. 1 in. 5 ft. 7 in. 5 ft. 9 in.

24. One weekend, Tony drove 298.3 miles and used 11.2 gallons of gasoline. How many miles did he average per gallon of gasoline? 24. ________

25. Find the cost of a $50.00 item after a 25% discount. 25. ________

CHAPTER

# 2 The Real Numbers and Their Properties

## 2.0 *Vocabulary Preview*

Define these words without using a dictionary. Formulate their meanings from past knowledge or experience.

**approximate**
**associative**
**commutative**
**distributive**
**even**
**identity**
**infinite**
**integer**
**inverse**
**graph**
**natural**
**negative**
**number line**
**odd**
**positive**
**real**
**set**

## 2.1 *The Set of Real Numbers*

The **set** of numbers which we first encounter in our lives is the set of **natural** numbers (the counting numbers). This set C = {1,2,3,4,5, . . . } is represented graphically with the number line:

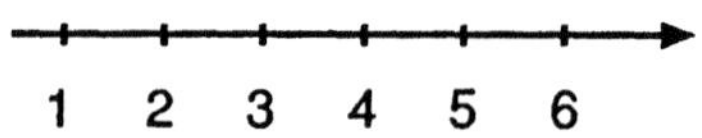


This "line" extends indefinitely to include all the counting numbers. This set of numbers is said to be **infinite;** that is, a non-ending set of numbers.

Using zero as our point of origin, we may extend the line in the opposite direction and mark off equal units of measure (designated as $-1$, $-2$, $-3$, etc.) to indicate that they are located to the left of zero. The line:

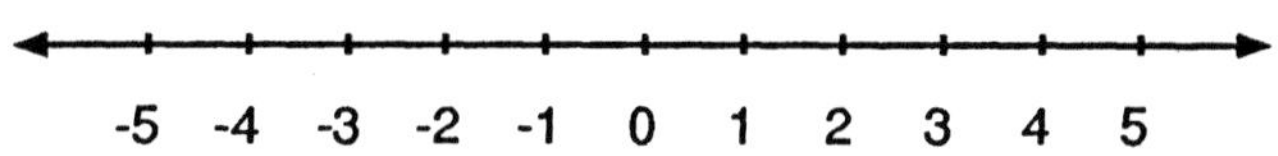


represents all the positive and negative **integers** (whole numbers) and all the numbers between them (non-integers).

This set of all integers and non-integers is called the set of **real** numbers.

*Example:* Locate the number 4 on the number line.

The number is positive so it is to the right of zero, four units:

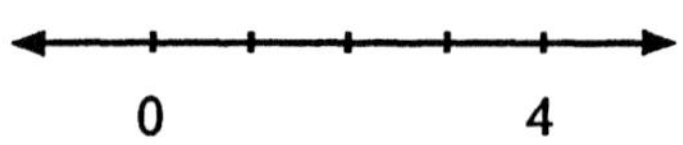

*Example:* Locate the number 3.5 on the number line.

Since the sign of the number is positive, it is located to the right of zero. 3.5 lies midway between 3 and 4 on the line. (See number line below.)

*Example:* Locate the number $-5$ on the number line.

The sign of the number is negative, so the number is five units LEFT of zero. (See number line below.)

*Example:* Locate $-2\frac{1}{4}$ on the number line.

Since the number is negative, it is to the left of zero. It is between $-2$ and $-3$, just $\frac{1}{4}$ of the way past $-2$.

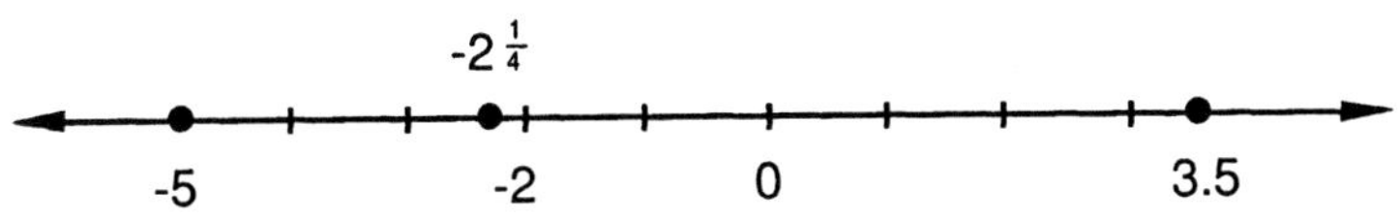


**Exercise Set 2.1**

Locate the following numbers on the real number line. In the case of non-integers, *approximate* the location of the number.

1. 13
2. 6.5
3. $-0.25$
4. $-5\frac{1}{2}$
5. 7.1
6. $-9$
7. 5.67
8. $2.\overline{3}$

## 2.2 *Comparing Real Numbers*

- Numbers to the right of zero are greater than zero.
- Numbers to the left of zero are less than zero.
- A number is greater than any number to its left.
- A number is less than any number to its right.
- The symbol for greater than is $>$
- The symbol for less than is $<$
- Two numbers which are equal share the same location on the real number line.

*Examples:*

| | | |
|---|---|---|
| $4 < 8$ | $8 > 4$ | $1.23 > 0$ |
| $1.20 < 1.23$ | $4.06 < 4.60$ | $4.60 > 4.06$ |
| $7.5 = 7\frac{1}{2}$ | $-4 > -6$ | $-2.06 < 2.06$ |
| $-0.1 < -0.01$ | $\sqrt{25} < \sqrt{64}$ | $-\sqrt{100} = -10$ |

**Exercise Set 2.2**

Insert the sign $<$, $>$, or $=$ to make the statements true:

| | | | |
|---|---|---|---|
| 1. 4.5 | 5 | 2. 8.90 | 8.99 |
| 3. $-3.33$ | $-3$ | 4. 4.57 | $-4.58$ |
| 5. 2.66 | 2.5 | 6. $1\frac{1}{4}$ | 1.25 |
| 7. 5.0 | $-5.3$ | 8. $-3.5$ | $-3.2$ |
| 9. $\sqrt{81}$ | 9 | 10. $-\sqrt{36}$ | $-6.2$ |
| 11. $-1$ | $-0.1$ | 12. 3.54 | 3.45 |

## 2.3 *Properties of Real Numbers*

The properties of real members are defined extensively in other college mathematics courses. In this text we will explain through examples, those properties of real numbers which are most important in the study of algebra; namely,

**the commutative property**
**the associative property**
**the distributive property**
**inverses and identities.**

## 2.4 *The Commutative Property*

This property of real numbers states that the sum and product of any two real numbers is the same, regardless of the "way" in which you add or multiply; that is:

$$5 + 4 = 4 + 5$$

and

$$5 \times 4 = 4 \times 5.$$

$$0.5 + 3.1 = 3.1 + 0.5$$

and

$$0.5 \times 3.1 = 3.1 \times 0.5$$

**In symbols, the commutative property states:**

**For any two real numbers a and b: $a + b = b + a$**
**and $a \times b = b \times a$**

*Examples:* $a + 6 = 6 + a$

$12 + (b - 2) = (b - 2) + 12$

$(7 + c) + (d - 3) = (d - 3) + (7 + c)$

*Examples:* $(a) \times 6 = 6 \times (a)$

$12 \times (b - 2) = (b - 2) \times 12$

$(d - 3) \times (7 + c) = (7 + c) \times (d - 3)$

**Exercise Set 2.4 A**

Use your calculator to perform the indicated operations which illustrate the commutative property of addition and multiplication:

(*NOTE:* Parentheses will be used instead of "×" to indicate multiplication.)

1. $3 + 122$ and $122 + 3$
2. $3(122)$ and $122(3)$
3. $5 + 1.3$ and $1.3 + 5$
4. $5(1.3)$ and $1.3(5)$
5. $-3 + 4$ and $4 + (-3)$
6. $-3(4)$ and $4(-3)$
7. $5.5 + 6$ and $6 + 5.5$
8. $5.5(6)$ and $6(5.5)$
9. $2 + 0.5$ and $0.5 + 2$
10. $2(0.5)$ and $0.5(2)$
11. $-2 + 2.3$ and $2.3 + (-2)$
12. $-2(2.3)$ and $2.3(-2)$
13. $.6 + (-8)$ and $(-8) + .6$
14. $(-8)(.6)$ and $(.6)(-8)$
15. $0 + 12.6$ and $12.6 + 0$
16. $0(12.6)$ and $12.6(0)$
17. $-3 + 3$ and $3 + (-3)$
18. $-3(3)$ and $3(-3)$
19. $.4 + 3.4$ and $3.4 + .4$
20. $.4(3.4)$ and $3.4(.4)$

**Exercise Set 2.4 B**

The following examples show that the commutative property DOES NOT hold for the operations of subtraction and division. Use your calculator to perform the indicated operations:

| | | | | | | |
|---|---|---|---|---|---|---|
| 1. $3 - 122$ | and | $122 - 3$ | 2. | $\frac{3}{122}$ | and | $\frac{122}{3}$ |
| 3. $5 - 1.3$ | and | $1.3 - 5$ | 4. | $\frac{5}{1.3}$ | and | $\frac{1.3}{5}$ |
| 5. $3 - 4$ | and | $4 - 3$ | 6. | $\frac{3}{4}$ | and | $\frac{4}{3}$ |
| 7. $5.5 - 6.1$ | and | $6.1 - 5.5$ | 8. | $\frac{5.5}{6.1}$ | and | $\frac{6.1}{5.5}$ |
| 9. $2 - 0.5$ | and | $0.5 - 2$ | 10. | $\frac{2}{0.5}$ | and | $\frac{0.5}{2}$ |
| 11. $2 - 2.3$ | and | $2.3 - 2$ | 12. | $\frac{2}{2.3}$ | and | $\frac{2.3}{2}$ |
| 13. $5.6 - 8.9$ | and | $8.9 - 5.6$ | 13. | $\frac{5.6}{8.9}$ | and | $\frac{8.9}{5.6}$ |
| 14. $0 - 12$ | and | $12 - 0$ | 15. | $\frac{0}{12}$ | and | $\frac{12}{0}$ |

## 2.5 *The Associative Property*

The **associative** property of real numbers states that the sum or product of any three real numbers is the same, regardless of the order in which you add or multiply.

**In symbols, the associative property states:**

**For any real numbers a, b and c:**

$$a + (b + c) = (a + b) + c \quad \textbf{and}$$
$$a \times (b \times c) = (a \times b) \times c$$

This latter expression may be written as

$a(bc) = (ab)c$ or $a \cdot (b \cdot c) = (a \cdot b) \cdot c$

Parentheses or a raised dot may replace the "×" to indicate multiplication.

*Example:* $23 + (12 + 5) = 23 + 17 = 40$

$(23 + 12) + 5 = 35 + 5 = 40$

$23 \times (12 \times 5) = 23 \times 60 = 1380$

$(23 \times 12) \times 5 = 276 \times 5 = 1380$

**Exercise Set 2.5**

Perform the indicated operations which illustrate the associative property of addition and multiplication:

1. $15 + (11.5 + 3.78)$
2. $(15 + 11.5) + 3.78$
3. $15 \times (11.5 \times 3.78)$
4. $(15 \times 11.5) \times 3.78$
5. $(21.2 + 5.0) + 7.1$
6. $21.2 + (5.0 + 7.1)$
7. $(21.2 \times 5.0) \times 7.1$
8. $21.2 \times (5.0 \times 7.1)$
9. $\left(\frac{1}{2} \times 6\right) \times 7$
10. $\frac{1}{2} \times (6 \times 7)$
11. $\left(\frac{1}{2} + 6\right) + 7$
12. $\frac{1}{2} + (6 + 7)$
13. $(5 + 0) + 12.4$
14. $5 + (0 + 12.4)$

## 2.6 *The Distributive Property*

This property actually involves the TWO operations of multiplication and addition. We might say that multiplication is "distributed" over a sum:

The product $3(9) = 27.$

Since $9 = 5 + 4,$
$3(5 + 4) = 27.$

Also, $3(5) + 3(4) = 15 + 12 = 27.27.$

We may say, then, that $3(5 + 4) = 3(5) + 3(4)$

**The above example illustrates the distributive property for real numbers, which may be symbolized:**

$$\mathbf{a(b + c) = ab + ac}$$

*Example:*
$$\begin{aligned} 3(45) &= 3(40) + 3(5) \\ &= 120 \quad + 15 \\ &= 135 \end{aligned}$$

*Example:*
$$\begin{aligned} 47(53) &= (40 + 7)(50 + 3) \\ &= (40 \cdot 50) + (40 \cdot 3) + (7 \cdot 50) + (7 \cdot 3) \end{aligned}$$

*Example:* $6(b + 4) = 6b + 24$

*Example:* $(a + b)(c + d) = ac + ad + bc + bd$

MORE ON THIS IN CHAPTER TEN

**Exercise Set 2.6 A**

Perform the indicated operations. Each pair illustrates the distributive property of real numbers:

1. 6(75) and 6(70) + 6(5)
2. 5.3(8) and 5(8) + .3(8)
3. 4.5(7) and 4(7) + .5(7)
4. 9(11) and 9(10) + 9(1)
5. 10(17) and 10(10) + 10(7)
6. 12(15) and 12(10) + 12(5)

## 2.7 *Identities and Inverses*

ADDITIVE **INVERSES** are numbers whose sum is zero.

MULTIPLICATIVE INVERSES (reciprocals) are numbers whose product is one.

**The number ZERO is called the IDENTITY ELEMENT of addition (or subtraction), since adding (or subtracting) zero does not change the identity of a number.**

*Examples:* $5 + 0 = 5$ $13 - 0 = 13$

**The number ONE is the IDENTITY ELEMENT for multiplication (or division), since multiplying (or dividing) by one does not change the identity of a number.**

*Examples:* $5(1) = 5$ $\frac{5}{1} = 5$

REMEMBER that the number **one** may be expressed as $\frac{3}{3}, \frac{6}{6}, \frac{1.4}{1.4}$

In general: $\frac{n}{n} = 1$

*Examples:* 6 and $-6$ are additive inverses since $6 + (-6) = 0$

6 and $\frac{1}{6}$ are multiplicative inverses since $6\left(\frac{1}{6}\right) = 1$.

In general:

$n + (-n) = 0$ and $n \cdot \frac{1}{n} = 1$

**Exercise Set 2.7 A**

Supply the missing number to make the following statements true:

1. $10 + (?) = 0$
2. $10(?) = 1$
3. $\frac{5}{5}(?) = 75$
4. $-75 + 75 = (?)$
5. $2.5(?) = 2.5$
6. $-2.5 + (?) = 0$
7. $3.75(?) = 1$
8. $3.75 + (?) = 0$
9. $-4.2(?) = 1$
10. $-4.2 + (?) = 0$

**Exercise Set 2.7 B**

Indicate the property of real numbers illustrated by each:

1. $3 + (-3) = (-3) + 3$
2. $3 + (-3) = 0$
3. $5(4 + 6) = 5(4) + 5(6)$
4. $5(4 + 6) = (4 + 6)5$
5. $7\left(\frac{1}{7}\right) = \left(\frac{1}{7}\right)(7)$
6. $7\left(\frac{1}{7}\right) = 1$
7. $16\left(\frac{4}{4}\right) = 16$
8. $8(12) = 8(10) + 8(2)$
9. $-8 + 8 = 0$
10. $-8\left(-\frac{1}{8}\right) = 1$
11. $-8 + 8 = 8 + (-8)$
12. $-8(8) = 8(-8)$
13. $7(a + b) = 7a + 7b$
14. $x + y = y + x$

**Exercise Set 2.7 C**

Use the indicated property of real numbers to fill the blanks:

| | | |
|---|---|---|
| 1. Commutative: | $a + (6 - b) = \underline{\ \ ?\ \ } + a$ | $a(6 - b) = \underline{\ \ ?\ \ }\,a$ |
| 2. Inverses: | $3 + \underline{\ \ ?\ \ } = 0$ | $3(\underline{\ \ ?\ \ }) = 1$ |
| 3. Distributive: | $6(c + 5) = 6c + \underline{\ \ ?\ \ }$ | |
| 4. Identity: | $\frac{6}{7}\left(\frac{5}{5}\right) = \underline{\ \ ?\ \ }$ | $\frac{6}{7} + \underline{\ \ ?\ \ } = \frac{6}{7}$ |
| 5. Associative: | $6 + (a + 4) = \underline{\ \ ?\ \ } + 4$ | $6(a \cdot 4) = (\underline{\ \ ?\ \ }) \cdot 4$ |

## CHAPTER EXERCISES

### Section 2.1

Locate the following numbers on the real number line:

| | |
|---|---|
| 1. 7.5 | 2. $-3.75$ |
| 3. 1.9 | 4. $-8$ |
| 5. $\frac{3}{4}$ | 6. $+.784$ |
| 7. 2.75 | 8. 7.66 |
| 9. $\sqrt{25}$ | 10. $-4\frac{1}{5}$ |
| 11. $-\sqrt{64}$ | 12. $\sqrt{37}$ |

### Section 2.2

Insert either sign, > or <, between the following numbers:

| | | | | | |
|---|---|---|---|---|---|
| 13. 10 | ____ | 12.7 | 14. 3.5 | ____ | 1.2 |
| 15. $-13$ | ____ | 5 | 16. $-12$ | ____ | $-23$ |
| 17. 2.25 | ____ | 0 | 18. 8.2 | ____ | $-17.1$ |
| 19. $2\frac{3}{4}$ | ____ | 1.93 | 20. $-4.7$ | ____ | $\sqrt{2}$ |
| 21. $\frac{3}{4}$ | ____ | $-.75$ | 22. $-\sqrt{36}$ | ____ | $\sqrt{4}$ |
| 23. .01 | ____ | .0001 | 24. 16.3 | ____ | 1.63 |

### Sections 2.4 to 2.6

State the property of real numbers (commutative, associative, distributive) illustrated by each of the following:

25. $23 + 47 = 47 + 23$

26. $(7 + 12) + 4 = 7 + (12 + 4)$

27. $2(5 + 3) = (2 \times 5) + (2 \times 3)$

28. $\frac{1}{2} \times 6 = 6 \times \frac{1}{2}$

29. $\sqrt{3} + \sqrt{2} = \sqrt{2} + \sqrt{3}$

30. $.5(11 + 3) = .5(11) + .5(3)$

31. $\left[\frac{6}{7}(2.1)\right](.5) = \frac{6}{7}[(2.1)(.5)]$

32. $51.2 + 7 = 7 + 51.2$

33. $\sqrt{5} \times (2.1 + 3.\overline{7}) = (\sqrt{5} \times 2.1) + (\sqrt{5} \times 3.\overline{7})$

**Section 2.7**

Find the numerical value for the following:

34. $7(8 + 2)$
35. $7.5 + (3.25 + 6.5)$
36. $.7 \times [11.1 \times 1.5]$
37. $\frac{1}{2}(50 + 30)$
38. $(7 \times 8) + (7 \times 5)$
39. $(.02 \times 7.4) + (.02 \times 3)$

**Section 2.7**

Find the correct inverse to make each statement true:

40. $6 + \underline{\quad ? \quad} = 0$
41. $\underline{\quad ? \quad} \times (-3) = 1$
42. $15 \times \underline{\quad ? \quad} = 1$
43. $\sqrt{5} + \underline{\quad ? \quad} = 0$
44. $\underline{\quad ? \quad} \times 3 = 1$
45. $4.3 \times \underline{\quad ? \quad} = 1$
46. $-2.67 + \underline{\quad ? \quad} = 0$
47. $-3 \times (\underline{\quad ? \quad}) = 1$

**Section 2.7**

Find additive AND multiplicative inverses of:

48. 33
49. $-10$
50. $\frac{2}{3}$
51. $-2.5$
52. 7.12
53. .8
54. 1
55. $-\sqrt{2}$
56. 0

**CHAPTER TEST** *Form 2A*

Locate the number that corresponds to the letter on the real number line:

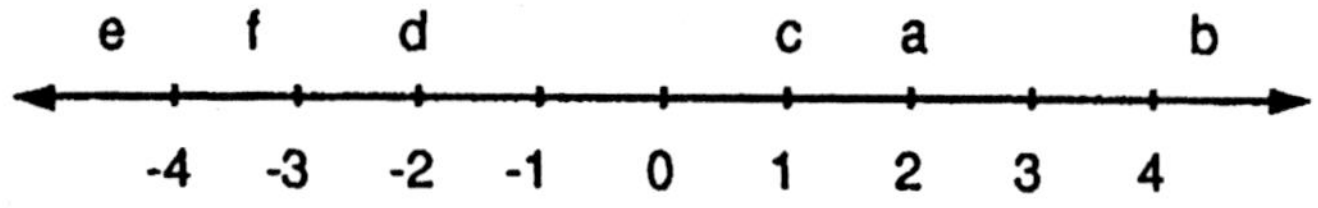


1. $-2$
2. $4.5$
3. $2$
4. $-4.5$
5. $1$
6. $-3.3$

Insert the symbol <, >, = which will make the statement true. **Hint**: Use your calculator to change each to decimal form; then compare the decimal forms.

| | | | **Answers** |
|---|---|---|---|
| 7. | $\frac{2}{3}$ | $0.666$ | 7.________ |
| 8. | $\frac{3}{4}$ | $0.075$ | 8.________ |
| 9. | $6$ | $\sqrt{36}$ | 9.________ |
| 10. | $-1.5$ | $-\frac{25}{3}$ | 10.________ |
| 11. | $50\%$ | $0.5$ | 11.________ |
| 12. | $\frac{2}{5}$ | $0.4$ | 12.________ |
| 13. | $8$ | $\sqrt{65}$ | 13.________ |
| 14. | $\frac{7}{10}$ | $\frac{14}{20}$ | 14.________ |
| 15. | $\frac{5}{8}$ | $\frac{3}{4}$ | 15.________ |
| 16. | $\frac{1}{3}$ | $0.\underline{3}$ | 16.________ |
| 17. | $\frac{11}{12}$ | $\frac{19}{20}$ | 17.________ |

Indicate the property (commutative, associative, distributive, identity or inverse) illustrated by each:

**Answers**

18. $5(3) + 5(7) = 5(3 + 7)$ — 18. ________

19. $\left(\frac{1}{3}\right)(3) = 1$ — 19. ________

20. $x + y = y + x$ — 20. ________

21. $(-8) + 0 = -8$ — 21. ________

22. $\left(\frac{3}{3}\right)(72) = 72$ — 22. ________

23. $(x + 5) + y = x + (5 + y)$ — 23. ________

24. $\left(\frac{11}{3}\right)\left(\frac{3}{11}\right) = 1$ — 24. ________

25. $(ab)(9) = (9)(ab)$ — 25. ________

State the additive AND multiplicative inverses of each:

**Answers**

26. 51 — 26. ________

27. –31 — 27. ________

28. $\frac{2}{5}$ — 28. ________

29. 3.5 — 29. ________

30. 0.15 — 30. ________

31. 1 — 31. ________

32. $-\frac{1}{2}$ — 32. ________

33. 3.6 — 33. ________

34. 0 — 34. ________

CHAPTER

# 3 Operations Involving Real Numbers in Integer Form

## 3.0 *Vocabulary Preview*

Define these words without using a dictionary. Formulate their meanings from past knowledge and experience:

| | | |
|---|---|---|
| **commutative** | **inverse** | **real** |
| **difference** | **multiple** | **repetitive** |
| **dividend** | **power** | **root** |
| **divisor** | **prime** | **sum** |
| **evaluate** | **product** | **unit** |
| **factor** | **quotient** | |

## 3.1 *Addition of Real Numbers in Integer Form*

The expression $(5) + (-5)$ indicates the operation of addition.

The signs associated with the numbers indicate "direction"; thus, we progress 5 units to the right of 0 on the number line, and then progress 5 units to the left.

We arrive at the point of origin; namely, 0.

To add $(10) + (2)$ move 10 units to the right of 0 and continue 2 units in the same direction.

We arrive at the point 12 units to the right of 0 which is at $+12$ on the number line.

When adding $(10) + (-2)$, we move 10 units to the right of 0, and then move 2 units left, arriving at a point 8 units to the right of 0; at $+8$.

The sum of $(-10) + (2)$ necessitates a move of 10 units to the left of 0 and then 2 units to the right. The result is the point 8 units to the left of 0; $(-8)$.

Similarly, $(-10) + (-2)$ is the result of moving 10 units left from 0 and continuing in the same direction 2 more units. The result is at the point located 12 units to the left of 0; namely, at $(-12)$.

*Example:* $6 + (-12) = -6$
$6 + (-12) + (-3) = -6 + (-3) = -9$

$-4 + (-7) = -11$
$-4 + (-7) + 2 = -11 + 2 = -9$

*Examples:* $17 + (-7) = 10$
$17 + (-7) + (-6) = 10 + (-6) = 4$

$-15 + 4 = -11$
$-15 + 4 + (-2) = -11 + (-2) = -13$

**Exercise Set 3.1**

Add the integers:

1. $(+7) + (+8)$
2. $(-14) + (9)$
3. $(9) + (-6)$
4. $4 + (-6)$
5. $-14 + 6$
6. $-13 + (-3)$
7. $-25 + 16$
8. $7 + (-8)$
9. $(-14) + (-9)$
10. $3 + (-13)$
11. $25 + (-16)$
12. $(-4) + (-6)$
13. $(-11) + (-5) + (5)$
14. $(-5) + (3) + (-12)$
15. $(-7) + (-1) + (-6)$
16. $8 + (-2) + (-11)$
17. $12 + (-5) + (-3)$
18. $(-6) + (6) + 0$
19. $5 + (-6) + (-4)$
20. $(-7) + (-5) + (-1)$
21. $(-2) + (-13) + 5$
22. $13 + (-7) + (-6)$
23. $(-15) + (-2) + 17$
24. $13 + 12 + (-6)$
25. $(-7) + 5 + (-3)$
26. $-13 + (-12) + 6$

## 3.2 *Subtracting Real Numbers in Integer Form*

Actually, subtraction is finding the **difference** between two numbers; that is, finding the distance between them on the number line.

*Example:* Subtract (5) − (3) and then subtract (3) − (5)

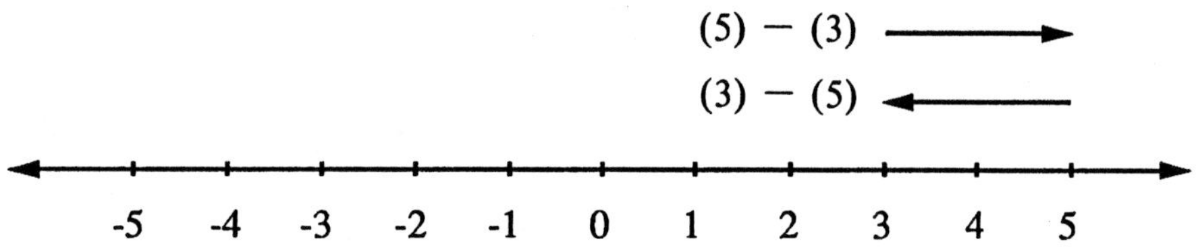


(5) − (3) = +2 because we move from 3 to 5 on the number line, or two units to the right.

(3) − (5) = −2 because we move from 5 to 3 on the number line, or 2 units to the left.

*Example:* Subtract (−5) − (3) and then (3) − (−5)

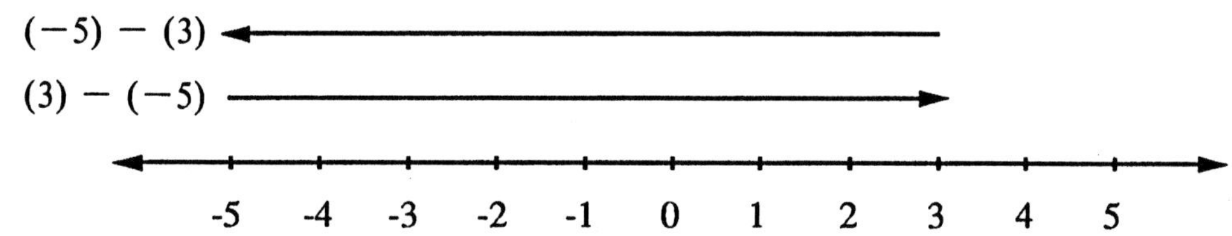


(−5) − (3) = −8 since we move from 3 to −5 on the number line, a move of 8 units to the left.

(3) − (−5) = +8 since moving from −5 to 3 is a move of 8 units to the right.

*Example:* Subtract (5) − (−3) and then (−3) − (5)

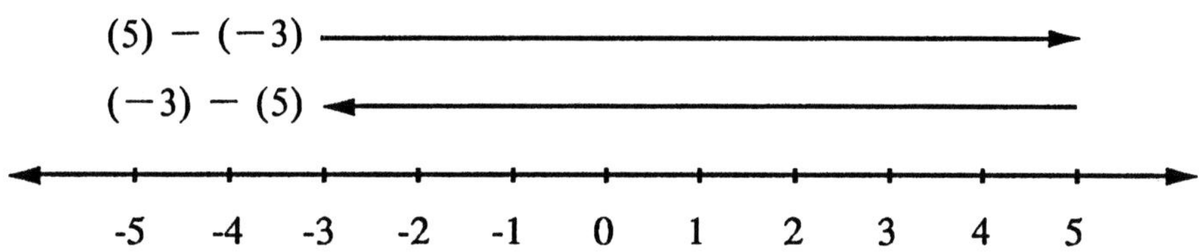


(5) − (−3) = +8 since we move from −3 to 5, which is 8 units to the right.

(−3) − (5) = −8 since the move is from 5 to −3 or 8 units to the left.

*Example:* Subtract (−5) − (−3) and (−3) − (−5)

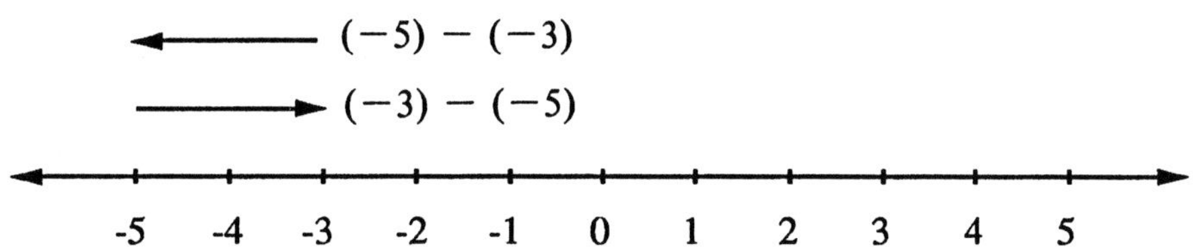


(−5) − (−3) = −2 since we move from −3 to −5 which is 2 units to the left.

(−3) − (−5) = +2 since moving from −5 to −3 is 2 units to the right.

NOTICE that the move always starts with the second number (the SUBTRAHEND) and goes toward the first number (the MINUEND).

To simplify matters, use the concept of inverses:

**addition may be replaced by subtraction if (and *only* if) the subtrahend is replaced by its additive inverse.**

**Remember from Chapter Two:**

ADDITIVE INVERSES ARE NUMBERS WHICH SUM TO ZERO.

$6 + (-6) = 0 \qquad -82 + 82 = 0$

*SUMMING UP:* The DIFFERENCE between the real numbers a and b, expressed (a − b), may be considered the SUM of a and −b, that is:
$(a - b) = (a) - (b) = (a) + (-b)$

*Examples:* $(+8) - (+6)$ is the same as $(+8) + (-6)$. The result is $+2$.

$(+6) - (+8)$ is the same as $(+6) + (-8)$. The result is $-2$.

$(-8) - (+6)$ is the same as $(-8) + (-6)$. The result is $-14$.

$(-6) - (-8)$ is the same as $(-6) + (+8)$. The result is $+2$.

**Exercise Set 3.2**

Add or Subtract as indicated.

1. $4 - 8$
2. $8 - (-3)$
3. $14 - 35$
4. $8 - 45$
5. $(-3) - 8$
6. $35 - 14$
7. $-15 - (-4)$
8. $(-16) - 19$
9. $33 - (-17)$
10. $-4 - (-15)$
11. $115 - 198 - (-16)$
12. $17 - 33$
13. $21 - (-4)$
14. $-4 - (21)$
15. $10 - (-10)$
16. $-7 - (-7)$
17. $187 - (-4)$
18. $-6 - (-6)$

## 3.3 *Multiplication of Real Numbers in Integer Form*

Usually the letter x is used to denote multiplication. In this text, we shall reserve the letter x for use as an algebraic symbol, and we'll use a raised dot, ·, or parentheses (or both) to indicate multiplication.

Multiplication is actually a quick way to do addition.

*Example:* When we write 3(18), we represent the **product** of 3 and 8, which is actually the sum of 3 eights.

Since $8 + 8 + 8 = 24$, $3(8) = 24$.

*Example:* 7 times 8 is the same as adding 7 eights:
$8 + 8 + 8 + 8 + 8 + 8 + 8 = 56$

*Example:* 10(3) is really like adding 10 threes.
$3 + 3 + 3 + 3 + 3 + 3 + 3 + 3 + 3 + 3 = 30$

It's faster to add 3 tens, and the answer will be the same, so:

$$10 + 10 + 10 = 30$$

This example illustrates the commutative property of multiplication which was discussed in Chapter Two.

$$10(3) = 3(10)$$

*Examples:*

| *Multiplication* | *Addition* | *Result* |
|---|---|---|
| 5(6) = | 6 + 6 + 6 + 6 + 6 | = 30 |
| 5(5) = | 5 + 5 + 5 + 5 + 5 | = 25 |
| 5(4) = | 4 + 4 + 4 + 4 + 4 | = 20 |
| 5(3) = | 3 + 3 + 3 + 3 + 3 | = 15 |
| 5(2) = | 2 + 2 + 2 + 2 + 2 | = 10 |
| 5(1) = | 1 + 1 + 1 + 1 + 1 | = 5 |
| 5(0) = | 0 + 0 + 0 + 0 + 0 | = 0 |

In each case, the **FACTOR** on the right has been decreased by 1. As a result, the **PRODUCT** was decreased by 5.

Continuing the pattern, then:

$$5(-1) = (-1) + (-1) + (-1) + (-1) + (-1) = -5$$
$$5(-2) = (-2) + (-2) + (-2) + (-2) + (-2) = -10$$
$$5(-3) = (-3) + (-3) + (-3) + (-3) + (-3) = -15$$

Multiplication of real numbers is commutative, that is, the product is the same regardless of the order in which you multiply:

so if $5(-1) = -5$, then
$(-1)(5) = -5$ and

$(-2)(5) = -10$
$(-3)(5) = -15$ and so forth.

Using the repetitive addition concept from above, and the commutative property for multiplication, we may write:

$$
\begin{aligned}
-3(4) &= 4(-3) = (-3) + (-3) + (-3) + (-3) = -12 \\
-3(3) &= 3(-3) = (-3) + (-3) + (-3) = -9 \\
-3(2) &= 2(-3) = (-3) + (-3) = -6 \\
-3(1) &= 1(-3) = -3 \\
-3(0) &= 0(-3) = 0
\end{aligned}
$$

While the FACTOR on the right decreases by 1, the PRODUCT is increased by 3; so continuing to reduce the factor by 1 and increase the product by 3, we find:

$$
\begin{aligned}
-3(-1) &= 3 \\
-3(-2) &= 6 \\
-3(-3) &= 9 \quad \text{etc.}
\end{aligned}
$$

***SUMMING UP:*** **The product of any two real numbers is POSITIVE if the signs of the factors are the SAME.**

**The product of any two real numbers is NEGATIVE if the signs of the factors are DIFFERENT.**

**Exercise Set 3.3 A**

Multiply

| | | |
|---|---|---|
| 1. $(4)(9)$ | 2. $(-4)(9)$ | 3. $(4)(-9)$ |
| 4. $(-4)(-9)$ | 5. $(10)(-2)$ | 6. $(-10)(-2)$ |
| 7. $(10)(2)$ | 8. $(-2)(10)$ | 9. $-7(3)$ |
| 10. $-7(-3)$ | 11. $-21(8)$ | 12. $21(-8)$ |
| 13. $-21(-8)$ | 14. $21(8)$ | 15. $(-12)(-12)$ |
| 16. $12(-12)$ | 17. $-12(12)$ | 18. $12(12)$ |

**Exercise Set 3.3 A:** ***(continued)***

19. $5(3)(2)$
20. $5(-3)(2)$
21. $-5(-3)(-2)$
22. $-5(-2)(-3)$
23. $-5(3)(-2)$
24. $5(3)(2)$
25. $(12)(-2)(-4)$
26. $(-12)(2)(-4)$

**Exercise Set 3.3 B**

Supply the missing number with its appropriate sign:

1. $-3(?) = 12$
2. $6(-2) = (?)$
3. $5(-4) = (?)$
4. $(?)(8) = -16$
5. $12(?) = -36$
6. $9(-5) = (?)$
7. $12(25) = (?)$
8. $15(?) = -45$
9. $(?)(16) = -32$
10. $16(-2) = (?)$
11. $4(-10) = (?)$
12. $-4(?) = 40$
13. $-7(?) = 14$
14. $2(-7) = (?)$
15. $11(-5) = (?)$
16. $5(?) = -55$
17. $-9(?) = 72$
18. $-9(8) = (?)$
19. $-1(13) = (?)$
20. $-13(?) = 13$
21. $-5(10) = (?)$
22. $15(?) = -15$
23. $25(-2) = (?)$
24. $78(-12) = (?)$

## 3.4 *Division of Real Numbers in Integer Form*

Since division and multiplication are essentially similar operations, the rules of signs which apply to multiplication also apply to division. That is:

The quotient of any two real numbers will be POSITIVE if the signs of the dividend and divisor are the same.

The quotient of any two real numbers will be NEGATIVE if the signs of the dividend and divisor are different.

*Examples:*

$$8 \div 2 = \text{one half of } 8 = 4$$
$$-8 \div 2 = \text{one half of } -8 = -4$$
$$8 \div -2 = -(\text{one half of } 8) = -4$$
$$-8 \div (-2) = -(\text{one half of } -8) = 4$$

**Remember the rules for multiplication and division:**
IF THE SIGNS ARE ALIKE, THE ANSWER IS POSITIVE.
IF THE SIGNS ARE DIFFERENT THE ANSWER IS NEGATIVE.

**Exercise Set 3.4 A**

Divide:

1. $54 \div 9$
2. $-54 \div 9$
3. $54 \div -9$
4. $-54 \div -9$
5. $10 \div -2$
6. $10 \div 2$
7. $-10 \div -2$
8. $-10 \div 2$
9. $-7 \div -3$
10. $7 \div -3$
11. $-7 \div -3$
12. $-7 \div 3$
13. $-21 \div 8$
14. $21 \div 8$
15. $-21 \div -8$
16. $21 \div -8$
17. $-12 \div -12$
18. $-12 \div 12$
19. $12 \div -12$
20. $12 \div 12$
21. $90 \div -10$
22. $-84 \div -4$
23. $35 \div -7$
24. $100 \div -20$
25. $-70 \div -7$
26. $-63 \div 3$

**Exercise Set 3.4 B**

Perform the indicated operations:

1. $3(-8) \div 4$
2. $-3(8) \div (-4)$
3. $5(12) \div 10$
4. $-5(12) \div (-10)$
5. $-2(-5) \div -10$
6. $-2(5) \div 10$
7. $2(-5) \div -10$
8. $6(-15) \div 2(9)$
9. $6(3) \div -2(9)$
10. $-6(-3) \div 2(9)$
11. $[2 \div (-6)](3)$
12. $[(-2) \div (-6)](-3)$
13. $[(-2) \div 6](3)$
14. $[15 \div (-3)] \div (-10)$

**Exercise Set 3.4 B *(continued)***

15. $[(-15) \div (-3)] \div (-10)$
16. $(15 \div 35)(114)$
17. $8(-3) \div -4$
18. $5(12) \div -10$
19. $-167 \div [(-24)(103)]$
20. $7 \div [(-6)(-7)]$

## 3.5 *Absolute Value*

The absolute value of a number is its distance from zero without regard to direction. Since 5 and $-5$ are equally distant from zero, their absolute values are equal.

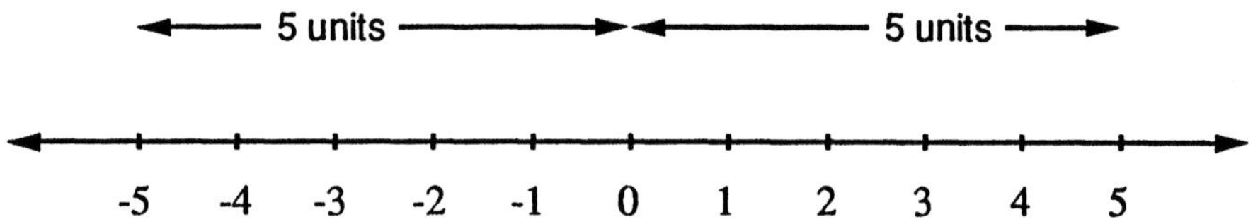


The absolute value of a number is expressed by placing the number between two bars.

*Examples:* $|-5| = 5 \qquad |3| = 3 \qquad -|-7| = -7$

$|6 - 8| = |-2| = 2$

$5 + |-3| = 5 + 3 = 8$

$3\,|-4| = 3(4) = 12$

**Exercise Set 3.5**

Evaluate:

1. $|-4|$
2. $-|-3|$
3. $-|9 - 4|$
4. $|2 - 6|$
5. $|8 + 3|$
6. $|4 - 6|$
7. $-|3 - 5|$
8. $|-7 + 5|$
9. $|-3|$
10. $6 - |-4|$
11. $|-2| + |8|$
12. $3 + |-9|$
13. $12 - |-7|$
14. $-5 + |-5|$

**Exercise Set 3.5** *(continued)*

15. $3\{5 - |-8|\}$
16. $-4[|-4|]$
17. $7(-2|-2|)$
18. $|9 - 13|$
19. $12(|-1|)$
20. $\{-2 + |-3|\}$
21. $(|-6|)^2$
22. $2[-|-9|]$
23. $-|-5| + 30$
24. $(|-3|)^2$
25. $-|-8| + 8$
26. $-[-4|-2|]$

## 3.6 *Raising an Integer to a Power*

In Chapter One, you investigated how to use your calculator to find the value of a number raised to a power. This is a review.

*Example:* $6^3 = 6 \cdot 6 \cdot 6 = 216$

*Example:* $7^4 = 7 \cdot 7 \cdot 7 \cdot 7 = 2401$

**Exercise Set 3.6**

Evaluate the following:

1. $12^4$
2. $8^3$
3. $5^6$
4. $4^7$
5. $2^8$
6. $10^6$

## 3.7 *Order of Operations for Simplifying Expressions*

It is important to know the proper order for simplifying mathematical expressions.

***The order of operations is as follows:***

| | |
|---|---|
| **First:** | **Simplify parentheses** |
| **Second:** | **Take roots or powers** |
| **Third:** | **Do multiplication and/or division** |
| **Finally:** | **Do addition and/or subtraction** |

*Example* $3(8-4)^2 + 2(7+3) =$
$3(4)^2 + 2(10) =$
$3(16) + 2(10) =$
$48 + 20 =$
which simplifies to 68

*Example* $6[2(8-12) + 5(3+2)] =$
$6[2(-4) + 5(5)] =$
$6[(-8) + 25] =$
$6[17] =$
which simplifies to $= 102$

*Example* $\frac{1}{5}\{6 - 3[7 - 13] - 4\} =$
$\frac{1}{5}\{6 - 3[-6] - 4\} =$
$\frac{1}{5}\{6 + 18 - 4\} =$
$\frac{1}{5}\{20\} =$
which simplifies to $= 4$

**Exercise Set 3.7 A**

Simplify:

1. $6 + 5\{7 - [2(5-8)]\}$
2. $3[5-7] + 2[8-2]$
3. $3\{6(3-2) - 8\}$
4. $(-5+8) - (16-6)$
5. $2\{16(5-7) - 4(8)\}$
6. $\{-(-5) + (-2)^2\}$
7. $3\{7 - 5[3-8] - 2\}$
8. $-14(3+2) - (3)^2$
9. $10\{3(8-10) + 6(2+4)\}$
10. $15 + 3\{8 - (2+5)\}$

**Exercise Set 3.7 B**

Follow the correct order of operations and evaluate the following:

1. $12 + [4\{3 - (7+2)\}]$
2. $5(|-8+3|)$
3. $2 + 3(|-2-3|)$
4. $(|-5| + |-16|) - (|14-7|)$
5. $(-4)(-4-12) - [-3(|-8|)]$

**Exercise Set 3.7 C**

Evaluate the following:

1. $3 + (-2) - (-6)$
2. $10(8) - 6(2)^3$
3. $12 - (-2) + 4$
4. $2[(3 - 5)^2]$
5. $15 + (-5) + 5$
6. $5 + (8 - 4)^2$
7. $-7 + (2 - 3)$
8. $7\{3 - (4 - 5) + |-7|\}$
9. $2[6(7 - 4)]$
10. $\{10 + |-10|\} + [3 - (-2)]$

## 3.8 *Word Problems*

Review Chapter One Section 1.8 before you try these word problems. Read them carefully before attempting them and check your work whenever possible.

**Exercise Set 3.8**

Solve the following problems:

1. The temperature at 2 A.M. at Mount Kateri point was $-3°$ F. By 7 A.M. it had dropped 25° F. What was the temperature at 7 A.M.?

2. The temperature in Garden City at 12 noon was 85°. If the temperature dropped to 78° by 8 P.M., express the drop in temperature as a signed number.

3. Louise made some purchases and charged the following amounts to her Master Card: $180.75, $22.34, $98.57 and $56.89. At the end of the month, she made a payment of $200. Express Louise's Master Card balance as a signed number.

4. Kay wrote the following checks to pay some of her monthly bills: $138.97, $35.00, $28.54, $156.00, and $140.00. At the time, she had $498.17 in her account. Express the balance as a signed number.

5. At the beginning of a business day, a share of a certain stock was valued at \$52.75. By noon it had risen in value \$2.35; then, later in the afternoon it dropped in value \$3.50. What was the value of a share of stock at the close of the business day?

6. Tom Jefferson's company gives him \$100.00 a day for travelling expenses. One day he spent \$65.00 for car rental, \$18.50 for a late lunch, and \$12.87 for gasoline. Did he have sufficient funds to cover his expenses? If so, how much was left?

7. John Thomas owns 120 shares of stock in a utility company. One business day the stock declined in value \$1.25 per share. Express his loss as a "signed number".

8. Joan Martin had a balance of \$335.00 due on one of her credit cards. She charged additional purchases of \$23.55, \$12.37, and \$17.99. Then she made a payment of \$250.00. What is her current balance due?

9. If the temperature in a certain area rose from 12° below zero to 15° above, what was the resulting increase in temperature?

10. If you were allowed \$75.00 a day for business expenses, how much would be left for lunch if you had spent \$43.50 on carfare, \$18.00 on supplies and \$2.35 on telephone calls?

11. A college freshman borrowed \$150.00 from his father to buy some computer supplies. He purchased the supplies for \$85.75, but decided to keep \$10.00 toward other college needs before giving the "change" to his father. How much could he give his Dad, and how much did he still owe?

## CHAPTER EXERCISES

### Section 3.1

Add the following:

1. $(+3) + (+7)$
2. $(-1) + (-4)$
3. $(+10) + (-2)$
4. $(-5) + (-2)$
5. $13 + (+3)$
6. $(-20) + 5$
7. $-7 + (-1)$
8. $-14 + 6$
9. $3 + (-8)$
10. $(+6) + (+2) + (+3)$
11. $(-1) + (+10) + (-1)$
12. $(+6) + (-7) + (-4)$
13. $(-3) + (-5) + (-4)$
14. $(+11) + (+12) + (-6)$
15. $20 + (-7) + (-5)$
16. $-18 + (+8) + (+2)$
17. $-1 + (14) + (-22)$
18. $(+16) + (+2) + 0$
19. $7 + 6 + (-5)$
20. $(-13) + (-5) + 6$

### Section 3.2

Perform the indicated operations:

21. $(17) - (5)$
22. $10 - (-2)$
23. $(-3) - (6)$
24. $12 - (-5)$
25. $27 - (-4)$
26. $(-3) - (-4)$
27. $(-9) - 1$
28. $(-5) - 10$
29. $(+8) - 4$
30. $52 - (-2)$
31. $-13 - (-10)$
32. $-11 - (20)$
33. $-7 - (-3) - (-5)$
34. $19 - (-5) - (-2)$
35. $15 + (6) - (3)$
36. $11 + (-3) + 6$
37. $-100 + 25 - 10$
38. $-17 - (-13) + (-2)$
39. $-400 - (-200) + (-10)$
40. $36 - (-7) + (-8) - (-5)$

**Section 3.3**

Multiply:

41. $(8)(3)$
42. $(6)(-2)$
43. $(-5)(6)$
44. $(-11)(-3)$
45. $(-7)(-5)$
46. $(10)(21)$
47. $(15)(-2)$
48. $(-12)(5)$
49. $200(-5)$
50. $(-3)(-11)$
51. $(-16)(-3)$
52. $(46)(-10)$
53. $7(-11)$
54. $-5(8)$
55. $3(201)$
56. $-4(-16)$
57. $-15(-3)(2)$
58. $14(3)(-1)$
59. $26(-3)(-10)$
60. $-7(-5)(-11)$

**Section 3.4**

Perform the indicated operations:

61. $27 \div 3$
62. $-26 \div 2$
63. $-82 \div 4$
64. $18 \div 9$
65. $63 \div (-3)$
66. $87 \div (-3)$
67. $-100 \div 25$
68. $-200 \div 5$
69. $77 \div (-7)$
70. $36 \div (-9)$
71. $(-15) \div (-5)$
72. $(-88) \div (-4)$
73. $25 \div (-2)$
74. $17 \div (-5)$
75. $108 \div 5$
76. $-450 \div 20$
77. $-82 \div (-5)$
78. $0 \div (-6)$
79. $7 \div (-6)$
80. $-150 \div (-25)$

### Section 3.4

Perform the indicated operations:

81. $23(5)$
82. $-24(10)$
83. $17(-3)$
84. $-11(-6)$
85. $-4(200)$
86. $105(-5)$
87. $2(8) \div 4$
88. $32(-1) \div 8$
89. $15 \div (-3)$
90. $(-18 \div 9)(5)$
91. $[(-120) \div (-6)](-2)$
92. $(10 \div 2)(-7)$
93. $-27(3) \div (-9)$
94. $26(-2) \div 13$
95. $-18 \div 9(2)$
96. $35 \div (-7)(5)$
97. $10(-5) \div 2$
98. $-48(-3) \div (-4)$
99. $-12(-5) \div 7$
100. $-105 \div (-5) \div (-3)$

### Section 3.5 and 3.6

Evaluate:

101. $|7|$
102. $|-12|$
103. $|7 - 5|$
104. $-|15|$
105. $-|-4|$
106. $|8 - 10|$
107. $3|-2|$
108. $|-5| + |10|$
109. $|15| - |-6|$
110. $-|4| - |-5|$
111. $-2(|8| - |-2|)$
112. $26 - |-5|$
113. $|-21| + 3(-11)$
114. $|-5|^2$
115. $|8| \div |-2|$
116. $|-4|^3$
117. $|121| + |-2|$
118. $|27| - |-3|$
119. $3|-7| - 2^3$
120. $2|-5| - 6|-3| + 7|-4|$

**Section 3.7**

Simplify:

121. $-5(2) + 2|-5|$

122. $3 + 2(4 - 10)$

123. $-8 + 7(-3) + 30$

124. $(-3 + 7) - (21 - 7)$

125. $-2(12 - 5) + 6(-3 + 10)$

126. $-26[8 + 2 - (6 - 4)]$

127. $-5(-6)^2$

128. $4(-25 + 2\{-5 - 6\})$

129. $3^2 - 4(25 - 27)^2$

130. $15(-16 - 4)$

131. $-100 + [83 - 6]$

132. $4\{3 + 5(8 - 2[-3])\}$

133. $|-7| - 3(2)^2$

134. $[-10 + 8] - [|-4| + 6]$

135. $5^2 - |18 - 4|$

136. $(-8)^2 + 5[15 - (3)^2]$

137. $144 - |-7|^2$

138. $\{-7 + [4 + |-8|] - 10\}$

139. $|-5|^2 + 6[15(-3) + 7]$

140. $10\{|-12| - 3[15(-2) - 3] + 4\}$

**Section 3.8**

Solve the following problems:

**Answers**

141. The temperature in Sunnyvale is usually 10 degrees lower than the temperature in the nearest city. If Sunnyvale temperatures average 70 degrees during a certain month, what is the average temperature in the nearest city that month?

141. ________

142. Each week Mr. Rivers deposits his pay check of $350.00 in a credit union account. This is after he has had payroll deductions amounting to $134.68. What is his weekly pay before deductions?

142. ________

143. Andrew Mann owns 150 shares of stocks in a certain company. One day each share was valued at $2.33, but by noon the value had dropped to $2.23 per share. How much loss did Mr. Mann suffer on his 150 shares that day?

143. ________

**CHAPTER TEST** *Form 3A*

Add the following:

1. $12 + 14$
2. $(-9) + (-6)$
3. $(-17) + 18$
4. $-15 + (-8) + (-91)$
5. $105 + (87) + (-134)$

Subtract the following:

6. $15 - 126$
7. $-19 - 14$
8. $-51 - 16 - 19$
9. $84 - (-16) - 100$
10. $-46 - (-46) - (-3)$

Multiply:

11. $(+7)(+12)$
12. $(-11)(5)$
13. $-4(-19)$
14. $(41)(78)(-6)(0)(-9)$
15. $42(-6)(-1)(2)$

Perform the indicated operation:

16. $42 \div (-6)$
17. $(-510) \div (-5)$
18. $(40)(7) \div (-10)$
19. $(125 \div 15) \div 5$
20. $(-120)(-40) \div (15)(-8)$

Find the absolute values and evaluate:

21. $|-36|$
22. $-|19 - 17|$
23. $3\,(|-8 - 5|)$
24. $-|35| - |-35| + 70$

Evaluate the following:

25. $16^2$
26. $|-2|^2$
27. $(-24)^2$
28. $(-3)^4$
29. $(-2)^5$
30. $4\,(3 + 7)^2 \div 5\,(-8 - 2)^2$

CHAPTER

# 4 Decimal Forms of Real Numbers

## 4.0 *Vocabulary Preview*

Define these words without using a dictionary. Formulate their meanings from past experience or knowledge.

| | | |
|---|---|---|
| **common** | **least** | **product** |
| **decimal** | **multiple** | **quotient** |
| **factor** | **non-integer** | **real** |
| **fraction** | **prime** | **subtrahend** |
| **inverse** | | |

## 4.1 *Adding Real Numbers in Decimal Form*

Since decimal fractions are real numbers, the same rules apply for addition of decimals as apply for integers; namely,

a. If the signs are alike, ADD and use the common sign for the answer.

b. If the signs are different, SUBTRACT and use the sign of the number with the higher absolute value for the answer.

*Examples:* $56.732 + 47.71 = 104.442$
signs are alike and the common sign is +

$(-56.732) + (-47.71) = -104.442$
signs are alike and the common sign is −

$56.732 + (-47.71) = 9.022$
signs are different
+56.732 has greater absolute value

$(-56.732) + (47.71) = -9.022$
signs are different
−56.732 has greater absolute value.

*Example:* Add $(-87.5) + 45.3 + (-65.7) + (22.6)$
First combine −87.5 and −65.7: the result is −153.2

Next combine 45.3 and 22.6; the result is 67.9

Combining those two results gives −85.3

*Example:* Add: $(-43.5) + 6.75 + (-33.4) + 4.75$

combine the positive numbers first:
$6.75 + 4.75 = +11.50$

next combine the negative numbers:
$-43.5 + (-33.4) = -76.9$

combining the two results:
$+11.50 + (-76.9) = -65.4$

**Exercise Set 4.1**

Add the following:

1. $56.32 + (36.124)$
2. $-56.32 + (-36.124)$
3. $-56.32 + (36.124)$
4. $56.32 + (-36.124)$
5. $142 + 35.6$
6. $-142 + (-35.6)$
7. $142 + (-35.6)$
8. $-142 + 35.6$
9. $0.098 + (-0.01)$
10. $-0.098 + (-0.01)$
11. $-0.098 + 0.01$
12. $0.098 + 0.01$
13. $-7.6 + -7.6$
14. $7.6 + (-7.6)$
15. $-7.6 + 7.6$
16. $7.6 + 7.6$

## 4.2 *Subtracting Real Numbers in Decimal Form*

**The rules for subtraction of non-integers are the same as those for subtracting integers, since they are both real numbers:**

**a. change the operation of subtraction to its inverse operation of addition**

**b. add the inverse of the subtrahend**

**that is**

**CHANGE THE SIGN OF THE SUBTRAHEND AND ADD.**

*Example:* $53.87 - 42.65 = 53.87 + (-42.65)$
$= 11.22$

*Example:* $53.87 - (-42.65) = 53.87 + 42.65$
$= 96.52$

*Example:* $-53.87 - 42.65 = -53.87 + (-42.65)$
$= -96.52$

*Example:* $-53.87 - (-42.65) = -53.87 + 42.65$
$= -11.22$

**Exercise Set 4.2 A**

Subtract the following:

1. $56.32 - 36.124$
2. $-56.32 - (-36.124)$
3. $-56.32 - 36.124$
4. $56.32 - (-36.124)$
5. $142 - 35.6$
6. $-142 - (-35.6)$
7. $142 - (-35.6)$
8. $-142 - 35.6$
9. $0.098 - (-0.01)$
10. $-0.098 - (-0.01)$
11. $-0.098 - 0.01$
12. $0.098 - 0.01$
13. $-7.6 - (-7.6)$
14. $7.6 - (-7.6)$
15. $-7.6 - 7.6$
16. $7.6 - 7.6$

COMPARE EACH OF THE ABOVE CALCULATIONS WITH THE CORRESPONDING ADDITION CALCULATIONS OF THE PREVIOUS EXERCISE.

DO YOU SEE THE INVERSE OPERATIONS AT WORK?

**Exercise Set 4.2 B**

Use your calculator to perform the indicated operations. Use the "change sign" key +/− when necessary.

1. $0.03 + 1.82 + (-1.01)$
2. $4 + 0.4 + 0.8$
3. $-2.7 + (-3.8) + (-0.1)$
4. $0.81 + (-2.86) + 1.95$
5. $3.68 + (-10.75)$

**Exercise Set 4.2 B** ***(continued)***

6. $76 + (-85.07) + 0.02$
7. $-21.76 + (-20.5) + (-1.26)$
8. $2.75 + 0.75 + 2.6$
9. $0.03 - (-1.82) - 1.01$
10. $4 - (-0.4) - (-0.8)$
11. $-2.7 - 3.8 - 0.1$
12. $0.81 - 2.86 - (-1.95)$
13. $3.68 - 10.75$
14. $76 - 85.07 - (-0.02)$
15. $-21.76 - 20.5 - 1.26$
16. $2.75 - (-0.75) - (-2.6)$

COMPARE YOUR ANSWERS FOR 1 THROUGH 8 WITH THOSE OF 9 THROUGH 16. WHY SHOULD THEY BE THE SAME?

## 4.3 *Multiplication and Division of Decimal Forms*

The rules for multiplication and division of non-integer decimal forms are the same as the rules for integers; namely,

**If the signs of the two numbers are the same, the product or quotient will be positive.**

**If the signs are different, the product or quotient will be negative.**

*Examples:*

$$(0.09)(-1.23) = -0.1107$$

$$(-23.5)(-3.1) = 72.85$$

$$(-0.1107) \div 0.09 = -1.23$$

$$72.85 \div -23.5 = -3.1$$

**Exercise Set 4.3**

Perform the indicated operations.

1. $(-7.4)(6.54)$
2. $-48.9 \div 2.5$
3. $(12.5)(-0.2)$
4. $6.3 \div -4.53$
5. $(-6.7)(-4.5)$
6. $14.78 \div -2.3$
7. $(-3.55)(-8.65)$
8. $-35.6 \div 4.56$
9. $0.3(-0.8)$
10. $-0.3 \div -0.8$
11. $2(-8.4) \div -2.5$
12. $0.6(-9.8) \div 2$
13. $0.5(-2.1)$
14. $0.5(-2.1 \div 2.1)$

## 4.4 *Powers and Square Root*

There is no difference between finding powers and roots of integers or non-integers in decimal form. Review the procedure from Chapter Three.

*Examples:* To evaluate $\sqrt{12.35}$, enter 12.35 into your calculator, then enter $\sqrt{\phantom{x}}$ followed by =. The result is 3.51.

To evaluate $(3.4)^2$, enter $3.4 \times 3.4 =$
The result is 11.56.

$$(0.09)^3 = 0.09 \times 0.09 \times 0.09 = 0.000729$$

$$\sqrt{0.025} = 0.158$$

$$(1.2)^5 = 1.2 \times 1.2 \times 1.2 \times 1.2 \times 1.2$$
$$= 2.48832$$

**Exercise Set 4.4**

Evaluate, rounding answers to two decimal places.

1. $\sqrt{54.17}$
2. $(7.5)^3$
3. $\sqrt{12.5}$
4. $(3.65)^2$
5. $\sqrt{98.1}$
6. $(4.69)^3$

## 4.5 *Order of Operations*

**Recall the order of operations for real numbers.**

| | |
|---|---|
| **Parentheses** | **—simplify** |
| **Exponents** | **—find powers and roots** |
| **Multiply or Divide** | **—from left to right** |
| **Add or Subtract** | **—from left to right** |

*Example:*

$$\begin{aligned} &3.2[5.1(3.4 - 1.5) - 5.6] = \\ &3.2[5.1(\ 1.9\ ) - 5.6] = \\ &3.2[\ 9.69 - 5.6] = \\ &3.2[\ 4.09\ ] = 13.088 \end{aligned}$$

*Example:*

$$\begin{aligned} &-2[3(8 - 5) + (6 - 8)] + 2[9 + (-2)] = \\ &-2[3(3) + (-2)] + 2[7] = \\ &-2[\ 9 + (-2)\ ] + 14 = \\ &-2[7] + 14 = \\ &-14 + 14 = 0 \end{aligned}$$

*Example:*

$$\begin{aligned} &3\{(.4)^2 - 2(.6)\} - \{5|-1.3|\} + (.4 - .7)^2 = \\ &3\{.16 - .12\} - \{5(1.3) + (-.3)^2\} = \\ &3\ (.04) - \{\ 6.5 + .09\ \} = \\ &.12 - 6.59 = -6.47 \end{aligned}$$

**Exercise Set 4.5**

Evaluate the following:

1. $3.1 + [4.2\{3.5 - (7.2 + 2.8)\}]$
2. $0.2\{6.5 - 3.4(7.8 - 5.4)\}$
3. $3.8(2.3)^2 - 5.2(1.6)^2$
4. $6.45[(2.20^2 - 8.4) + (6.1)^2]$
5. $1.25 - 3.46(1.5)^2 + 6.1(2.3)^2$
6. $6.5 - 2.1\{|-3.4| - 6.2(3.4)$

## CHAPTER EXERCISES

### Section 4.1

Add:

1. $17.2 + 3.6$
2. $-15.75 + 5.25$
3. $3.52 + 11.25$
4. $84.50 + (-25.25)$
5. $-8.4 + (-7.6)$
6. $-6.83 + 3.5$
7. $-17.1 + (-1.3)$
8. $-.5 + 11.2$
9. $7.36 + (1.01)$
10. $-23.6 + (.06)$
11. $-22.3 + 22.3 + .06$
12. $4.3 + (-2.14) + 3$
13. $-21.5 + (-5.2) + 3.0$
14. $48.0 + .02$
15. $-11.99 + 3.16 - 4.7$
16. $-.001 + .01 + (-.1)$
17. $-34.7 + (-2.1) + (-1.6)$
18. $127.6 + (-30.5)$
19. $-347.83 + (-45.95)$
20. $56.1 + (-1.05)$

### Section 4.2

Subtract the following:

21. $18.56 - 8.24$
22. $-31.5 - (3.4)$
23. $-7.1 - (2.6)$
24. $-11.34 - (-6.14)$
25. $93.7 - 50.1$
26. $-39.6 - 8.5$
27. $123.99 - 4.79$
28. $-6.01 - (-1.70)$
29. $6.01 - 3.49$
30. $-401.7 - (84.9)$
31. $.07 - .52$
32. $-.008 - (.049)$
33. $-69.7 - (-3.84)$
34. $11.1 - 5.64$
35. $-30.4 - 8.96$
36. $-15.62 - (-5.98)$
37. $-15.703 - (-15.703)$
38. $-22.7 - (4.0)$
39. $134.6 - (-79.48)$
40. $35.01 - (6.3)$
41. $4.6 - 2.73 - 8.24$
42. $-12.84 - (-7.1) - 8.2$
43. $-11.01 - (-7.5) + 8.79$
44. $-11.273 + 8.71 - (-7.9)$
45. $-12.4 - 41 - 8.92$
46. $7.6 - 4.03 - (-1.045)$

## Section 4.3

Perform the indicated operations:

47. $(3.2)(1.3)$
48. $(-8.7)(1.2)$
49. $(5.12)(+3.7)$
50. $(-4.1)(7.9)$
51. $(8.31)(2.1)$
52. $(16.3)(-6.1)$
53. $(9.2)(-4.2)$
54. $(-5.1)(-.6)$
55. $(-8.1)(-.3)$
56. $(.05)(-.092)$
57. $(14.50) \div (2.5)$
58. $-56.1 \div 3.9$
59. $132.7 \div (-5.1)$
60. $-37.6 \div (-3)$
61. $(.4) \div (.2)$
62. $-27.9 \div 9.1$
63. $(-11.6)(.5)$
64. $8.75 \div (-.9)$
65. $-70.5 \div (-.51)$
66. $(17.1)(-.003)$

## Section 4.4

Find the roots to two decimal places.

67. $\sqrt{17.45}$
68. $\sqrt{6.72}$
69. $\sqrt{.79}$
70. $\sqrt{4.5}$
71. $\sqrt{104.63}$
72. $\sqrt{11.072}$
73. $\sqrt{476.4}$
74. $\sqrt{2271.3}$
75. $\sqrt{8.10}$
76. $\sqrt{.404}$
77. $\sqrt{-12.5}$
78. $\sqrt{-65.8}$

Your calculator displayed ERROR on the last two problems. Why?

## Section 4.4

Evaluate. Round your answers to two decimal places.

79. $(2.71)^2$
80. $(8.64)^3$
81. $(15.04)^3$
82. $(.71)^2$
83. $(1.032)^2$
84. $(41.7)^3$
85. $(1.07)^5$
86. $(.08)^3$

## Section 4.5

Evaluate the following:

87. $1.5 + (2.1 + 7.6)$
88. $2(7.8) - 6.2$
89. $21.7 - 2(11.6)$
90. $-8.09 - 3(4.5) + 7$
91. $(-52.12)(-4.3 + 7.91)$
92. $3.1\{26.7(8.04)\}$
93. $(6.1)^2 - 4.7(8.11)$
94. $(3.2)^2 - \sqrt{4}.6 + 12$
95. $-11.45 - (6.4)^2 + \sqrt{3}$
96. $(-7.1)(18.0) - (62.1)(2.7)$
97. $(.005)^2 - (1.1)(-4.2)$
98. $-67.3(.03)^2 \div (.38)$
99. $(-32.1)^2 \div (-4.1) + 2.76$
100. $\{(-19.7) + .07\}^2 \div [(41.5) - (-6.2)]$

## CHAPTER TEST *Form 4A*

Perform the indicated operations. Round your answers to three decimal places.

1. $2.54 + 3.67$
2. $8.14 + 9.0102$
3. $-175 + 128.6$
4. $-79.2 + (-8.9)$
5. $-6.13 + (-2.54)$
6. $-0.92 - 0.92$
7. $78.6 - 38.9$
8. $-52.5 - 7.2$
9. $11.1 - 0.22$
10. $175.6 - (-5.1)$
11. $-10.01 - (-21.3)$
12. $-23.6 - 0.002$
13. $-1.96 + 3.64 - (-1.92)$
14. $-1.84 + 1.62 + (-3.12)$
15. $-79 - 9.1 - 6.2$
16. $-84 - 8.1 - 0.63$
17. $9.002 - 0.0002 + (-9.001)$
18. $-11.076 - 0.076 + 3.1$
19. $(14.1)(1.4)$
20. $(6.75)(7.13)$
21. $(1.75)(-12.4)$
22. $(-0.6)(107.1)$
23. $(-5)(1.72)(-0.6)$
24. $(112.6)(-1.3)(-0.09)$
25. $6.75 \div 45.6$
26. $-1.091 \div 3.01$
27. $-48.6 \div -0.01$
28. $-34.6 \div (-6.01)$
29. $14(-6.2) \div 3.01$
30. $-6(4.1) \div 2(-6.31)$
31. $18.14 \div 2(-1.3)$
32. $(-8.14) \div (-3.6)(-7.2)$

Evaluate. Round your answers to three decimal places.

33. $(13.2)^2$
34. $\sqrt{13.2}$
35. $(114.19)^3$
36. $(-1.64)^3$
37. $\sqrt{12.11}$
38. $(-0.101)^3$

Evaluate. Round your answers to three decimal places.

39. $-3(4.1)^2$

40. $2(-1.76)^2$

41. $-6.1\sqrt{3.87}$

42. $(12.54)^2(\sqrt{6})$

43. $(-7.1)^2(\sqrt{1.01})$

44. $(0.16)^3(\sqrt{1024})$

Use the correct order of operations to simplify the following:

45. $5 + 6[3 + 4(2 - 11)$

46. $-2[1 + 397 + 5)] + 9$

47. $12 + 4[1 - (6 + 11)]$

48. $-100 - [7 - (-8 - 9)]$

49. $-3(7) + [-4 + 16(-1 + 21) + 5]$

50. $3[11 - \{9 - 2(17 - 8)\}]$

**CUMULATIVE EXERCISES Chapters 1–4 *Set A***

Name the property of real numbers illustrated by the following:

**Answers**

1. $11 + 23 = 23 + 11$ 1. ________
2. $3(15 + 6 = 3(15) + 3(6)$ 2. ________
3. $(10 + 12) + 20 = 10 + (12 + 20)$ 3. ________
4. $(25)\left(\frac{1}{25}\right) = 1$ 4. ________
5. $-18 + 18 = 0$ 5. ________

Combine and simplify:

**Answers**

6. $14 + (-5) + 6$ 6. ________
7. $12(4) + 16 - (-8)$ 7. ________
8. $3^2 + |-7| - (-24)$ 8. ________
9. $-8\{2(4.3) - 5\}$ 9. ________
10. $\{-105 \div (-5)\} \div 3$ 10. ________
11. $7.1 - [(2)^3 - 4.3 + 3(5)]$ 11. ________
12. $(1.05)^2 + 3|12 - 9| + (-3.1) - 8.4$ 12. ________
13. $-6.22 - 4.3 + (-7.7)$ 13. ________
14. $(2.41 - 6.7) - 2(-8.4 - .5)$ 14. ________
15. $1.2[16.4(2.2 - 8.7)]$ 15. ________

Perform the indicated operations:

**Answers**

16. $12.1(-4.5)$ 16. ________
17. $-32.6 \div 2$ 17. ________
18. $(1.1)^2 - 6.3 + 7.4$ 18. ________
19. $22.5 - (-8.2) - (-4.5)$ 19. ________
20. $(-.6)(-3.0)$ 20. ________
21. $(-3.6) \div (-9.0)$ 21. ________
22. $\sqrt{15.7}$ 22. ________
23. $(.4)^3$ 23. ________
24. $(1.4)^2 - .7(-2) + \{8.4 \div 3\}$ 24. ________
25. $\sqrt{.58}$ 25. ________

## CUMULATIVE TEST Chapters 1–4 *Form A*

Match each of the expressions in Column A with the property in Column B that it illustrates.
NOTE: A property may be used more than once.

| Column A | Column B |
|---|---|
| 1. $18 \times (5 \times 24) = (18 \times 5) \times 24$ | a. Commutative Property |
| 2. $-45 + 45 = 0$ | b. Property of Inverses |
| 3. $6 \times (83 + 16) = (6 \times 83) + (6 \times 16)$ | c. Distributive Property |
| 4. $73 + 9 = 9 + 73$ | d. Associative Property |
| 5. $-7 \times \left(-\frac{1}{7}\right) = 1$ | e. Property of Identity |

Perform the indicated operations and simplify your answer.

**Answers**

6. $-50 - (-5) + 6$ 6. ________
7. $-13(-3) + 3(5)$ 7. ________
8. $-200 \div (-50)$ 8. ________
9. $2(16) - 7^2 + 3$ 9. ________
10. $-5\{-6(5)\}$ 10. ________
11. $-17(15)$ 11. ________
12. $-2\sqrt{25} + |-7|$ 12. ________
13. $-2.71 + (-7.9)$ 13. ________
14. $(.7)^2$ 14. ________
15. $17.6 \div (-2)$ 15. ________
16. $-.3(4.71)$ 16. ________
17. $(25.1 - 6.2) - (8.7 + 2.3)$ 17. ________
18. $(-18.4 \div 4.2) \div 3.1$ 18. ________
19. $-7.3(-4.6)$ 19. ________
20. $17.3 - 8.2 - (-1.1)^2$ 20. ________
21. $|-7.9| + 4(-3.8)$ 21. ________
22. $-8.93 + 3|-4|$ 22. ________
23. $[(-1.4)^2 + 9.6 - 1.6]$ 23. ________
24. $+2(-10.1) - 3.4 \div .1$ 24. ________
25. $8.2 - 10 + .4$ 25. ________

CHAPTER

# 5 Fraction Forms of Real Numbers

## 5.0 *Vocabulary Preview*

Define these words without using a dictionary. Formulate their meaning from past experience or knowledge.

| | | |
|---|---|---|
| **common** | **fraction** | **non-integer** |
| **denominator** | **infinite** | **numerator** |
| **divisible** | **integer** | **prime** |
| **factor** | **inverse** | **reciprocal** |

## 5.1 *Prime Factors*

Any positive integer may be written as a product of prime numbers. A **prime** number is a number which has only itself and 1 as **factors;** that is, a number which is not **divisible** by any other number except itself.

$2 = 2 \cdot 1$ is a prime number.

*NOTE:* 2 is the only prime number which is even.

$5 = 5 \cdot 1$ is a prime number.

$8 = 8 \cdot 1$ or $4 \cdot 2$ is **not** a prime number.

The list of prime numbers is infinite; that is, the list of prime numbers is a never ending sequence of numbers:

2 3 5 7 11 13 17 19 23 29 31 37 etc.

**Exercise Set 5.1 A**

Write the factors of those numbers which are **not** prime:

| | | |
|---|---|---|
| 1. 36 | 2. 54 | 3. 101 |
| 4. 43 | 5. 81 | 6. 49 |
| 7. 72 | 8. 39 | 9. 121 |

*Example:* Write the number 32 as a product of prime numbers.

$$\begin{aligned} 32 &= 2(16) \\ &= 2(2)(8) \\ &= 2(2)(2)(4) \\ &= 2(2)(2)(2)(2) = 2^5 \end{aligned}$$

*Example:* Write 35 and 42 as products of primes.

$35 = 5 \cdot 7 \qquad 42 = 2 \cdot 21 = 2 \cdot 3 \cdot 7$

*Example:* Write 36, 54 and 90 as products of primes.

$36 = 2(18) = 2(2)(9) = 2(2)(3)(3)$

$54 = 2(27) = 2(3)(9) = 2(3)(3)(3)$

$90 = 2(45) = 2(3)(15) = 2(3)(3)(5)$

*Example:* Find the greatest **common** factor of 35 and 42.

From the example above, we see that their only common factor is 7. It is, therefore, their greatest common factor.

*Example:* Find the greatest common factor of 36, 54 and 90.

From the example above, we see that EACH of the numbers contains the factors 2 and 3. In fact, 36, 54 and 90 each contains $2 \cdot 3 \cdot 3$ so we may say that their **greatest common factor** is $2 \cdot 3^2$ or $2 \cdot 9$ or 18.

**Exercise Set 5.1 B**

Write each number of the group as a product of primes. Then find their greatest common factor.

1. 22 55
2. 70 98
3. 60 99
4. 143 169
5. 68 42
6. 130 190 350
7. 63 126 189
8. 36 40 80 180
9. 4 15 21 24
10. 17 34 51 68

## 5.2 *Common Factors and Reducing Fractions*

**To reduce a fraction to simplest terms:**

**1. write the numerator and denominator as products of primes**

**2. divide off the fators which are common (alike) in numerator and denominator**

**3. multiply the factors which remain.**

*Example:* Using the numbers from examples in 5.2:

$$\frac{35}{42} = \frac{5 \cdot 7}{2 \cdot 3 \cdot 7} = \frac{5}{2 \cdot 3} = \frac{5}{6}$$

$$\frac{54}{90} = \frac{3 \cdot 18}{5 \cdot 18} = \frac{3}{5}$$ (There was no need to break the 18 down any further.)

$$\frac{75}{125} = \frac{3 \cdot 25}{5 \cdot 25} = \frac{3}{5}$$ (Again, there was no need to break down the 25.)

**Exercise Set 5.2 A**

Write the fractions in their simplest (reduced) form:

1. $\frac{7}{140}$ 2. $\frac{15}{60}$ 3. $\frac{9}{108}$

4. $\frac{15}{10}$ 5. $\frac{34}{186}$ 6. $\frac{115}{30}$

7. $\frac{6}{2}$ 8. $\frac{12}{172}$ 9. $\frac{10}{860}$

10. $\frac{8}{136}$ 11. $\frac{11}{121}$ 12. $\frac{54}{36}$

*Example:* Show that $\frac{7}{140}$ and $\frac{1}{20}$ are equal fractions by changing them to decimal form.

$7 \div 140 = 0.05$ $1 \div 20 = 0.05$

*Example:* Show that $\frac{35}{60}$ and $\frac{7}{12}$ are equal fractions by changing them to percents.

$(35 \div 60)100 = 58.\overline{3}\%$ $(7 \div 12)100 = 58.\overline{3}\%$

**Exercise Set 5.2 B**

Use the problems from Exercise Set 5.2 A and change each from fraction form to decimal form and then to percent form.

## 5.3 *Multiplication of Fractions*

The product of two fractions is actually the product of the numerators' factors divided by the product of the denominators' factors.

*Example:* $\frac{3}{4} \cdot \frac{7}{8} = \frac{3 \cdot 7}{4 \cdot 8} = \frac{21}{32}$

*Example:* $\frac{6}{7} \cdot \frac{7}{12} = \frac{6 \cdot 7}{7 \cdot 12} = \frac{42}{84}$

In the second example, the answer may be reduced to simpler form.

This may be avoided by dividing off the common factor 7 from the numerator and denominator, and likewise dividing off the factor 6 which is common in the numerator and denominator. $(12 = 2 \cdot 6)$

$$\frac{6}{7} \cdot \frac{7}{12} = \frac{6}{12} \cdot \frac{7}{7} = \frac{6 \cdot 7}{2 \cdot 6 \cdot 7} = \frac{1}{2}$$

*Example:* To multiply $\frac{7}{8} \cdot \frac{24}{9} \cdot \frac{11}{14}$

think of 24 as 3 times 8
think of 9 as 3 times 3
think of 14 as 2 times 7, so that the common factors in numerator and denominator may be divided off:

$$\frac{7}{8} \cdot \frac{24}{9} \cdot \frac{11}{14} = \frac{7}{8} \cdot \frac{3 \cdot 8}{3 \cdot 3} \cdot \frac{11}{2 \cdot 7} = \frac{11}{6}$$

The common factors of 3, 8 and 7 were divided off, leaving 11 in the numerator and $3 \cdot 2$ in the denominator.

*REMEMBER:* Non-integers in decimal or fraction form are REAL NUMBERS and the same rules for signs apply:

**If the signs of the factors are alike, the product will be positive.**
**If the signs of the factors are different, the product will be negative.**

**Exercise Set 5.3**

Wherever possible, divide off the common factors before multiplying. Answers should be in reduced form.

1. $\frac{2}{5} \cdot \frac{25}{40}$
2. $\frac{8}{3} \cdot \frac{9}{4}$
3. $\frac{6}{7} \cdot \frac{2}{3}$
4. $\frac{32}{36} \cdot \frac{15}{16}$
5. $\frac{-7}{9} \cdot \frac{63}{21}$
6. $\frac{-11}{15} \cdot \frac{5}{110}$
7. $\frac{12}{7} \cdot \frac{14}{-4}$
8. $\frac{-13}{21} \cdot \frac{21}{-13}$
9. $\frac{1}{-39} \cdot \frac{13}{4}$
10. $\frac{10}{12} \cdot \frac{3}{5}$

## 5.4 *Division of Fractions*

Remember from Chapter Two, that **multiplicative inverses** (reciprocals) are numbers whose product is 1.

$\frac{1}{2}$ and 2 are multiplicative inverses.

$\frac{2}{5}$ and $\frac{5}{2}$ are multiplicative inverses.

$-8$ and $-\frac{1}{8}$ are multiplicative inverses.

Now consider:

**Since division is the inverse operation of multiplication, we may replace division by multiplication, if (and only if) we replace the divisor by its multiplicative inverse (reciprocal).**

*Example:* $8 \div 2 = 4$ or

$$8 \times \frac{1}{2} = 4$$

*Example:* $\frac{3}{4} \div \frac{1}{2} = .75 \div .5 = 1.5$ or

$$\frac{3}{4} \div \frac{1}{2} = \frac{3}{4} \times \frac{2}{1} = .75 \times 2 = 1.5$$

*Example:* $\frac{5}{4} \div \frac{1}{4} = 1.25 \div 0.25 = 5$ or

$$\frac{5}{4} \div \frac{1}{4} = 1.25 \times \frac{4}{1} = 5$$

*Example:* $\frac{1}{8} \div \frac{4}{5} = 0.125 \div 0.8 = 0.15625$ or

$$\frac{1}{8} \div \frac{4}{5} = 0.125 \times 1.25 = 0.15625$$

**RULE: When dividing fractions:**

1. **change the operation from division to multiplication**
2. **change the divisor to its reciprocal**
3. **multiply. (FOLLOW RULES OF SIGNS!)**

*Example:* $\frac{7}{5} \div \frac{14}{15} = \frac{7}{5} \times \frac{15}{14} = \frac{7}{5} \times \frac{3 \times 5}{7 \times 2} = \frac{3}{2}$

*Example:* $\frac{2}{7} \div \frac{8}{14} = \frac{2}{7} \times \frac{14}{8} = \frac{2}{7} \times \frac{2 \times 7}{2 \times 2 \times 2} = \frac{1}{2}$

*Example:* $\frac{1}{9} \div \frac{7}{27} = \frac{1}{9} \times \frac{27}{7} = \frac{3}{7}$

**Exercise Set 5.4**

Divide the fractions:

1. $\frac{3}{4} \div \frac{6}{20}$
2. $\frac{5}{6} \div \frac{25}{24}$
3. $\frac{-14}{12} \div \frac{35}{12}$
4. $\frac{-16}{7} \div \frac{-40}{49}$
5. $\frac{12}{7} \div \frac{2}{11}$
6. $\frac{16}{3} \div \frac{4}{3}$

7. $36 \div \frac{1}{2}$

8. $\frac{-3}{7} \div 18$

9. $\frac{-3}{5} \div -27$

10. $\frac{-3}{32} \div \frac{-1}{16}$

## 5.5 *Least Common Multiple*

Since 12, 24 and 36 are each divisible by 3 and 4, they are **common multiples** of 3 and 4. However, since 12 is the smallest of these numbers, it is called their **least common multiple.**

Common multiples of 4 and 5 include 20, 40 and 60 (and an infinite list of others) since they are divisible by both 4 and 5; however, 20 is their LEAST common multiple.

**The least common multiple of a number is the smallest number which is divisible by each of them.**

To find the least common multiple of a group of numbers:

1. factor each number into a product of primes
2. use the highest degree of each of the combined factors and multiply.

*Example:* Find the least common multiple of 7, 14 and 35.

$7 = 1 \cdot 7, \quad 14 = 2 \cdot 7 \quad \text{and} \quad 35 = 5 \cdot 7$

The least common multiple is $2 \cdot 5 \cdot 7$ or 70.

*Example:* Find the least common multiple of 3, 4 and 8.

$3 = 1 \cdot 3 \quad 4 = 2 \cdot 2 = 2^2 \quad 8 = 2 \cdot 2 \cdot 2 = 2^3$

The least common multiple is $3 \cdot 2^3 = 3 \cdot 8 = 24$

*Example:* Find the least common multiple of 4, 6 and 16.

$4 = 2 \cdot 2 = 2^2 \quad 6 = 2 \cdot 3 \quad 16 = 2 \cdot 2 \cdot 2 \cdot 2 = 2^4.$

The least common multiple is $2^4 \cdot 3$ or $16 \cdot 3$ or 48.

*Example:* Find the least common multiple of 15, 24 and 35.

$15 = 3 \cdot 5 \qquad 24 = 3 \cdot 8 = 3 \cdot 2^3 \qquad 35 = 5 \cdot 7$

The least common multiple is $3 \cdot 5 \cdot 2^3 \cdot 7$

The LCM is $3 \cdot 5 \cdot 8 \cdot 7$ which is 840.

**Exercise Set 5.5**

Find the least common multiples of the following groups of numbers:

1. 2, 16 and 30
2. 3, 11 and 33
3. 3, 15 and 35
4. 6, 9 and 12
5. 3, 13 and 15
6. 5, 15 and 20
7. 2, 8 and 18
8. 4, 9 and 27
9. 9, 15 and 25
10. 8, 18 and 45

## 5.6 *Addition and Subtraction of Fractions*

It is only possible to add fractions which are alike. We can add halves to halves, quarters to quarters, thirds to thirds.

Likewise, we can subtract tenths from tenths, fifths from fifths, sevenths from sevenths, etc.

**RULE: To add or subtract fractions when the denominators are alike:**

**1. add or subtract the numerators**

**2. place the result over the common denominator**

*Example:* $\frac{1}{2} + \frac{7}{2} = \frac{8}{2} = 4 \qquad \frac{1}{2} - \frac{7}{2} = \frac{-6}{2} = 3$

*Example:* $\frac{6}{13} + \frac{11}{13} = \frac{17}{13} \qquad \frac{6}{13} - \frac{11}{13} = \frac{-5}{13}$

*Example:* $\frac{2}{11} + \frac{9}{11} - \frac{4}{11} = \frac{11}{11} - \frac{4}{11} = \frac{7}{11}$

**RULE: To add or subtract fractions when the denominators are different:**

1. **find the common denominator (LCM)**
2. **change the fractions to equivalent ones with the LCM as the denominator**
3. **add or subtract as indicated**

*Example:* $\frac{4}{3} + \frac{7}{6} = \frac{8}{6} + \frac{7}{6} = \frac{15}{6}$

*Example:* $6 + \frac{5}{2} = \frac{12}{2} + \frac{5}{2} = \frac{17}{2}$

*Example:* $\frac{4}{3} + \frac{3}{5} - 2 = \frac{20}{15} + \frac{9}{15} - \frac{30}{15} = \frac{-1}{15}$

**RULE: Since fractions are real numbers, the same laws of signs apply:**

**IF THE SIGNS ARE ALIKE, add absolute values and give the answer the common sign.**

**IF THE SIGNS ARE DIFFERENT, subtract absolute values and give the answer the sign of the number with the greater absolute value.**

**Exercise Set 5.6**

Add or subtract as indicated:

1. $\frac{2}{3} + \frac{8}{3}$
2. $\frac{7}{13} - \frac{5}{13}$
3. $\frac{3}{5} - \frac{7}{5} + \frac{2}{5}$
4. $\frac{5}{7} - \frac{3}{7} + \frac{6}{7}$
5. $\frac{2}{3} + \frac{5}{9}$
6. $\frac{3}{4} - \frac{3}{16}$
7. $\frac{3}{5} - \frac{2}{10} + \frac{7}{5}$
8. $\frac{5}{6} - \frac{5}{18}$
9. $\frac{2}{3} - \frac{3}{4} + \frac{1}{12}$
10. $\frac{3}{15} - \frac{5}{3} + \frac{2}{5}$
11. $\frac{8}{15} + \frac{5}{6} - \frac{7}{10}$
12. $\frac{4}{11} - \frac{6}{22} + 1$

## 5.7 *Raising Fractions to Powers*

When raising a fraction to a power, the numerator and the denominator are each raised to the power.

*Example:* The fraction $\left(\frac{4}{5}\right)^3$ is the same as $\frac{4}{5} \cdot \frac{4}{5} \cdot \frac{4}{5}$ which is $\frac{4^3}{5^3}$ or $\frac{64}{125}$

*Example:* The expression $\left(\frac{2}{3}\right)^2$ is equal to $\frac{2}{3} \cdot \frac{2}{3}$

This is the same as $\frac{2^2}{3^2} = \frac{4}{9}$

*Example:* $\left(\frac{-4}{5}\right)^2 = \frac{-4}{5} \cdot \frac{-4}{5}$ or $\frac{(-4)^2}{5^2} = \frac{16}{25}$

*Example:* $\frac{(-3)^3}{2} = \frac{-3}{2} \cdot \frac{-3}{2} \cdot \frac{-3}{2} = \frac{-3^3}{2^3} = \frac{-27}{8}$

*Example:* $\left(\left|\frac{-5}{-6}\right|\right)^2 = \left|\frac{-5}{-6}\right| \cdot \left|\frac{-5}{-6}\right| = \frac{5}{6} \cdot \frac{5}{6} = \frac{5^2}{6^2} = \frac{25}{36}$

**Exercise Set 5.7**

Raise the fractions to the indicated power:

1. $\left(\frac{1}{2}\right)^3$
2. $\left(\frac{2}{3}\right)^3$
3. $\left(\frac{7}{5}\right)^2$
4. $\left(\frac{8}{3}\right)^3$
5. $\left(\frac{7}{10}\right)^4$
6. $\left(\frac{3}{5}\right)^4$
7. $\left(\frac{2}{11}\right)^3$
8. $\left(\frac{3}{12}\right)^3$
9. $\left(\frac{3}{4}\right)^5$
10. $\left(\frac{1}{2}\right)^6$
11. $\left(\frac{12}{5}\right)^4$
12. $\left(\frac{13}{7}\right)^3$
13. $\left(\frac{-2}{3}\right)^2$
14. $\left(\frac{-5}{2}\right)^3$
15. $\left(\left|\frac{-4}{7}\right|\right)^2$
16. $\left(\left|\frac{-1}{5}\right|\right)^3$

## 5.8 *Order of Operations*

Remember that the order of operations is very important in finding a solution to algebra problems:

| | |
|---|---|
| First: | Simplify within PARENTHESES |
| Second: | Take roots or powers; that is, simplify EXPONENTS |
| Third: | Do MULTIPLICATION and DIVISION |
| Fourth: | Do ADDITION and SUBTRACTION |

*Example:* Simplify $\left(\frac{15}{3} \div \frac{5}{4}\right) - \frac{1}{2}$

First $\frac{15}{3} \div \frac{5}{4} = \frac{15}{3} \times \frac{4}{5} = 4$

Second $4 - \frac{1}{2} = 3\frac{1}{2}$

*Example:* Simplify $\frac{1}{2}\left(8 - \frac{1}{5}\right) + \frac{3}{10}$

First $8 - \frac{1}{5} = \frac{40}{5} - \frac{1}{5} = \frac{39}{5}$

Second $\frac{1}{2}\left(\frac{39}{5}\right) = \frac{39}{10}$

Finally $\frac{39}{10} + \frac{3}{10} = \frac{42}{10}$

**Exercise Set 5.8**

Use the correct order of operations to simplify the following:

1. $\left(\frac{4}{11}\right)(2) \div \left(\frac{1}{8}\right)$
2. $\frac{26}{3} + \left[\frac{1}{2}(6)\right]$
3. $\frac{2}{3}[3^2 - 6]$
4. $\left(\frac{1}{2}\right)\left(\frac{2}{5}\right) - \frac{1}{5}$
5. $\frac{3}{5} + \frac{1}{2}\left(\frac{3}{5} + \frac{3}{5}\right)$
6. $\left(\frac{1}{4} - \frac{3}{5}\right)^2$
7. $4\left[\frac{1}{2}(8 - 12)\right]$
8. $\left(\frac{5}{7} - \frac{4}{7}\right)(-7)$
9. $\left[\left(\frac{5}{6}\right)\left(\frac{3}{2}\right)\right] \div \left[\left(\frac{1}{4}\right)\left(\frac{10}{3}\right)\right]$
10. $\left(\frac{4}{3}\right)^2 - \left(\frac{5}{9} + 1\right)$
11. $\left(\frac{7}{11}\right)\left(\frac{2}{7}\right) \div \frac{11}{12}$
12. $\left[\frac{1}{7} - \frac{4}{7}\right]^2$
13. $\left[\frac{12}{7} \div \frac{3}{14}\right]^3$
14. $\left[\frac{-1}{9} - \frac{2}{27}\right]^2$
15. $\frac{21}{4} - \left(\frac{5}{2}\right)^3$
16. $-\left[12 + \frac{1}{3}\right] \div 4$

## CHAPTER EXERCISES

### Section 5.1

Factor each of the following into prime factors:

1. 15
2. 7
3. 11
4. 25
5. 4
6. 10
7. 12
8. 8
9. 100
10. 70
11. 56
12. 250
13. 78
14. 96
15. 35
16. 41
17. 156
18. 84
19. 500
20. 68

### Section 5.1

Find the greatest common factor for each group of numbers:

21. 12 and 33
22. 35 and 60
23. 6 and 15
24. 81 and 45
25. 16 and 32
26. 8 and 12
27. 27, 81 and 45
28. 26 and 39
29. 28, 196 and 70
30. 72 and 24
31. 16 and 35
32. 64, 40 and 124

### Section 5.2

Reduce each fraction to its simplest form:

33. $\frac{6}{8}$
34. $\frac{12}{15}$
35. $\frac{26}{50}$
36. $\frac{21}{24}$
37. $\frac{35}{55}$
38. $\frac{14}{35}$
39. $\frac{45}{225}$
40. $\frac{48}{32}$
41. $\frac{7}{11}$
42. $\frac{28}{36}$
43. $\frac{48}{8}$
44. $\frac{84}{28}$
45. $\frac{121}{44}$
46. $\frac{50}{125}$

### Section 5.3

Show that the following pairs of fractions are equal by converting each fraction in the pair to a decimal:

47. $\frac{3}{4} = \frac{75}{100}$
48. $\frac{1}{5} = \frac{3}{15}$
49. $\frac{2}{3} = \frac{12}{18}$
50. $\frac{8}{20} = \frac{4}{10}$
51. $\frac{25}{30} = \frac{10}{12}$
52. $\frac{20}{15} = \frac{16}{12}$

53. $\frac{3}{96} = \frac{1}{32}$

54. $\frac{30}{24} = \frac{35}{28}$

55. $\frac{15}{40} = \frac{21}{56}$

56. $\frac{35}{40} = \frac{21}{24}$

**Section 5.3**

Multiply the following:

57. $\frac{2}{3} \cdot \frac{5}{7}$

58. $\frac{3}{5} \cdot \frac{15}{27}$

59. $\frac{-4}{16} \cdot \frac{20}{30}$

60. $\frac{-22}{33} \cdot \frac{-36}{8}$

61. $\frac{-10}{15} \cdot \frac{-27}{4}$

62. $\frac{12}{7} \cdot \frac{-35}{16}$

63. $\frac{14}{24} \cdot \frac{40}{35}$

64. $\frac{18}{9} \cdot \frac{21}{42}$

65. $\frac{-15}{20} \cdot \frac{16}{12}$

66. $\frac{28}{105} \cdot \frac{35}{-42}$

67. $\frac{-13}{8} \cdot \frac{36}{24}$

68. $\frac{30}{18} \cdot \frac{-6}{20}$

**Section 5.4**

Divide the following:

69. $\frac{2}{3} \div \frac{16}{27}$

70. $\frac{100}{60} \div \frac{8}{27}$

71. $\frac{10}{16} \div \frac{5}{7}$

72. $\frac{21}{11} \div \frac{3}{5}$

73. $\frac{30}{33} \div \frac{12}{3}$

74. $\frac{210}{56} \div \frac{3}{8}$

75. $\frac{-15}{7} \div \frac{12}{18}$

76. $\frac{-8}{56} \div \frac{-2}{3}$

77. $-14 \div \frac{1}{2}$

78. $\frac{16}{100} \div -80$

**Section 5.5**

Find the least common multiple of the following groups of numbers:

79. 8 and 9

80. 25 and 15

81. 100 and 30

82. 16 and 20

83. 12, 15 and 20

84. 21, 15 and 9

85. 32, 14 and 8

86. 21 and 45

87. 3, 2 and 4

88. 27, 18 and 8

89. 6, 26 and 52

90. 7, 9 and 14

**Section 5.6**

Combine:

91. $\frac{12}{13} + \frac{2}{13}$

92. $\frac{3}{11} - \frac{6}{11} + \frac{5}{11}$

93. $\frac{3}{8} - \frac{7}{8} + \frac{6}{8}$

94. $\frac{5}{9} - \frac{7}{9}$

95. $\frac{2}{15} - \frac{7}{15} - \frac{3}{15}$

96. $\frac{8}{7} + \frac{2}{7} - \frac{11}{7}$

97. $\frac{-4}{17} - \frac{6}{17} - \frac{2}{17}$

98. $\frac{-11}{21} - \frac{3}{21} + \frac{5}{21}$

99. $\frac{3}{4} + \frac{5}{2}$

100. $\frac{8}{5} + \frac{7}{15}$

101. $\frac{5}{26} - \frac{2}{13}$

102. $\frac{5}{8} + \frac{1}{4} + \frac{7}{2}$

103. $\frac{2}{3} - \frac{15}{27} + \frac{7}{9}$

104. $\frac{3}{2} + \frac{5}{6} - \frac{2}{3}$

105. $\frac{3}{4} + \frac{4}{3} + \frac{3}{2}$

106. $\frac{7}{15} + \frac{5}{6}$

107. $\frac{3}{4} + \frac{7}{12} + \frac{3}{5}$

108. $\frac{1}{2} - \frac{3}{4} + \frac{4}{5}$

109. $\frac{7}{15} + \frac{8}{35}$

110. $\frac{13}{14} - \frac{7}{10}$

111. $\frac{3}{18} + \frac{7}{45}$

112. $\frac{5}{21} - \frac{8}{27}$

113. $\frac{1}{2} + \frac{4}{18} + \frac{8}{27}$

114. $\frac{20}{33} - \frac{7}{15}$

115. $\frac{17}{30} + \frac{7}{18} - \frac{4}{15}$

116. $\frac{9}{22} + \frac{7}{6} + \frac{5}{33}$

117. $\frac{19}{20} - \frac{11}{12} - \frac{2}{15}$

118. $\frac{5}{14} + \frac{11}{35} + \frac{5}{18}$

**Section 5.7**

Simplify:

119. $\left(\frac{2}{3}\right)^2$

120. $\left(\frac{4}{5}\right)^3$

121. $\left(\frac{7}{5}\right)^2$

122. $\left(\frac{2}{7}\right)^4$

123. $\left(\frac{5}{3}\right)^4$

124. $\left(\frac{7}{11}\right)^2$

125. $\left(\frac{13}{4}\right)^4$

126. $\left(\frac{4}{7}\right)^5$

127. $\left(\frac{8}{3}\right)^6$

128. $\left(\frac{17}{21}\right)^0$

129. $\left(\frac{3}{8}\right)^5$

130. $\left(\frac{7}{6}\right)^0$

## CHAPTER TEST *Form 5A*

1. List the prime numbers between 12 and 30.

For questions 2–4, list the prime factors of the given numbers:

2. 74 3. 120 4. 512

For numbers 5–7, list the greatest common factor for each group of numbers:

5. 42 and 98 6. 275 and 440 7. 36, 72 and 144

For questions 8–10, reduce each fraction to simplest form:

8. $\frac{64}{128}$ 9. $\frac{32}{256}$ 10. $\frac{121}{44}$

For questions 11–14, multiply the given fractions and express the answers in simplest form:

11. $\frac{2}{9} \cdot \frac{3}{5}$

12. $\frac{16}{81} \cdot \frac{9}{8}$

13. $\frac{11}{7} \cdot \frac{5}{44} \cdot \frac{49}{2}$

14. $5 \cdot \frac{1}{3} \cdot 9 \cdot \frac{1}{5}$

For numbers 15–18 divide the fractions. Reduce answers to simplest terms.

15. $\frac{11}{7} \div \frac{55}{14}$

16. $\frac{2}{9} \div \frac{16}{45}$

17. $\frac{23}{5} \div \frac{23}{10}$

18. $\frac{5}{6} \div \frac{30}{6}$

Add or subtract as indicated:

19. $\frac{7}{12} + \frac{3}{20}$

20. $\frac{2}{3} - \frac{5}{6} + \frac{3}{4}$

21. $\frac{3}{7} - \frac{8}{14} + \frac{9}{2}$

22. $\frac{1}{3} + \frac{4}{5} + \frac{6}{7}$

## CUMULATIVE EXERCISES Chapters 1–5 *Set A*

1. Evaluate: $(10^2)(10) - 1000$

2. Evaluate: $16(55) - 34(7.3)$

3. If Margo had 16 pennies, 37 nickels, 43 dimes and 17 quarters in a change bank, how much was in the bank?

4. What is the additive inverse of $\frac{79}{3}$?

5. What property of real numbers is illustrated by: $16(a + b) = 16a + 16b$

6. True or False? $(8 \cdot a)b = 8(a \cdot b)$ illustrates the commutative property of real numbers.

7. Evaluate: $|-14| \div |-7|$

Perform the indicated operation and round numbers to three decimal places:

8. $17.1 \div (5.4)^2$

9. $-19.2 - 16.3 + 10.7$

10. $\sqrt{17.6} \div -3.4$

11. $(5)(-6) - (1.2)(-3)$

Express each as a product of prime numbers:

12. 360

13. 1240

14. 720

15. 1648

Simplify:

16. $\frac{105}{495}$

17. $\frac{344}{473}$

18. $\frac{-300}{475}$

19. $\frac{-220}{312}$

Find the LCM of the groups of numbers:

20. 10, 12 and 15

21. 21, 27 and 49

Perform the indicated operation and simplify:

22. $\frac{18}{-13} \cdot \frac{39}{27}$

23. $\frac{-189}{15} \div \frac{-34}{45}$

24. $\left(\frac{6}{5}\right)\left(\frac{35}{4}\right)\left(\frac{-8}{7}\right)$

25. $-14\left(\frac{33}{21}\right) \div 11$

26. $\frac{1}{5} + \frac{2}{7} - \frac{1}{6}$

27. $\frac{-11}{12} - \frac{7}{8} + \frac{3}{6}$

28. $\frac{7}{8} - \frac{4}{5} + \frac{4}{15}$

29. $\frac{7}{16} + 7 - \frac{7}{8} - 14$

Evaluate:

30. $(14.2)^2(3.7) - 2.5(-4)^3$

31. $121 - 7(4)^2 + 10^3$

32. $(25)^2$

33. $(176)^1$

34. $(-147)^0$

35. $7^2 - (-7) - 7^3$

## CUMULATIVE TEST Chapters 1–5 *Form A*

1. Evaluate: $(25)^2(175) - 90$

2. Mark made monthly payments of \$275.35 to pay off a five year loan. What was the amount of the loan?

3. What is the reciprocal of $\frac{165}{4}$?

4. Evaluate: $|-36| \div -9$

5. Evaluate: $15(-3)^2$

Simplify. Round answers to 3 decimal places:

6. $1.7(3.6)^2 - 2.71$

7. $-18(\sqrt{3.17})$

8. $-4.7 \div 0.06$

9. $(-5)^2 - (5)^2$

Express as a product of prime factors:

10. 450

11. 1720

Reduce the fractions to simplest terms:

12. $\frac{38}{190}$

13. $\frac{48}{256}$

14. Find the LCM of 9, 20 and 25.

Perform the indicated operation and simplify:

15. $\left(\frac{-7}{10}\right) \cdot \left(\frac{25}{14}\right)$

16. $\frac{81}{(-16)} \div \frac{-27}{8}$

17. $\frac{1}{3} + \frac{1}{7} + \frac{1}{5}$

18. $\frac{1}{8} - \frac{2}{3} - \frac{5}{6}$

19. $\left(\frac{-7}{8}\right)^2$

20. $-(-3)^3$

Evaluate:

21. $(14.1)(-3)^2 - 22(-4)^2$

22. $108 - 3\{5 - [6 + 2(3 - 5)^2]\}$

23. $\left(\frac{5}{7}\right)^2$

24. $(23)^3$

25. $(472)^1$

CHAPTER

# 6 Properties of Exponents

## 6.0 *Vocabulary Preview*

Define these words without using a dictionary. Formulate their meanings from past experience and knowledge.

| | | |
|---|---|---|
| **base** | **exponent** | **power** |
| **coefficient** | **factor** | **product** |

## 6.1 *Powers of Ten*

The number system which we use is based on groups of ten. We know (and can easily check with a calculator) that:

$$100{,}000 = 10 \cdot 10 \cdot 10 \cdot 10 \cdot 10 = 10^5$$
$$10{,}000 = 10 \cdot 10 \cdot 10 \cdot 10 = 10^4$$
$$1{,}000 = 10 \cdot 10 \cdot 10 = 10^3$$
$$100 = 10 \cdot 10 = 10^2$$

Notice that each number may be written as a POWER of ten.

The number 10 is the BASE;

the EXPONENT (the small number written above and to the right of the base) tells the POWER to which the base is raised; that is, how many times 10 is used as a FACTOR.

As the exponent was decreased by 1, the result became $\frac{1}{10}$ of the previous answer; that is, the previous product was divided by 10.

It follows, then, that:

$$10^1 = 100 \cdot \frac{1}{10} = 10$$
$$10^0 = 10 \cdot \frac{1}{10} = 1$$
$$10^{-1} = 1 \cdot \frac{1}{10} = \frac{1}{10}$$
$$10^{-2} = \frac{1}{10} \cdot \frac{1}{10} = \frac{1}{100}$$
$$10^{-3} = \frac{1}{100} \cdot \frac{1}{10} = \frac{1}{1000}\text{, and so forth.}$$

*NOTE:* Any number may be used as a base; however, base 10 is used here because it is the basis for our number system.

*Example:* Evaluate $10^6$.

$$\begin{aligned} 10^6 &= (10)(10)(10)(10)(10)(10) \\ &= 1000000 \end{aligned}$$

*Example:* Evaluate $10^{-4}$.

$$\begin{aligned} 10^{-4} &= \left(\frac{1}{10}\right)\left(\frac{1}{10}\right)\left(\frac{1}{10}\right)\left(\frac{1}{10}\right) \\ &= \frac{1}{10000} \end{aligned}$$

**Exercise Set 6.1**

Evaluate the following:

1. $10^4$
2. $10^0$
3. $10^8$
4. $10^{-2}$
5. $10^3$
6. $10^{-3}$
7. $10^9$
8. $10^{-8}$
9. $10^{-7}$

## 6.2 *Powers of Bases Other Than Ten*

Using 2 as the base, and proceeding as above:

$$\begin{aligned} 2^5 &= 2 \cdot 2 \cdot 2 \cdot 2 \cdot 2 = 32 \\ 2^4 &= 2 \cdot 2 \cdot 2 \cdot 2 \quad = 16 \\ 2^3 &= 2 \cdot 2 \cdot 2 \quad\quad = 8 \end{aligned}$$

Again notice the pattern. As the exponent is decreased by 1, the new product becomes $\frac{1}{2}$ the previous one; that is, the previous product is divided by 2.

So, continuing the pattern above:

$$\begin{aligned} 2^2 &= 8 \cdot \frac{1}{2} = 4 \\ 2^1 &= 4 \cdot \frac{1}{2} = 2 \\ 2^0 &= 2 \cdot \frac{1}{2} = 1 \\ 2^{-1} &= 1 \cdot \frac{1}{2} = \frac{1}{2} \\ 2^{-2} &= \frac{1}{2} \cdot \frac{1}{2} = \frac{1}{4} \text{ and so on.} \end{aligned}$$

Consider base five:

$$5^4 = 5 \cdot 5 \cdot 5 \cdot 5 = 625$$
$$5^3 = 5 \cdot 5 \cdot 5 = 125$$
$$5^2 = 5 \cdot 5 = 25$$

As the exponent decreases by 1, the product is $\frac{1}{5}$ the previous one.

$$5^1 = 25 \cdot \frac{1}{5} = 5$$
$$5^0 = 5 \cdot \frac{1}{5} = 1$$
$$5^{-1} = 1 \cdot \frac{1}{5} = \frac{1}{5}$$
$$5^{-2} = \frac{1}{5} \cdot \frac{1}{5} = \frac{1}{25}, \text{ etc.}$$

**Exercise Set 6.2 A**

Write the number represented by each of the following:

1. $3^3$
2. $6^2$
3. $8^1$
4. $10^{-3}$
5. $5^4$
6. $7^{-2}$
7. $9^0$
8. $2^8$
9. $4^{-2}$
10. $5^0$
11. $6^3$
12. $3^{-2}$
13. $10^{-1}$
14. $4^{-1}$
15. $7^1$

**Exercise Set 6.2 B**

Supply the exponent which will make the statement true:

1. $2^x = 64$
2. $8^x = 64$
3. $3^x = 243$
4. $10^x = 0.1$
5. $7^x = 343$
6. $8^x = 8$
7. $6^x = 216$
8. $4^x = 1$
9. $9^x = \frac{1}{9}$
10. $2^x = \frac{1}{8}$
11. $5^x = \frac{1}{25}$
12. $4^x = \frac{1}{16}$
13. $16^x = 16$
14. $12^x = 1$
15. $2^x = \frac{1}{128}$

## 6.3 *Negative Exponents and the Exponent Zero*

From the previous examples note:

**a. any non-zero number raised to the zero power is 1**

**b. a number raised to the first power is unchanged**

**c. when the exponent is negative, the result is the same as raising its reciprocal to that same positive power.**

*Examples:*

$$6^1 = 6 \qquad 6^{-1} = \frac{1}{6}$$

$$\left(\frac{1}{6}\right)^1 = \frac{1}{6} \qquad \left(\frac{1}{6}\right)^{-1} = 6$$

$$(-3)^4 = 81 \qquad (-3)^{-4} = \frac{1}{81}$$

$$\left(\frac{1}{3}\right)^4 = \frac{1}{81} \qquad \left(\frac{1}{3}\right)^{-4} = 81$$

$$12^2 = 144 \qquad 12^{-2} = \frac{1}{144}$$

### Exercise Set 6.3 A

Write the numbers represented by the following:

1. $8^2$
2. $\left(\frac{2}{5}\right)^3$
3. $\left(\frac{1}{7}\right)^1$
4. $\left(\frac{1}{8}\right)^2$
5. $\left(\frac{1}{8}\right)^{-2}$
6. $\left(\frac{1}{7}\right)^3$
7. $\left(\frac{1}{7}\right)^{-3}$
8. $6^3$
9. $4^4$
10. $4^{-4}$
11. $\left(\frac{1}{6}\right)^3$
12. $\left(\frac{1}{6}\right)^{-3}$
13. $\left(\frac{1}{4}\right)^4$
14. $\left(\frac{1}{4}\right)^{-4}$
15. $7^1$
16. $7^{-1}$
17. $\left(\frac{1}{4}\right)^0$

### Exercise Set 6.3 B

Write the numbers represented by the following:

1. $\left(\frac{1}{2}\right)^5$
2. $\left(\frac{2}{5}\right)^3$
3. $\left(\frac{2}{3}\right)^4$
4. $\left(\frac{7}{8}\right)^0$
5. $\left(\frac{5}{6}\right)^{-2}$
6. $\left(\frac{1}{4}\right)^{-1}$
7. $\left(\frac{3}{5}\right)^{-2}$
8. $\left(\frac{1}{3}\right)^0$
9. $\left(\frac{1}{5}\right)^{-1}$
10. $\left(\frac{1}{2}\right)^{-5}$
11. $\left(\frac{2}{3}\right)^{-2}$
12. $\left(\frac{7}{8}\right)^{-1}$
13. $\left(\frac{1}{3}\right)^{-2}$
14. $\left(\frac{1}{6}\right)^2$
15. $\left(\frac{1}{6}\right)^{-2}$

## 6.4 *Addition and Subtraction of Like Terms*

Recall from Chapter Three that multiplication may be viewed as repetitive addition.

Since $4(2^3) = 2^3 + 2^3 + 2^3 + 2^3$

and $3(2^3) = 2^3 + 2^3 + 2^3$,

the sum $4(2^3) + 3(2^3) =$

$(2^3 + 2^3 + 2^3 + 2^3) + (2^3 + 2^3 + 2^3) =$

$7(2^3)$

Likewise, $4(2^3) - 3(2^3) = 1(2^3)$

since $(2^3 + 2^3 + 2^3 + 2^3) - (2^3 + 2^3 + 2^3) =$

$2^3 + 2^3 + 2^3 + 2^3 - 2^3 - 2^3 - 2^3 =$

$2^3$ or $1(2^3)$

*Example:* $4(y^4) + 3(y^4) = (4 + 3)(y^4)$
$= 7y^4$

*Example:* $8(c^5) - 2(c^5) = (8 - 2)(c^5)$
$= 6c^5$

*Example:* $12(5^6) + 8(5^6) - 4(5^2) =$
$(12 + 8)(5^6) - 4(5^2) =$
$20(5^6) - 4(5^2)$

*Example:* $2c^3 + 5b^3 - 6c^3 + 8b^3 =$
$(2 - 6)c^3 + (5 + 8)b^3 =$
$-4c^3 + 13b^3$

**IMPORTANT NOTICE:** It is only possible to **add** or **subtract** LIKE TERMS; that is, terms which have the SAME BASE and the SAME EXPONENT.

*Example:* $(8^2)$ and $(8^3)$ cannot be added or subtracted since the **exponents are different.**

*Example:* $(2^5)$ and $(3^5)$ cannot be added or subtracted since the bases are different.

*Example:* Find the sum and difference of $6x^2$ and $2x^2$.

$$6x^2 + 2x^2 =$$
$$(6 + 2)x^2 = 8(x^2)$$

$$6x^2 - 2x^2 =$$
$$(6 - 2)x^2 = 4(x^2)$$

Notice that in each case, only the COEFFICIENTS of the terms were added or subtracted; the exponential part remained unchanged.

In effect, this is the distributive law in reverse.

**Exercise Set 6.4**

Simplify by combining like terms. DO NOT MULTIPLY.

1. $2(4^3) + 3(4^5) - (4^3) + 6(4^5)$
2. $10x^5 + 8y^3 - 2z + 3x^5 - 4y^3$
3. $5(4^4) + 6(4^2) + 2(4^4) - 3(4^2)$
4. $2y^3 + 3y^5 - y^3 + 6y^5$
5. $10(2^5) + 8(3^3) - 2(5^2) + 3(2^5) - 4(3^3)$
6. $10(8^{-2}) - 4(8^{-2})$
7. $12k^3 - 6g^5 + 4k^3 - 3g^5 + k^3 - g^5$
8. $5(12^0) + 2(12^0)$
9. $2x^2 + 6x^2$
10. $16t^5 - 8t^5$
11. $7y^4 + 8w^3 - 3y^4$
12. $a^2 + b^2 - 3a^2 + 7b^2$
13. $p^5 + q^2 - p^5 + q^2$
14. $6r^3 + 12s^2 - 9r^3 - 3t^5 + 5s^2 + 4t^5$
15. $d^3 - g^5 + 3d^3 - 3g^5$

16. $r^2 + s^4 + t^2 + s^3 + t^2 + r^3$

17. $4(3^{-2}) + 6(3^{-2}) - 7(3^2) + 2(3^2)$

18. $10 - k + 2k - 2k^2 - 2 + 3k^3$

19. $6y^2 - 5y - 6y^2 + 8y - 5$

20. $3x^2 - 4y^2 + 7y - 2xy + 2y^2$

## 6.5 *Multiplication of Terms Involving Exponents*

Since $3^2 = 3 \cdot 3$ and $3^5 = 3 \cdot 3 \cdot 3 \cdot 3 \cdot 3$, it follows that

$3^2 \cdot 3^5 = (3 \cdot 3)(3 \cdot 3 \cdot 3 \cdot 3 \cdot 3)$ or $3^7$

Likewise,

$2^4 \cdot 2^1 = (2 \cdot 2 \cdot 2 \cdot 2)(2) = 2^5$ and

$6^2 \cdot 6^4 = (6 \cdot 6)(6 \cdot 6 \cdot 6 \cdot 6) = 6^6$

Also: $\left(\frac{1}{3}\right)^2 \cdot \left(\frac{1}{3}\right)^3 =$

$$\left[\left(\frac{1}{3}\right)\left(\frac{1}{3}\right)\right]\left[\left(\frac{1}{3}\right)\left(\frac{1}{3}\right)\left(\frac{1}{3}\right)\right] = \left(\frac{1}{3}\right)^5$$

**WHEN NEGATIVE EXPONENTS ARE INVOLVED, BE CAREFUL!**

$(4^5)(4^{-2}) =$

$(4)(4)(4)(4)(4)\left(\frac{1}{4}\right)\left(\frac{1}{4}\right) = (4)(4)(4) = 4^3$

**RULE: When multiplying numbers with like bases, ADD the exponents.**

*Examples:*

$(16^2)(16^3) = 16^{2+3} = 16^5$

$(8^2)(8^{-2}) = 8^{2+(-2)} = 8^0 = 1$

$(x^3)(x^5) = x^{3+5} = x^8$

$(t^5)(t^{-3}) = t^{5-3} = t^2$

$(18^4)(18^0) = 18^{4+0} = 18^4$

$(r^3)(r^{-3}) = r^{3-3} = r^0 = 1$

**Exercise Set 6.5**

Simplify the following without evaluating. Leave the answer in exponential form.

1. $3^3 \cdot 3^5$
2. $5^2 \cdot 5^3$
3. $2^5 \cdot 2^0$
4. $\left(\frac{1}{3}\right)^4\left(\frac{1}{3}\right)$
5. $(.7)^2(.7)^6$
6. $6^2 \cdot 6^{-2}$
7. $4^3 \cdot 4^6$
8. $\left(\frac{1}{10}\right)^2\left(\frac{1}{10}\right)$
9. $16^2 \cdot 16^7$
10. $6^{10} \cdot 6^2$
11. $4^{-3} \cdot 4^3$
12. $2^5 \cdot 2^3$
13. $12^0 \cdot 12^0$
14. $(.9^3)(.9)$

## 6.6 *Division of Terms Involving Exponents*

Since $5^3 = (5)(5)(5)$

and $5^2 = (5)(5)$,

$$\frac{5^3}{5^2} = \frac{(5)(5)(5)}{(5)(5)} = 5$$

*NOTE:* In section 6.1 and 6.2 we developed the concept that:

$$\frac{1}{x} = x^{-1}$$
$$\frac{1}{x^2} = x^{-2} \quad \text{etc.}$$

With that in mind, consider $8^3 \div 8^7$:

Since $8^3 = (8)(8)(8)$ and $8^7 = (8)(8)(8)(8)(8)(8)(8)$,

$$\frac{8^3}{8^7} = \frac{(8)(8)(8)}{(8)(8)(8)(8)(8)(8)(8)} = \frac{1}{8^4} = 8^{-4}$$

**RULE: When dividing numbers with LIKE bases, SUBTRACT the exponents.**

*Examples:* $7^6 \div 7^2 = 7^{6-2} = 7^4$

$9^2 \div 9^6 = 9^{2-6} = 9^{-4} = \left(\frac{1}{9}\right)^4$

$17^{10} \div 17^3 = 17^{10-3} = 17^7$

$8^3 \div 8^{10} = 8^{3-10} = 8^{-7} = \left(\frac{1}{8}\right)^7$

$5^8 \div 5^{-3} = 5^{8-(-3)} = 5^{8+3} = 5^{11}$

$4^{13} \div 4^{13} = 4^{13-13} = 4^0 = 1$

**Exercise Set 6.6**

Divide the following without evaluating. Leave the answer in exponential form.

1. $10^{12} \div 10^6$
2. $5^3 \div 5^4$
3. $6^2 \div 6^2$
4. $11^5 \div 11^2$
5. $3^7 \div 3^9$
6. $3^2 \div 3^{-2}$
7. $12^3 \div 12^3$
8. $9^2 \div 9^5$
9. $4^5 \div 4^{-2}$
10. $2^5 \div 2^{-2}$
11. $7^4 \div 7^3$
12. $8^0 \div 8^{-2}$
13. $10^{-2} \div 10^2$
14. $12^2 \div 12^0$
15. $5^5 \div 5^{10}$

## 6.7 *Raising a Power to a Power*

Since $(15)^3 = 15 \cdot 15 \cdot 15$

$[(15^3)]^2 = (15 \cdot 15 \cdot 15)(15 \cdot 15 \cdot 15)$
$= 15^6$

Also $[3^4]^3$
$= [3 \cdot 3 \cdot 3 \cdot 3]^3$
$= [(3)(3)(3)(3)][(3)(3)(3)(3)][(3)(3)(3)(3)]$
$= 3^{12}$

*Examples:*

$$[7^6]^2 = (7)^{6\cdot 2} = 7^{12}$$
$$[9^2]^6 = (9)^{2\cdot 6} = 9^{12}$$
$$[x^8]^3 = (x)^{8\cdot 3} = x^{24}$$
$$[s^3]^{-3} = (s)^{3(-3)} = s^{-9}$$
$$[14^3]^3 = (14)^{3\cdot 3} = 14^9$$

**RULE: In general, when a number which is already raised to a power is being raised to another power, MULTIPLY EXPONENTS.**

**Exercise Set 6.7**

Perform the indicated operations. Leave answers in exponential form.

1. $(2^4)^3$
2. $[20^2]^5$
3. $(7^0)^3$
4. $\left[\left(\frac{1}{10}\right)^2\right]^7$
5. $\left[\left(\frac{1}{10}\right)^7\right]^2$
6. $[20^5]^2$
7. $[(.3)^2]^{10}$
8. $[(.3)^{10}]^2$
9. $[6^2]^2$
10. $(2^3)^4$
11. $(8^5)^3$
12. $(8^3)^5$
13. $\left[\left(\frac{1}{2}\right)^4\right]^2$
14. $(13^6)^0$
15. $[11^3]^{-3}$
16. $[11^{-3}]^3$

## CHAPTER EXERCISES

### Section 6.1

Evaluate:

1. $10^3$
2. $10^6$
3. $10^{-3}$
4. $10^{-1}$
5. $10^5$
6. $10^{-6}$
7. $5^3$
8. $3^{-2}$
9. $2^{-4}$
10. $7^{-2}$
11. $6^{-1}$
12. $13^{-2}$

### Section 6.2 and 6.3

Find the value of the exponent x which makes each statement true. If no value for x will satisfy the equation, write "no solution".

13. $2^x = 16$
14. $7^x = 49$
15. $5^x = 125$
16. $3^x = \frac{1}{9}$
17. $14^x = \frac{1}{196}$
18. $18^x = 18$
19. $2^x = 128$
20. $25^x = 1$
21. $3^x = 81$
22. $10^x = .01$
23. $.25^x = 0$
24. $.3^x = 0$
25. $.2^x = .04$
26. $\left(\frac{1}{2}\right)^x = \frac{1}{32}$
27. $10^x = 1000$

### Section 6.4

Simplify by combining like terms. **Do not multiply.**

28. $2(7^2) + 3(7^2)$
29. $17(4^2) - 6(4^3) + 2(4^2)$
30. $5(2^3) - 3(2^3)$
31. $6(5^2) + 2(5^2) - 3(5^2)$
32. $10(4^3) - 3(4^3) + (4^3)$
33. $7(3^5) - 5(3^5)$
34. $12(8^3) - 15(8^3)$
35. $15(9^2) - 6(9^2) + 3(7^2)$
36. $3(7^3) + 2(8^3) - 5(7^3)$
37. $11(2^3) + 6(2^4)$
38. $6(10^2) - 8(5^2) + 2(10^2)$
39. $8(5^3) + 2(5^3) - 2(5^2)$
40. $7\left(\frac{1}{2}\right)^4 - 3(6^2) + 2\left(\frac{1}{2}\right)^4$
41. $9(8^2) - 6(8^2) + 2(8^4)$
42. $6x^2 - 7x^2$
43. $15x^3 - 7x^3$
44. $18y^2 - 8z^2 + 2y^2$
45. $7c^0 + 6d^3 - 8c^0$
46. $13a^5 - 2a^5 + 6b^2 + 2b^2$
47. $9s^5 + 3t^5 - 7s^5 + 2t^5$

48. $6b^2 - 8b^3 + 4b^2 - 6b^3$

49. $17x^7 + 8x^5 - 6x^2$

50. $3d^6 - 2a^4 - 3b^3 + 2a^4$

51. $7x^3 + 5y^2 - 10x^3 + 2y^2$

52. $5a^2 - 3c + 5a^2 - 4c^2$

53. $16x - 4y + 8xy - 10x$

54. $8x^2 - 16y^2 + 4xy + 16y^2$

55. $6s^2 - 8s + 16s - 2st$

**Section 6.5**

Simplify each of the following. Leave your answer in exponential form.

56. $2^3 \cdot 2^4$

57. $(-2)^3 \cdot (-2)^5$

58. $3^2 \cdot 3^5$

59. $7^2 \cdot 7^3 \cdot 7^5$

60. $6^2 \cdot 6^5 \cdot 6^0$

61. $3^8 \cdot 3^3 \cdot 3^4$

62. $25^2 \cdot 25^3 \cdot 25^4$

63. $16^3 \cdot 16^1 \cdot 16^0$

64. $(.2)^5 \cdot (.2)^7$

65. $\left(\frac{1}{3}\right)^3 \cdot \left(\frac{1}{3}\right)^5 \cdot \left(\frac{1}{3}\right)$

66. $8^2 \cdot 8^1$

67. $5^3 \cdot 5^2 \cdot 5^{-1}$

68. $9^3 \cdot 9^{-2} \cdot 9^2$

69. $10^4 \cdot 10^{-3} \cdot 10^0$

70. $\left(\frac{1}{3}\right)^5 \cdot \left(\frac{1}{3}\right)^{-2}$

71. $(.7)^6 \cdot (.7)^8 \cdot (.7)^9$

72. $(.25)^3 \cdot (.25)^{-2}$

73. $a^3 \cdot a^2 \cdot a^5$

74. $x^3 \cdot x^5 \cdot x^7$

75. $y^5 \cdot y^{-3}$

76. $z^3 \cdot z^6 \cdot z^{-4}$

77. $s^3 \cdot s^{-2} \cdot s^0 \cdot s^{-1}$

78. $t^{12} \cdot t^{-10} \cdot t^3$

79. $a^3 \cdot b^3 \cdot a^5$

80. $x^2 \cdot x^3 \cdot x^5 \cdot t^3$

81. $(xy)^2 \cdot (xy)^3 \cdot (xy)^{-1}$

82. $(.7x)^2 \cdot (.7x)^3$

83. $x^2 \cdot z^3 \cdot x^{-3} \cdot z^2$

84. $(ab)^4 \cdot (ab)^3 \cdot c^2 \cdot ab$

85. $(abc)^3 \cdot (abc)^{-6}$

**Section 6.6**

Simplify the following. Leave your answers in exponential form.

86. $4^4 \div 4^2$
87. $7^5 \div 7^3$
88. $15^3 \div 15^1$
89. $(.5)^6 \div (.5)^5$
90. $(.2)^3 \div (.2)^1$
91. $x^7 \div x^3$
92. $s^8 \div s^5$
93. $t^9 \div t^5$
94. $w^3 \div w^4$
95. $y^8 \div y^{10}$
96. $z^7 \div z^{10}$
97. $a^0 \div a^2$
98. $x^3 \div x^0$
99. $(.3)^3 \div (.3)^0$
100. $(3a)^4 \div (3a)^5$
101. $x^3 \div x^6$
102. $y^2 \div y^4$
103. $t^3 \div t^{10}$
104. $(xy)^3 \div (xy)^3$
105. $(abc)^3 \div (abc)^6$

**Section 6.7**

Perform the indicated operations. Leave your answer in exponential form.

106. $(2^2)^3$
107. $(5^2)^2$
108. $(10^2)^5$
109. $(.2^2)^4$
110. $(3^4)^2$
111. $(.27^2)^0$
112. $(x^3)^4$
113. $(y^5)^3$
114. $(s^6)^2$
115. $(t^4)^3$
116. $(x^2y^3)^2$
117. $(w^3z^5)^4$
118. $(xy^2)^5$
119. $(tw^6)^0$
120. $(s^2t^0)^6$
121. $(a^3)^3$
122. $[(x^2)^3]^4$
123. $(3^2y^3)^5$
124. $(2^3x^3)^4$
125. $(6a^3)^6$
126. $(.4y^5)^3$
127. $(5x^2y)^4$

## CHAPTER TEST *Form 6A*

Evaluate the following:

1. $10^5$
2. $10^{-4}$
3. $12^2$
4. $100^{-1}$
5. $3^{-3}$
6. $11^3$
7. $728^{-1}$
8. $(-3)^2$

Supply the value for x which makes each statement true:

9. $2^x = \frac{1}{2}$
10. $74^x = 1$
11. $25^x = 25$
12. $4^x = 64$

Rewrite the following using POSITIVE exponents:

13. $\left(\frac{1}{8}\right)^{-1}$
14. $(175)^{-8}$
15. $\left(\frac{3}{7}\right)^{-4}$
16. $(y)^{-2}$

Evaluate:

17. $(5)^{-2}$
18. $(-8)^{-2}$
19. $1^{-1}$
20. $(18)^{-1}$
21. $\left(\frac{4}{7}\right)^{-2}$
22. $\left(\frac{2}{5}\right)^{-3}$

Simplify by combining like terms. **Do not multiply.**

23. $5(6^2) + 18(6^2)$
24. $13(x^3) - 24(x^3) + 2(x^3)$
25. $8x^2 + 9x^2 - 3$
26. $11(9^5) - 3(9^5) + 4(2^7) + 5(2^7)$

Simplify the following without evaluating:

27. $(17^3)(17^5)$
28. $58^0 \cdot 58^{-7}$
29. $x^5 \cdot x^{12}$
30. $175^5 \div 175^2$
31. $5^0 \div 5^{-6}$
32. $77^7 \div 77^{-7}$

Simplify the following. Leave answers in exponential form.

33. $(121^3)^5$
34. $[14^0]^{12}$
35. $(9^{-6})^2$

## CUMULATIVE EXERCISES Chapters 1–6 *Set A*

1. Evaluate: $(10)^3(10)^2 - (10)^5$

2. Evaluate: $12(10) - 14(2.5)$

3. Jake spent \$14.00 a day on meals, \$25.00 a day for transportation, and \$17.50 a day for supplies while he was on a three day business tour. If his company allowed him \$300.00 for expenses, how much (if anything) was left unspent?

4. What is the reciprocal of $\frac{-1}{7}$?

Complete the statements:

5. The distributive property for real numbers states that

   $7(a + b) =$ ____ + ____.

6. $8 + a + b = a + 8 + b$ illustrates the ____________ property of real numbers.

7. $|-6| \cdot |2| =$ ________

Perform the indicated operations and round answers to three decimal places:

8. $18.5 \div (1.2)^2$

9. $-10.3 - 15.6 + 12.9$

10. $\frac{\sqrt{15.4}}{-2.2}$

11. $(3)(-2) - (2.3)(-1)$

Express each as a product of prime numbers:

12. 250

13. 1000

14. 240

15. 125

Simplify:

16. $\frac{10}{45}$

17. $\frac{340}{180}$

18. $\frac{155}{35}$

19. $\frac{-200}{48}$

Find the LCM of the groups of numbers:

20. 8, 36 and 42

21. 13, 26 and 39

Perform the indicated operation and simplify:

22. $\frac{13}{-18} \cdot \frac{27}{39}$

23. $\frac{15}{189} \div \frac{-45}{34}$

24. $\left(\frac{-35}{7}\right)\left(\frac{6}{4}\right)\left(\frac{8}{5}\right)$

25. $-21\left(\frac{33}{14}\right) \div 7$

26. $\frac{1}{4} + \frac{2}{3} - \frac{1}{12}$

27. $\frac{-1}{5} - \frac{3}{4} + \frac{1}{2}$

28. $\frac{7}{3} + \frac{5}{9} - \frac{7}{6}$

29. $\frac{-3}{5} + 5 + \frac{1}{3}$

Evaluate:

30. $(10)^2(4.5) - 2(1.8)$

31. $12 - 4(2^2) + 10^3$

32. $\left(\frac{2}{3}\right)^{-2}$

33. $(100)^{-1}$

Combine like terms to simplify:

34. $5xy + 4x^2 + 6y^2 - 3xy - 6x^2$

35. $4 + 4y^5 - 7 + 6xy^5 - 8xy$

36. $2(7^2) - 5(7^2) + 2(3^2)$

37. $a^3 - 6a + a^2 - 4a + 5a^3$

Simplify. Write your answers using POSITIVE exponents:

38. $(5^5)(5^6)$

39. $s^2 \div s^5$

40. $\left(\frac{3}{2}\right)^{-2}$

41. $(6xy)^0$

## CUMULATIVE TEST Chapters 1–6 *Form A*

**Answers**

1. Evaluate: $(5)^3(3)^2 - 10^3$ — 1. ________

2. The average amount spent weekly for food by the Smith family is \$123.20. At that rate what is the family's food bill for a month? — 2. ________

3. What is the additive inverse of $\frac{-13}{2}$? — 3. ________

4. Evaluate: $|-36| - 6^2$ — 4. ________

5. Evaluate $25(-2)^2$ — 5. ________

Simplify. Round answers to two decimal places:

**Answers**

6. $1.5(3.2)^2 - 2.53$ — 6. ________

7. $-2\sqrt{49}$ — 7. ________

8. $-2.3 \div 0.01$ — 8. ________

9. $(-3)^2 - 3^2$ — 9. ________

Express as a product of prime numbers:

**Answers**

10. 245 — 10. ________

11. 1025 — 11. ________

Reduce the fractions to simplest terms:

**Answers**

12. $\frac{30}{180}$ — 12. ________

13. $\frac{48}{108}$ — 13. ________

14. Find the LCM for 12, 14 and 28. — 14. ________

Perform the indicated operations and simplify:

**Answers**

15. $\left(\frac{-3}{5}\right)\left(\frac{15}{27}\right)$ 15.________

16. $\frac{75}{-12} \div \frac{15}{2}$ 16.________

17. $\left(\frac{-3}{4}\right)^2$ 17.________

18. $\frac{1}{7} - \frac{2}{3} + \frac{6}{21}$ 18.________

Evaluate

**Answers**

19. $10\,(2.5)^2 - 2\,(-4)^2$ 19.________

20. $12 - 3\,\{-1\,(5 + 9) + 2^2\}$ 20.________

21. $\left(\frac{-1}{4}\right)^2$ 21.________

22. $(435)^0$ 22.________

Combine like terms:

**Answers**

23. $4\,(5^6) - 3\,(5^6) + 2\,(5^6)$ 23.________

24. $14xy + 8x^2 - 3xy - 7y^2 - 2x^2$ 24.________

Rewrite using only positive exponents:

**Answers**

25. $\left(\frac{a}{b}\right)^{-3}$ 25.________

Perform the indicated operations, using only positive exponents for your answers:

**Answers**

26. $(5^3)\,(5^{-2})$ 26.________

27. $c^3 \div c^4$ 27.________

28. $(z^4)^5$ 28.________

CHAPTER

# 7 Algebraic Expressions

## 7.0 *Vocabulary Preview*

Define these words without using a dictionary. Formulate their meaning from past experience and knowledge.

| | | |
|---|---|---|
| **algebraic** | **descending** | **polynomial** |
| **ascending** | **literal** | **term** |
| **binomial** | **monomial** | **trinomial** |
| **constant** | **numeric** | **variable** |

## 7.1 *Types of Algebraic Expressions*

The content of Chapter Six, and the exercises which were part of it, actually involved what are called **algebraic expressions.**

An ALGEBRAIC EXPRESSION is a statement which contains any or all of the following:

a. literal part(s)
b. numeric part(s)
c. signs indicating operations
d. grouping symbols.

The LITERAL part of an algebraic expression may change in value; thus it is called a VARIABLE.

The NUMERIC part remains unchanged; (for example, 3 always remains 3) thus, it is called a CONSTANT.

*Example:* The TERM $3x^2$ contains the variable x and the constants 3 and 2.

This SINGLE TERMED EXPRESSION is called a *MONOMIAL.*

*Example:* The expression $3x^2 - 4$ contains the variable x, and the constants 2 (the exponent), 3 and 4. It also contains the operational sign $-$.

This TWO TERMED EXPRESSION is called a *BINOMIAL.*

*Example:* The expression $3x^4 - (x^3 + 2)$ contains the variable x, the constants 3 and 2, the exponent which is the constant 4, the operational sign, $-$, and the grouping symbol ( ).

Remembering that subtraction is the inverse of addition and $(-x^3 - 2)$ is the additive inverse of $x^3 + 2$, we may write:

$$3x^4 - (x^3 + 2) \text{ as } 3x^4 - x^3 - 2$$

This THREE TERMED EXPRESSION is called a *TRINOMIAL.*

Any algebraic expression which contains MORE THAN ONE TERM is called a POLYNOMIAL.

## 7.2 *Combining Algebraic Expressions*

When combining algebraic expressions, you add or subtract the numerical coefficients of the terms which are alike; that is, terms with the same variable portion.

REVIEW THE RULES FOR ADDING AND SUBTRACTING SIGNED NUMBERS, section 3.1 and 3.2.

*Examples:* $3x + 7x = (3 + 7)x = 10x$

$$12y^3 - 16y^3 = (12 - 16)y^3 = -4y^3$$

$$6ab + 7ab = (6 + 7)ab = 13ab$$

$$-14a^2b + 8a^2b = (-14 + 8)a^2b = -6a^2b$$

$16a^2b^2 + 2a^2b$ cannot be combined since their variable portions are not alike.

$$-5x^2 + 3x^2 + 12x^2 = (-5 + 3 + 12)x^2 = 10x^2$$

$$19y + 14y + 7y^2 = (19 + 14)y + 7y^2 = 33y + 7y^2$$

**Exercise Set 7.2**

Simplify by combining like terms:

1. $2v + 41v + 19$
2. $3x - 7x + 2$
3. $5x + 3y + 2x - y$
4. $3 + 5a - 4b + 6a - 3b + 2$
5. $5s + 2r - 3s - 3r + 2t$
6. $2y^2y^3 - 9x^2 + 4 + 5y^3$
7. $3cd^2 - 4c^2d^2 + 5cd^2 + cd^2$
8. $-6 + 4x^4 - 3x + 5x^4 - x^3 + 3x^2$
9. $2t^2 - 6 + t^2 - 9t^3 + 5t^2 + 5$
10. $-3s^2t^2 - 4s^2t^2 + 7st^2$
11. $m^2 + o^3 - 3m^2 - 4o^5 + 2m^2$
12. $9g^2h^2 - 6g^2h^2 - 5g^2h^3$
13. $rs^5 - 6rs^3 + 7rs^5 + 9r^3s$
14. $7x^3y + 4x^2y^2 + 2x^2y^2$
15. $3uv^3 - 4u^3v^3 + 5uv^3$
16. $-14k - 12k + 3k^2$
17. $7x^2 + 5x - 2 - 3x^2 - 8x + 7$
18. $15x + 3y + 2z + 3x - 3y + 3z$
19. $14a^2b - 8ab^2 + 6a^2b$
20. $-12y^3 + 16y^2 - 3y^3$

## 7.3 *Adding Polynomials*

When adding polynomials, it helps to arrange the terms in **descending** order; that is with the highest power of the variable first.

*REMINDER:* COMBINE ONLY THOSE TERMS WHICH ARE ALIKE.

*Examples:* a. $(3x + 5) + (8x^2 + 3x + 2) =$
$8x^2 + (3x + 3x) + (5 + 2) = 8x^2 + 6x + 7$

b. $(7x^3 - 3x^2 + 5) + (6 - 12x^2) + 3x^3 =$
$(7x^3 + 3x^3) + (-3x^2 - 12x^2) + (5 + 6) =$
$10x^3 - 15x^2 + 11$

c. $(2 - a^3 + 2a) + (a^5 - 2 + 6a^4 - 3a^3) =$
$a^5 + 6a^4 + (-a^3 - 3a^3) + 2a + (2 - 2) =$
$a^5 + 6a^4 - 4a^3 + 2a$

d. $(a^2 + 2ab - c^2) + (3a^2 + 2ab - 5c^2) =$
$(a^2 + 3a^2) + (2ab + 2ab) + (-c^2 - 5c^2) =$
$4a^2 + 4ab - 6c^2$

e. $(7a^5 + 4a^6 - 2a^2) + (4a^2 - 3a^5 + 2a^6 - 3) =$
$(4a^6 + 2a^6) + (7a^5 - 3a^5) + (-2a^2 + 4a^2) + (-3) =$
$6a^6 + 4a^5 + 2a^2 - 3$

**Exercise Set 7.3**

Add the polynomials:

1. $(3x + 12) + (-3x - 2)$
2. $(6m - 2n) + (3m + 10n)$
3. $(8c^3 + 5d^2 - 10a) + (3c^3 + 2d^2 - 5a)$
4. $(x^2 - 5x + 15) + (-2x^2 + 10x - 5)$
5. $(6a^2 - 7b) + (3a^2 - 2b) + (2a^2 + 3b)$
6. $(y^2 - y - 1) + (3y - 2y^2 + 1) + (y^2 - y + 1)$
7. $(5y - 3x + 6) + (-3 - 5y + 2x)$
8. $(15z^2 + 3z + 2) + (3z^2 - 3z + 2)$
9. $(-8a^2b + 2b^2 - 6a^2b^2) + (8a^2b - 2b^2)$
10. $(4x^3 + 5x + 7 - 3x^2) + (3x^3 + 2x^2 + 3x - 4)$
11. $(3a^2b + 6a + 7) + (4a - 2a^2b - 5)$
12. $(xy^2 + 10x^2y + 2z) + (3z + 5x^2y + 2xy^2)$

13. $(9c^5 + 8b^4 - 2bc + 4) + (8bc + 6b^4 - 16b + 2c^5)$

14. $(2r + 3t^2 + 12s) + (2t^2 + 3s) + (75 - 2s + 5r)$

15. $(15x + 3y + 2z) + (3x - 3y + 3z) + (2x + 3y - z)$

## 7.4 *Subtracting Polynomials*

Remember from Chapter Three that subtraction of real numbers, a and b, is defined as follows:

$$a - b = (+a) - (+b) = (+a) + (-b).$$

So, when subtracting polynomials:
(the first polynomial) − (the second polynomial) =
(the first polynomial) + (the negative of the second)

*Examples:*

a. $(6x + 7y) - (8x + 2y) = (6x + 7y) + (-8x - 2y)$
$= (6x - 8x) + (7y - 2y)$
$= -2x + 5y$

b. $(3x^2 - 2xy + z^2) - (x^2 + 3xy - 6z^2) =$
$(3x^2 - 2xy + z^2) + (-x^2 - 3xy + 6z^2) =$
$(3x^2 - x^2) + (-2xy - 3xy) + (z^2 + 6z^2) =$
$2x^2 - 5xy + 7z^2$

c. $(7a + 2b + c) - (6a - 2b + c) =$
$(7a + 2b + c) + (-6a + 2b - c) =$
$(7a - 6a) + (2b + 2b) + (c - c) =$
$a + 4b$

d. $(a^2 + b^2 + c^2) - (2a + c^2) =$
$(a^2 + b^2 + c^2) + (-2a - c^2) =$
$a^2 + b^2 + (c^2 - c^2) + (-2a) =$
$a^2 + b^2 - 2a$

e. $(5a - 5b) - (6a - 18b) = (5a - 5b) + (-6a + 18b)$
$= -a + 13b$

f. $(x^2 + y^2 + z^2) - (3x^2 - 2y^2 - z) =$
$(x^2 + y^2 + z^2) + (-3x^2 + 2y^2 + z) =$
$-2x^2 + 3y^2 + z^2 + z$

g. $(2m^2 + 6) - (6m^2 - 4) =$
$(2m^2 + 6) + (-6m^2 + 4) =$
$-4m^2 + 10$

**Exercise Set 7.4**

Subtract the polynomials

1. $(2a - 3b + 2) - (a + b - 7)$
2. $(m^2 - 4n) - (n + 2)$
3. $(5c^2 + 2) - (3 + 2c^2)$
4. $(11b^2 + 2b - 1) - (6b^2 - 3b + 5)$
5. $(c^3 - 10c) - (8c^3 + 2c)$
6. $3ab - (4a - 2ab)$
7. $(5y^2 + 3y - 6) - (2y^2 + 2y - 1)$
8. $(8r^3 + 2r - 1) - (3r^3 - 2r + 5)$
9. $(4mn + 2n^2 - 1) - (3n^2 - 2mn + m^2)$
10. $(7x^3 + 2x^2 - 3x + 2) - (3x^3 - 5x^2 + 4x - 6)$
11. $(5x^3 - 6x^2 + 2x - 7) - (2x^3 + 6x + 1)$
12. $(-10xy^2 + 10y^3) - (7x^2y - 3xy^2 - 2y^3)$
13. $(10y^2 - 3y - 2) - (7y^2 + 2y + 5)$
14. $(13a^3 + 5a^2 - 2a + 3) - (6a^2 - 5a + 2)$
15. $(y - 7x^4y^2 + 3x^2) - (2y + 6x^4 - 3x^5y)$

## 7.5 *Multiplying Monomials by Monomials*

To multiply $(7xy)(-2x^2y^3)$:

1. multiply the coefficients: $(7)(-2) = -14$
2. since bases are alike, multiply $(x)(x^2)$: and $y(y^3)$
3. $x(x^2) = x^3$ and $y(y^3) = y^4$

The result of $(7xy)(-2x^2y^3)$ is $-14x^3y^4$.

To evaluate $(2x^3y^4z^2)^2$:

1. rewrite as $(2x^3y^4z^2)(2x^3y^4z^2)$
2. multiply coefficients and terms with like bases.

The result is $4x^6y^8z^4$.

*Examples:* a. $(3a^2)(6ab) = (3)(6)(a^2)(a)(b) = 18a^3b$

b. $2a^5(4a^6) = (2)(4)(a^5)(a^6) = 8a^{11}$

c. $\frac{1}{2}ab(36a^4b) = \left(\frac{1}{2}\right)(36)(a)(a^4)(b)(b) = 18a^5b^2$

d. $(3xy)^2 = (3xy)(3xy) = 9x^2y^2$

e. $(-2x)^2 = (-2x)(-2x) = 4x^2$

f. $(xyz)^3 = (xyz)(xyz)(xyz) = x^3y^3z^3$

g. $(7x)(7y) = 49xy$

h. $(-2a^6)(10a^2) = -20a^8$

i. $(.3a^2bc^3)(.5ab^3) = .15a^3b^4c^3$

**Exercise Set 7.5**

Multiply the monomials:

1. $(70x^3yz)(2x^5y^2z)$
2. $7b(a^2b^2)$
3. $3t^2(4st)$
4. $(-2x^2y)^2$
5. $.8c(.2c^2)$
6. $mn(m^2n^2)$
7. $3ab(-2ab^2)$
8. $.3t^2(.6t^3r)$
9. $5m^2(-2m^2n)$
10. $6(-2ab)$
11. $p^2(pqr)$
12. $\frac{3}{4}bc(8b)$
13. $\frac{1}{4}a(16ab)$
14. $\left[\frac{1}{2}(r^2)\right]^2$
15. $-13a(-2ab)$
16. $6a^2(-2ab)$
17. $2abc(-abc)$
18. $3ab(-3ab)$
19. $a^2cd(a^3b^2cd)$
20. $2(.2cd^2e^3)$

## 7.6 *Multiplying Polynomials by Monomials*

When multiplying a polynomial by a monomial, use the distributive property as defined in Chapter Two.

*Examples:*

$$c(a + b) = ca + cb$$
$$8(x + y) = 8x + 8y$$
$$-3x(x^2 + y^2) = -3x^3 - 3xy^2$$
$$-2xy(3xy^2 - 2) = -6x^2y^3 + 4xy$$
$$3x^3(2x^2 - 3x + 2) = 6x^5 - 9x^4 + 6x^3$$
$$p(q^4 + pq - p^3) = pq^4 + p^2q - p^4$$

**Exercise Set 7.6**

Multiply the following:

1. $5(a + b)$
2. $2(x^2 - y + 3z)$
3. $cd(c - 2d)$
4. $12x(4xy - 3y)$
5. $5y(-8y^3z)$
6. $5x^3(4xy + 8x)$
7. $7y(4y^3 - 6y + 2)$
8. $-5abc(-2ab + 2bc - c^2)$
9. $6x(x^2 + 3x + 2)$
10. $7y(x^3 - 2y + z)$
11. $2r(3 + 7r)$
12. $-4(r^2 + 2t + s^3)$
13. $-ab^2(2ab + c)$
14. $-3g(4gh - 2h^2 + 2)$
15. $3a^3(2a^2b - 6a)$

## 7.7 *Multiplying Binomials*

Remember from Chapter Three, that the **distributive law** for real numbers states:

$$a(b + c) = ab + ac$$

*Example:* $6(5 + 4) = 6(5) + 6(4) = 30 + 24 = 54$

This is easily verified, since $5 + 4 = 9$ and $6(9) = 54$

*Example:* $6(x + 4) = 6(x) + 6(4) = 6x + 24$

*Example:* $10(2 + 8) = 10(2) + 10(8) = 20 + 80 = 100$

This also is easily verified.

*Example:* $10(y + 8) = 10(y) + 10(8) = 10y + 80$

*Example:* $x(y + x - z) = x(y) + x(x) - x(z) = xy + x^2 - xz$

*Example:*
$$\begin{aligned}(2 + 5)(6 - 4) &= 2(6 - 4) + 5(6 - 4)\\ &= 2(6) + 2(-4) + 5(6) + 5(-4)\\ &= 12 - 8 + 30 - 20\\ &= 14\end{aligned}$$

*Example:* To multiply the binomials $(a + b)(c + d)$, use the distributive property:

$a(c + d) + b(c + d) = ac + ad + bc + bd$

*Example:* The product of
$$\begin{aligned}(x + 5)(x + 5) &= x(x + 5) + 5(x + 5)\\ &= x^2 + 5x + 5x + 25\\ &= x^2 + 10x + 25\end{aligned}$$

*Example:* The product of
$$\begin{aligned}(x - 5)(x - 5) &= x(x - 5) + [-5(x - 5)]\\ &= x^2 - 5x + [-5x + 25]\\ &= x^2 - 10x + 25\end{aligned}$$

*Example:*
$$\begin{aligned}(x + 5)(x - 5) &= x(x - 5) + 5(x - 5)\\ &= x^2 - 5x + 5x - 25\\ &= x^2 - 25\end{aligned}$$

*Example:*
$$\begin{aligned}(x - 5)(x + 5) &= x(x + 5) + [-5(x + 5)]\\ &= x^2 + 5x - 5x - 25\\ &= x^2 - 25\end{aligned}$$

*Example:*
$$\begin{aligned}(2x - 1)^2 &= (2x - 1)(2x - 1)\\ &= 2x(2x - 1) + [-1(2x - 1)]\\ &= 4x^2 - 2x - 2x + 1\\ &= 4x^2 - 4x + 1\end{aligned}$$

**Exercise Set 7.7**

Multiply the binomials and simplify:

1. $(3x - 4)(x + 2)$
2. $(3x - 4)(x - 2)$

**Exercise Set 7.7 *(continued)***

3. $(x + 4)^2$
4. $(6a + 2)(a - 3)$
5. $(6a - 2)(a - 3)$
6. $(s + 5)(s + 3)$
7. $(a + b)^2$
8. $(a + b)(a - b)$
9. $(a + 4)(a + 2)$
10. $(2r - s)(2r + s)$
11. $(2r + s)(2r + s)$
12. $(3y - 1)^2$
13. $(b - c)(b + 2c)$
14. $(b + 2c)(b + c)$
15. $(4x + 2)(x - 5)$
16. $(x + 10)(x - 1)$
17. $(3x + 4)(x - 2)$
18. $(x - 4)^2$
19. $(x - 4)(x + 4)$
20. $(6a - 2)(a + 3)$
21. $(s - 5)(s - 3)$
22. $(s - 5)(s + 3)$
23. $(a - b)^2$
24. $(a + 4)(a - 2)$
25. $(a - 4)(a + 2)$
26. $(2r - s)(2r - s)$
27. $(3y + 1)^2$
28. $(3y + 1)(3y - 1)$
29. $(b + c)(b - 2c)$
30. $(4x + 2)(x + 5)$
31. $(4x - 2)(x - 5)$

## 7.8 *Multiplying Trinomials and Other Polynomials*

By virtue of the distributive property of real numbers, the product of $3x(4x^3 - 8x^2 + 3x - 1)$ is equal to $3x(4x^3) + (3x)(-8x^2) + 3x(3x) + (3x)(-1)$ which equals $12x^4 - 24x^3 + 9x^2 - 3x$.

Likewise, the product of $(3x - 2)(4x^3 - 8x^2 - 1)$ is the sum of the two products:

$3x(4x^3 - 8x^2 + 3x - 1)$ and $-2(4x^3 - 8x^2 + 3x - 1)$, which becomes:

$(12x^4 - 24x^3 + 9x^2 - 3x) + (-8x^3 + 16x^2 - 6x + 2)$.

This simplifies to $12x^4 - 32x^3 + 25x^2 - 9x + 2$.

**IN GENERAL: When multiplying polynomials, multiply EACH term of the second expression by EACH term of the first. Then combine like terms to simplify.**

*Example:* $(5x - 1)(2x^2 - 3x + 2) =$
$5x(2x^2 - 3x + 2) + [-1(2x^2 - 3x + 2)] =$
$(10x^3 - 15x^2 + 10x) + (-2x^2 + 3x - 2) =$
$10x^3 - 17x^2 + 13x - 2$

**Exercise Set 7.8 A**

Multiply:

1. $3(x + 4y - 6z)$
2. $2x(3y - 8z)$
3. $2a(b + 5c - 7d)$
4. $6t(-8u + 9v - 5w)$
5. $(a + 3)(b - 4)$
6. $(x + 5)(x - 3)$
7. $(3a + 2b)(a - 3b)$
8. $(3s - t)(2s + 5)$
9. $(y - 9)(7y + 1)$
10. $(p + 2q)(p + 5q)$
11. $(m + 2q)(m - q)$
12. $(4b - 3d)(2b + d)$

**Exercise Set 7.8 B**

Multiply:

1. $(a + b)(a + b)$
2. $(a - b)(a - b)$
3. $(2a + c)(2a + c)$
4. $(2a - c)(2a - c)$
5. $(d + 3e)(d + 3e)$
6. $(d - 3e)(d - 3e)$
7. $(5s + t)(5s + t)$
8. $(5s - t)(5s - t)$
9. $(3q + 4t)(3q + 4t)$
10. $(3q - 4t)(3q - 4t)$
11. $(7p + 2r)(7p + 2r)$
12. $(7p - 2r)(7p - 2r)$

**Exercise Set 7.8 C**

Multiply:

1. $(a + b)(a - b)$
2. $(c - d)(c + d)$
3. $(2s + 3t)(2s - 3t)$
4. $(4 + z)(4 - z)$
5. $(t - 4r)(t + 4r)$
6. $(2q + r)(2q - r)$
7. $(u - 7v)(u + 7v)$
8. $(h + 9i)(h - 9i)$

**Exercise Set 7.8 D**

Multiply:

1. $(a + b)(a - b + c)$
2. $(2a + b)(3a + 4b - c)$
3. $(3d + e)(4e - 5f + g)$
4. $(g + 2h)(g + h - j)$
5. $(5s + t)(s + t - u)$
6. $(k + 3L)(L - m + 4n)$

## 7.9 *Division of Polynomials by Monomials*

To divide $\frac{8 + 4}{2}$ either add 8 + 4 first and then divide by 2, or divide **each** number in the numerator by 2 and then add.

In the first instance, you get $12 \div 2 = 6$, and in the second instance, you get $4 + 2 = 6$.

Suppose you want to divide the sum $x + y$ by 2.

Since you cannot add x and y, you must express $(x + y) \div 2$ as $\frac{x}{2} + \frac{y}{2}$.

*Example:* To divide $6x^2 + 12y^3 + 48$ by 6, divide each term by 6.

$$\frac{6x^2}{6} + \frac{12y^3}{6} + \frac{48}{6} = x^2 + 2y^3 + 8$$

*Example:* To divide $5xy + 15x^2y^2 + 25z$ by 5, divide each term by 5.

$$\frac{5xy}{5} + \frac{15x^2y^2}{5} + \frac{25z}{5} = xy + 3x^2y^2 + 5z$$

*Example:* To divide $7y^3 - 14y + 28y^5$ by 7y, divide each term by 7y.

$$\frac{7y^3}{7y} - \frac{14y}{7y} + \frac{28y^5}{7y} = y^2 - 2 + 4y^4$$

**Reminder:** When dividing expressions with like bases, subtract exponents.

*Example:* Divide $5x^3 + 10x + 25$ by $5x$

$$\frac{5x^3}{5x} + \frac{10x}{5x} + \frac{25}{5x} = x^2 + 2 + \frac{5}{x}$$

**Exercise Set 7.9**

Divide each of the following. Leave your answers in simplest form.

1. $\dfrac{25x^2 - 30x}{5}$

2. $\dfrac{6x^3 - 3x^2 - 15x + 12}{3}$

3. $\dfrac{8x^2 + 6x}{x}$

4. $\dfrac{-7x^3 - 6x^2}{2x}$

5. $\dfrac{2x^3 - 6x}{2x}$

6. $\dfrac{1.6x^3 - 1.8x^2}{0.2}$

7. $\dfrac{21a^2 - 18a}{-3}$

8. $\dfrac{ax^2 + ax + a}{a}$

9. $\dfrac{ax^2 + bx}{x}$

10. $\dfrac{-28x^3 + 16x^2 - 6x}{2x}$

11. $\dfrac{17c + 16c^2 - 1.5c^3}{-c}$

12. $\dfrac{3d^3 - 4d^2 + 8d}{-d}$

13. $\dfrac{8.8w^4 - 3.3w^3 - 2.2w}{1.1w^2}$

14. $\dfrac{7xy + 8xz}{2xy}$

15. $\dfrac{3abc + 6abd}{-2ab}$

## CHAPTER EXERCISES

### Section 7.2

Combine like terms. Leave answers in simplest form.

1. $3x + 7x$
2. $12x^2 - 3x^2$
3. $8y^2 - 6y^2 - 7$
4. $5s^2 + 6s^2 + 12$
5. $15 - 5z + 4$
6. $3a^2 + 6a^2 - 17$
7. $4b^2 - 6b^2 + 4$
8. $2y^3 - 8y^3 - 6$
9. $18 - 4x + 3 - 8x$
10. $2t - 7 + 6t - 3$
11. $4xy + 6xy - 1$
12. $3s^2t + 8s^2t + 4$
13. $2x - 6b + a - 4b$
14. $15xy - 4x^2 - 4xy$
15. $12qt + 6q + 2t + 3q$
16. $4abc - 6a^2b$
17. $100x^2 - 4y^2 + 6x^2 - 7$
18. $s^4t^2 + 4t + 5s^4t^2$
19. $4.2t - 6.3s + 3.6t$
20. $4u^2v - 7u + 6u^2 - 8u$
21. $3t + 4t^2 - 6 + 2t^2 - 7t$
22. $18.6y^3 - 2.4y^3 - 6.7$
23. $g^2h - 4 + 6g^2h - 7.1$
24. $3x^3 - 2x^2 - 6x^3 - 5x^2 + 3$

### Section 7.3 and 7.4

Simplify the following by combining like terms:

25. $(2x + 5) + (8x + 3)$
26. $(8y - 3) + (7y - 8)$
27. $(4xy - 6) + (5 - 6xy)$
28. $(7s^2 - 2) + (6 - 15s^2)$
29. $(5y^2 - 6y) + (8y^2 - 4y - 1)$
30. $(25x^2 - 16) + (100 - 81x^2)$
31. $(3st - 8) + (4st - 6t)$
32. $(2x + 3y - 7) + (4y - 10x)$
33. $(4x^2 - 7x + 5) + (8x^2 - 17x + 6)$
34. $(-s^2 - 2s - 3) + (-5s^2 - 8s - 3)$
35. $(5a^2 - ab - b^2) + (8a^2 - 4ab + b^2)$
36. $(3.2y^2 - 4y + 6.1) + (1.3y^2 - 2.1y)$
37. $(-1.5 + 2.2x - x^2) + (7.2 + .8x)$
38. $(3a^2 + 4) + (5a^2 + 7) + (12a^2 - 9)$
39. $(-10z - 5y) + (-2z + 11y) + (-3 - 4y)$
40. $(-3x + 4y) + (7y - 3) + (4x - 11)$

**Section 7.3 and 7.4**

Simplify by combining like terms:

41. $(15b^2 - 6b) + (-3b + 5) + (b^2 - 1)$
42. $(25x^2 - 4y^2) + (-8y^2 - x^2)$
43. $(42a - 7.3) + (3a^2 + 1.4)$
44. $(18x + 7) - (8x + 4)$
45. $(24y^2 + 6) - (3y^2 - 5)$
46. $(4z - 17) - (3 - 5z)$
47. $(s^2 - 3s + 2) - (2 - 8s)$
48. $(7x^2 + 4xy - 3) - (2x^2 + 3xy + 4)$
49. $(-5b^2 - 6b) - (-3b^2 + 10b)$
50. $(-10x^2 + 15) - (-25 - 3x^2)$
51. $(2.1s^2 - 3) - (6.2s^2 + 4)$
52. $(7y^2 - 4y + 6) - (2y^2 + 5y + 7)$
53. $(7.2r^2 + 4r - 1) - (1.6r^2 - 2r + .8)$
54. $(4w^2 - 6w + 5) - (3w^2 - 5w - 10)$
55. $(2 - 3t + 4t^2) - (-6t^2 - 4t + 5)$
56. $(18.7 + z^2) - (-6.1z^2 - 1.3)$
57. $(2a - 5) - (6a + 4) - (7a - 4)$
58. $(2b - 3c + 4) - (4 - 3c + b^2)$
59. $(3k - 41 - 2) - (111 - 4 + 8k)$
60. $(-20w^3 - 6w + 4) - (4 - 3c + b^2)$
61. $(-32st + 6s^2) - (7t^2 - 7st)$
62. $(-26.4 - 4x - 6y) - (-2y - 17.3)$

**Section 7.5**

Multiply the monomials and combine like terms to simplify your answers:

63. $2(8y)$
64. $(4x)(3x)$
65. $(-7y)(2y)$
66. $(-8a)(-6a)$
67. $(14b)(-3b)$
68. $(3x)(5x^3)$
69. $(10y^2)(-6y^2)$
70. $(4x)(-2x^2)$
71. $(-6x)(+3x)(-4x)$
72. $(-2x)(-3x)(-5x)$
73. $(+3xy)(-4xy)$
74. $(2x)(-7xy)(+6y)$
75. $(10x^2)(-x)(-3xy)$
76. $(2.1a)(3a^3)$
77. $(3x)^2$
78. $2x(3y)$
79. $(3x)^2$
80. $(-4y)^2$
81. $(-5z^2)^2$
82. $(a^2bc)^3$

83. $-7x(-5x^2)$

84. $(-4ab)(.2a)$

85. $(-2x^2y^2)^4$

86. $(0.1x^5)^3$

87. $(-3x^4)^3$

88. $(4x^2)(2x)$

89. $(-14xy)(-2x)$

90. $(-16ab)(2a^2b)$

91. $(.2x)(10x^3)$

92. $(0.4x)(0.2x^2)$

93. $\left(\frac{2y}{3}\right)(9y^2)$

94. $(-3a)(4a^3b)(-b^2)$

95. $(4a)^2(3a)$

## Section 7.6

Multiply. Leave your answers in simplest form.

96. $3(x + y)$

97. $10(3x - 4z)$

98. $-5(2a^2 - 3b)$

99. $-7(6w - 2w^2)$

100. $-4(3a^2 - 6a + 4)$

101. $2x(4x + 3)$

102. $-3c(5c^2 - 2c - 1)$

103. $15x(-4 - 3z + z^2)$

104. $4xy(6xy - 4)$

105. $-3ab(7a^2b - 6)$

106. $14sw(s^2w^2 - 2s + 3)$

107. $.2xy(.4x^2y - 3y^2)$

108. $\frac{2}{3}x(6x^2 - 15)$

109. $3w^2(10 - 3w + 5w^2)$

110. $-5x^2yz(-7xyz + 2x^2y)$

111. $-7(4c - 6d)$

112. $-5(5k + 71 - 4m)$

113. $-2(-10r^2 - 3r + 4)$

114. $-(6x^2 - 16x - 3)$

115. $6a(-7a^2 + 3a - 11)$

116. $0.3x(0.4x^2 - 0.9x + 5)$

117. $\frac{3}{5}y(25y^2 - 15y - 5)$

118. $st(4s^2t - 6st^2 + 2t)$

119. $-2wx^2y(-3wxy^2 + 1)$

120. $(-4d^2 + 6d)(-3d)$

121. $5x^2y^3(7 - 2x^3y^3)$

122. $1.2ab^2(0.3a^2b^2 - 0.5ab)$

123. $\frac{21}{4}s^2y(8sy - 24sy^2)$

124. $(-6x + 5 + 3z)(-4a)$

125. $-14x^2y(3y + 2x^2 - 5xy^2)$

**Section 7.7**

Multiply. Combine like terms to simplify.

126. $(x + 3)(x + 2)$
127. $(y - 4)(y - 3)$
128. $(z - 5)(z + 7)$
129. $(w + 12)(w - 4)$
130. $(a + 8)(2a + 1)$
131. $(2d + 1)(5d - 2)$
132. $(12s - 1)(3s - 2)$
133. $(a - 8)(a + 8)$
134. $(x + 7)(x - 7)$
135. $(3x - 2)(x + 10)$
136. $(r - 9)(r + 9)$
137. $(c - d)(2c + d)$
138. $(x + 4)^2$
139. $(y - 7)^2$
140. $(c + d)^2$
141. $(x - y)^2$
142. $(x - a)(x + a)$
143. $(w + z)(w - z)$
144. $(2a - 4)(3a + 3)$
145. $(12x - 7)(x + 4)$
146. $(3x + 10)(x - 1)$
147. $(x - 1)(x + 1)$
148. $(z - w)(2z + 3w)$
149. $(b + 7)(-2b - 3)$
150. $(-5 - t)(t + 3)$
151. $(-2 - k)(2k + 11)$
152. $(a + b)(a - 4)$
153. $(3x - 4)(y + 5)$
154. $(3x + 2y)(3y - 2x)$
155. $(2w + 3)^2$
156. $(11 - 4y)^2$
157. $(x^2 - 3)(2x^2 + 1)$
158. $(2y^3 - 4)(3y^2 - 1)$
159. $(x + y^2)^2$
160. $(3a - b)(3a + b)$
161. $(2d + 6)(2d - 6)$
162. $(x - y)(y + x)$
163. $(2x - 0.4)(3x + 0.5)$
164. $(6y - 15)\left(\frac{2}{3}y + 1\right)$
165. $(0.8 + x^2)^2$

## Section 7.8

Multiply. Simplify your results.

166. $(a + b)(2a + 3b)$
167. $(a + 3)(3a^2 + 2a + 4)$
168. $(x - 4)(2x^2 + 7x + 1)$
169. $(x + 5)(9 - 3x + 5x^2)$
170. $(2x - 1)(10x^2 - 6x - 3)$
171. $(r + t)(2r^2 + rt - t^2)$
172. $(y - 3)(8y^3 - 6y - 4)$
173. $(a + b)(a^2 - 6b)$
174. $(2z - 3)(4z^2 - z + 7)$
175. $(w - 3)(5w^3 - 2w^2 - 7w + 12)$
176. $(2a - 4)(6a^2 - 7a + 7)$
177. $(-3b - 5)(7b^2 - 2b - 4)$
178. $(2x - 9)(5 - 3x + x^2)$
179. $(4 - y)(3 - 5y + 6y^2)$
180. $(3 + t)(11 - 6t - 3t^2)$
181. $(2 + t^2)(-4 - 6t + 8t^2)$
182. $(2x^2 - 3x - 5)(4x - 7)$

Extras for experts:

183. $(a - b)(3a^2 + 4ab + 6b^2)$
184. $(x + y + z)(3x^2 - 4xy - 6y^2)$
185. $(2a - 3y + 4c)(5a - 6y - 7c)$

## Section 7.9

Divide the following:

186. $\dfrac{12b^2c + 6bc}{bc}$
187. $\dfrac{6a^2bc^2 + 12abc - 2a}{2a}$
188. $\dfrac{x^4 + 27x^2 + 4x}{x}$
189. $\dfrac{9d^2 + 6d^3 - 12}{3d}$

## CHAPTER TEST *Form 7A*

Combine each of the following. Classify each answer as monomial, binomial, or polynomial.

1. $6w + 7 - 12w$
2. $8x^2 + 7y - 10y + 8x^2$
3. $-3a^2 + 2ab + b^2 - 4a^2 - 2ab$

Combine the following. Leave answers in simplest form.

4. $(4x + 12) + (8x - 11)$
5. $(16a^2 - 11) - (10a^2 - 4)$
6. $(8a^2 - 6a + 2) + (-2a^2 - 6a - 2)$
7. $(5w^2 + 3w - 2) - (-2w^2 - w - 3)$
8. $15b^2 + (6b^2 - 2b) - (11b^2 - 4)$

Perform the indicated operations. Simplify results.

9. $(6x^2y)(2xy)$
10. $(-3ab)^2$
11. $4xy^2(-2xy^2)$
12. $6x(2x^2 - 5x + 2)$
13. $-11y(y^2 - 3y - 2)$
14. $4ab^2(-3a^2b - 2ab)$
15. $(x - 7)(x + 3)$
16. $(x + 6)^2$
17. $(x - 5)(x + 2)$
18. $(x + 8)(x - 8)$
19. $(2x + 3)(3x - 1)$
20. $(2x + 3)(3x - 1)$
21. $(2x + 3)(3x^2 - x + 4)$
22. $(2x - 3y) - (4x - 6y) - (x + 2y)$
23. $(2a + b)(a - b)^2$
24. $(2x + 5)(2x - 5)$
25. $(a^2 + 9a + b)(a - b)$

Divide the polynomials by the monomials:

26. $\dfrac{4 - 12y + 8y^2}{4}$
27. $\dfrac{6y^2 + 3y - 9}{3y}$
28. $\dfrac{7k + 49k^2}{7k}$
29. $\dfrac{5z^2 + 10zx}{5z}$
30. $\dfrac{x^2 - 10x + 5}{x}$

## CUMULATIVE EXERCISES **Chapters 1–7** *Set A*

1. If you purchased 100 stocks valued at \$11.75, and a month later sold them for \$12.70, how much would you have gained in the process?

2. What must be added to $-87$ for the result to be zero?

3. What is the multiplicative inverse of 75?

4. What property of real numbers is represented by the statement $2a + 3b = 3b + 2a$?

5. Evaluate $(5.2)^2 - (2.2)^2$

6. Evaluate $(\sqrt{16})^2$

7. Simplify: $25^2 \div 10^2$

8. Simplify: $\dfrac{\sqrt{25}}{|-5|}$

9. Simplify: $(-5)^2(.5)^2$

Express the following as products of prime numbers:

10. 220

11. 36

Simplify:

12. $\dfrac{-121}{11}$

13. $\dfrac{550}{45}$

Find the LCM of the groups of numbers:

14. 4, 16 and 24

15. 5, 10 and 100

Perform the indicated operations and simplify:

16. $\frac{3}{4} - \frac{2}{5} + \frac{1}{10}$

17. $\frac{4}{15} \div \frac{8}{3}$

18. $\frac{1}{3} \cdot \frac{3}{7} + \frac{1}{2} \cdot \frac{8}{9}$

19. $\frac{3}{8} - 7 + \frac{1}{8} + 3$

Evaluate:

20. $(1.2)^2(.2) - (2.5)^2$

21. $13 - 3\{(2 - 3)^5\}$

Evaluate the following without multiplying. Write answers with positive exponents only.

22. $15(100)^0$

23. $\left[\frac{2}{5}\right]^2$

24. $[20^{-2}]^2$

25. $(23^4)(23^{-4})$

26. $27^{-3} \div 27^5$

27. $x^2 \div x^5$

28. $(-1^2)^4$

Simplify by combining like terms:

29. $4t^3 - 6t + 7t^3 + 7$

30. $2x^2y^3 + 7xy^3 - 5x^2y^3$

31. $15a^2bc^2 + 8a^2b^2 - 8a^2bc^2$

32. $-5g^4h + 9g^2h^2 + 3g^4h - 5g^2$

Perform the indicated operations. Simplify answers.

33. $(3rs + 6t^2s - 7tr) + (8tr - 3t^2s) - (tr + 3t^2s + 3rs)$

34. $(2a + 9b + 1) - 3(a + b + c) + (a - 4c)$

35. $2a(b + 4d^2) - 6(c + 5d^2) - 3ab(1 + d^2)$

36. $\dfrac{3abc + 6b^2 - 12}{3b}$

37. $\dfrac{14xy^2z^2 + 7yz^3 - 21z}{7z}$

## CUMULATIVE TEST Chapters 1–7 *Form A*

What properties of real numbers are illustrated by the following:

**Answers**

1. $a + (3b + c) = (a + 3b) + c$ 1. ________
2. $2a(b + 3c) = 2ab + 6ac$ 2. ________
3. $4b + 7c = 7c + 4b$ 3. ________
4. $\frac{-1}{3} \cdot -3 = 1$ 4. ________
5. $a + 0 = a$ 5. ________

Insert the symbol, =, >, or <, which will make the statements #6 through #9 true:

**Answers**

6. 0.63 ______ 0.633 6. ________
7. –0.5 ______ –0.51 7. ________
8. $|-8|$ ______ $|7|$ 8. ________
9. $6.\overline{3}$ ______ 6.3 9. ________

For questions 10 and 11, express each number as a product of prime numbers:

**Answers**

10. 35 10. ________
11. 185 11. ________
12. Simplify: $\frac{35}{185}$ 12. ________

Perform the indicated operations and simplify answers:

**Answers**

13. $5x(x + 1) - (x^2 + 8x - 3)$ 13. ________
14. $-3x(5x - 9)$ 14. ________
15. $(x + y)^2 - 2(xy + 6)$ 15. ________
16. $(5x + 2)(5x - 2)$ 16. ________
17. $(x + 1)^2 - (x - 1)^2$ 17. ________
18. $-(17c^2d + 14cd^2 - 3)$ 18. ________
19. $18xy(-2xz + 5) - (16x^2yz + 45xy)$ 19. ________
20. $(w - 1)(w + 3) - w(w + 3)$ 20. ________
21. $(xy^2 + y^3 + 5z^2) - (6y^3 + 7) + (3xy^2 - 5z^2)$ 21. ________

Divide the following:

**Answers**

22. $\frac{12cd^2 + 14cd^2 - 6c}{2c}$ 22. ________
23. $\frac{xy^2 + x^2y^2 - xy}{xy}$ 23. ________

CHAPTER

# 8 Writing and Solving Linear Algebraic Equations

## 8.0 *Vocabulary Preview*

Define these words without using a dictionary. Formulate their meanings from past experience or knowledge.

| | | |
|---|---|---|
| **algebraic** | **equation** | **quantity** |
| **constant** | **increased** | **substitute** |
| **decrease** | **linear** | **variable** |
| **diminished** | **literal** | |

## 8.1 *Expressing Word Problems Algebraically*

From Chapters six and seven, you know that **algebraic expressions** are made up of:
numbers (numeric parts),
letters (literal parts),
signs indicating operations, and grouping symbols (parenthesis).

The **literal** part usually indicates an unknown quantity, while the other symbols show the relationship between that quantity and another (or others).

When dealing with NUMBER RELATIONSHIPS, n is generally used to represent the unknown quantity, called a VARIABLE.

When two algebraic expressions have the same numeric value, the two expressions are connected with the sign of equality, =, and this is called an **algebraic equation.**

| *Examples:* | *Expression* | *Equation* |
|---|---|---|
| | $n + 4$ | $n + 4 = 8$ |
| | $n - 6$ | $n - 6 = 12$ |
| | $3n$ | $3n = 24$ |
| | $2(n - 3)$ | $2(n - 3) = 0$ |

*REMINDER:* When indicating the operation of multiplication, the × sign is not used, since × is used (like n) to represent a variable.

Simply write the expression without an operational sign, or use parentheses as: (3)(n) or 3(n) or 3n.

*Example:* "One third of a number is 10" is written: $\left(\frac{1}{3}\right)(n) = 10.$

*Example:* "A number divided by 3 is 10" is written: $\frac{n}{3} = 10$

Note here that multiplying by $\frac{1}{3}$ and dividing by 3 are the same.

$\frac{1}{3}$ of 30 is 10 and 30 divided by 3 is 10.

*Example:* "Twice a number diminished by 2" is written $2n - 2$

*Example:* "Twice a number diminished by 2 is 26 is written $2n - 2 = 26$

*Example:* "Three times the sum of a number and 2" is $3(n + 2)$

*NOTE:* In the last expression the parenthesis show the ENTIRE SUM of $(n + 2)$ is doubled.

Without the parenthesis, $3n + 2$ would read: "the sum of three times a number plus 2".

### Exercise Set 8.1

Write the following as algebraic expressions or equations, using n to represent the variable.

1. Twice a number decreased by 4
2. Twice a number decreased by 4 is 8.
3. The sum of twice a number and 16
4. Twice the sum of a number and 16
5. Twice the sum of a number and 16 is 34.
6. The sum of twice a number and 16 is 26.
7. Half a number
8. Half a number is 10.
9. Half a number increased by 4
10. Half a number increased by 4 is 12.
11. A number decreased by 3 times itself
12. A number is twice the sum of 10 and 5.
13. Twice a number increased by 7 times itself
14. One-half a number decreased by 2 is 6.
15. Two-thirds a number increased by the number is 15.

## 8.2 *Equations Involving Sums and Differences*

Suppose we want to find the number represented by an algebraic equation:

*Example:* A number increased by 4 is 8. Find the number.

*Solution:* Let n represent the number, and

| | |
|---|---|
| translate the statement to: | $n + 4 = 8$ |
| subtract 4 from each side: | $-4 = -4$ |
| The result is: | $n = 4$ |

*Example*: A number decreased by 6 is 12. What is the number?

*Solution*: Let n represent the number and

| | |
|---|---|
| translate the statement to: | $n - 6 = 12$ |
| add 6 to each side: | $+6 = +6$ |
| The result is: | $n = 18$ |

*Example*: A number decreased by 10 is 2. Find the number.

*Solution*: Using n to represent the number,

| | |
|---|---|
| translate the statement to: | $n - 10 = 2$ |
| add 10 to each side: | $+ 10 = +10$ |
| The result is: | $n = 12$ |

*Example*: What number increased by 12 is 37?

*Solution*: Let n represent the number,

| | |
|---|---|
| translate the statement to | $n + 12 = 37$ |
| subtract 12 from each side: | $- 12 = -12$ |
| The result is: | $n = 25$ |

*Example*: If the result of decreasing a number by 13 is 89, what is the number?

*Solution*: Using n for the original number,

| | |
|---|---|
| translate the statement to; | $n - 13 = 89$ |
| add 13 to each side: | $+ 13 = + 13$ |
| The result is: | $n = 102.$ |

NOTE that we ADDED to "undo" subtraction, and we SUBTRACTED to "undo" addition.

Remember from Chapters Two and Four, that addition is the inverse of subtraction and subtraction is the inverse of addition.

**Exercise Set 8.2**

Write as algebraic equations and solve:

1. A number increased by 17 is 35. Find the number.

2. Thirty five decreased by a number is 17. Find the number.

3. What number is 2 more than 18?

4. A number decreased by 2 is 18. Find the number.

5. Twenty diminished by what number results in 18?

6. A number is diminished by 6. The result is 13. What is the number?

7. Thirteen increased by what number is 19?

8. What number diminished by 13 is 6?

9. When 14 is subtracted from a number, the result is 20. Find the number.

10. What number added to 14 results in 34?

11. What number decreased by 8 is 7?

12. What number added to 7 gives 15?

13. What number subtracted from 15 is 8?

14. What number must be added to 6 to get 20?

15. What number is 14 less than 20?

16. What number is 14 more than 6?

17. If 3 is added to a number the result is 33. What is the number?

18. Thirty less than what number is 3?

19. Thirty-three diminished by what number is 30?

20. Three more than a number is 17. What is the number?

21. Three less than a number is 17. What is the number?

22. 21 decreased by what number is 3?

23. 14 increased by what number is 17?

24. 17 is three less than what number?

25. 14 is three more than what number?

26. Eighteen is how much less than 35?

## 8.3 *Equations Involving Products and Quotients*

*Example*: Three times a number is 24. What is the number?

*Solution*: Let n represent the number, and write

$$3n = 24$$

divide each side by 3

$$\frac{3n}{3} = \frac{24}{3}$$

$$n = 8$$

*Example*: One third a number is 25. Find the number.

*Solution*: $\frac{1}{3}n = 25$

multiply each side by 3 since $3 \cdot \frac{1}{3} = 1$

$$3 \cdot \frac{1}{3}n = 25 \cdot 3$$

$$n = 75$$

NOTE: multiplication was used to "undo" division, and division was used to "undo" multiplication.

Remember from Chapters Two and Four, that multiplication is the inverse of division, and division is the inverse of multiplication.

*Example:* What number divided by 3 is 10?

*Solution:*

$$\frac{n}{3} = 10$$

multiply each side by 3:

$$3 \cdot \frac{n}{3} = 10 \cdot 3$$

$$n = 30$$

*Example:* Three fourths of what number is 60?

*Solution:*

$$\frac{3}{4}n = 60$$

divide each side by $\frac{3}{4}$

$\left(\text{or multiply each side by } \frac{4}{3} \text{ since } \frac{4}{3} \cdot \frac{3}{4} = 1\right)$

$$\frac{4}{3} \cdot \frac{3}{4}n = 60 \cdot \frac{4}{3}$$

$$n = 80$$

*Example:* Half a number is 10. What is the number?

*Solution:*

$$\frac{1}{2}n = 10$$

Multiply each side by 2: $2 \cdot \frac{1}{2}n = 10 \cdot 2$

$$n = 20$$

*Example:* Three fourths of what number is 12?

Translate the statement to:

$$\frac{3}{4}n = 12$$

divide each side by $\frac{3}{4}$

$\left(\text{that is, multiply by } \frac{4}{3}\right)$

$$\frac{4}{3} \cdot \frac{3}{4}n = (12)\frac{4}{3}$$

$$n = 16$$

*Example:* Forty two is $\frac{3}{5}$ of what number?

$$42 = \frac{3}{5}n$$

divide each side by $\frac{3}{5}$ $\left(\text{multiply by } \frac{5}{3}\right)$

$$\frac{5}{3}(42) = \frac{5}{3} \cdot \frac{3}{5}n$$

$$70 = n$$

**Exercise Set 8.3**

Write as algebraic equations and solve for the variable n:

1. Twice a number is 36. What is the number?
2. One-half a number is 36. Find the number.
3. What number divided by 2 is 36?
4. Three times a number is 39. Find the number.
5. A number is tripled and the result is 39. Find the number.
6. One-third of what number is 13?
7. Two-thirds of a number is 4. Find the number.
8. Four is three-halves of what number?
9. If you double a number the result is 100. What is the number?
10. Half of what number is 50?
11. Three times a number is thirty. What is the number?
12. What number divided by 3 is 10?
13. One-fourth of a number is 120. What is the number?
14. Four times a number is 120. What is the number?
15. Thirty is one-quarter of what number?
16. Twenty is two-fifths of what number?
17. Twelve is one-third of what number?

18. Three-fifths of 45 is what number?

19. Take two-thirds of a number, the result is 24. What is the number?

20. One-half of what number is 37?

21. Eight-tenths of 100 is what number?

22. One-fifth of 30 is what number?

23. Seven-tenths of what number is 70?

24. Five-ninths of what number is 50?

25. Three-elevenths of a number is 3. What is the number?

## 8.4 *Expressions Involving Several Operations*

When solving equations which involve several operations, such as: TWICE A NUMBER DECREASED BY 4 IS 8,

*it is important to perform the operations in order*:

| | | |
|---|---|---|
| first: | write | $2n - 4 = 8$ |
| second: | add 4 to each side | $2n = 12$ |
| then: | divide each side by 2 | $n = 6$ |

TO CHECK YOUR ANSWER: substitute 6 for n in the *original* equation:

$$2n - 4 = 8$$
$$2(6) - 4 = 8$$
$$12 - 4 = 8$$
$$8 = 8$$

*Example*: The sum of twice a number and 10 is 40. Find the number.

*Solution*:

| | |
|---|---|
| write | $2n + 10 = 40$ |
| subtract 10 from each side | $2n = 30$ |
| divide each side by 2 | $n = 15$ |

*Check*:

$$2n + 10 = 40$$
$$2(15) + 10 = 40$$
$$30 + 10 = 40$$
$$40 = 40$$

*Example*: Twice the sum of a number and 16 is 34. What is the number?

*Solution*:

| | |
|---|---|
| write | $2(n + 16) = 34$ |
| multiply through the ( ) | $2n + 32 = 34$ |
| subtract 32 from each side | $2n = 2$ |
| divide each side by 2 | $n = 1$ |

*Check*:

$$2(n + 16) = 34$$
$$2(1 + 16) = 34$$
$$2(17) = 34$$
$$34 = 34$$

*Example*: One-half the sum of a number and 6 is 8. Find the number.

write: $\frac{1}{2}(n+6) = 8$

multiply each side by 2: $2\left(\frac{1}{2}\right)(n+6) = 8(2)$

$n + 6 = 16$

subtract 6 from each side: $n = 10$

*Check*:

$$\frac{1}{2}(n+6) = 8$$
$$\frac{1}{2}(10+6) = 8$$
$$\frac{1}{2}(16) = 8$$
$$8 = 8$$

**Exercise Set 8.4**

Write each of the following as an algebraic equation and solve for the variable n. Check the results.

1. One-half a number decreased by 2 is 6.

2. Twice a number increased by 2 is 8.

3. One-third a number increased by 4 is 10.

4. Three times a number is decreased by 9. What is the number if the result is 18?

5. Ten is increased by 4 times a number. The result is 22.

6. Ten times what number must be subtracted from 60 to give 20?

7. Five is $\frac{1}{3}$ of what number increased by 2?

8. Subtract 15 from four times a number. The result is 5.

9. The sum of twice a number and 12 is 28.

10. Twice the sum of a number and 12 is 28.

11. The difference between 3 times a number and 10 is 20.

12. The sum of $\frac{1}{2}$ a number and 3 is 9.

13. The difference between twice a number and 5 is 15.

14. The sum of 5 times a number and 21 is 56.

15. Three times a number plus 9 is 21.

16. The sum of twice a number and 5 is 29.

17. Five is added to 4 times a number. The result is 37.

18. Add 12 to twice a number. The result is 26.

19. Five times a number decreased by 13 is 12.

20. Ten is $\frac{2}{3}$ of what number decreased by 2?

## 8.5 *Multiplying by –1 to Find n*

Sometimes the solution for an equation is found to be in terms of –n rather than n. This is easy to resolve, if you remember that **–n is a "reflection" of +n on the number line,** just as –10 is a "reflection" of +10 on the number line.

So: if n = 10, it makes sense to say than –n = –10; and if –n = 5, it is true that n = –5.

This same result is obtained by multiplying each side of the equations by –1.

*Example*: Five is decreased by 3 times a number. The result is 2.

*Solution*:

| | |
|---|---|
| write | $5 - 3n = 2$ |
| subtract 5 from each side | |
| (or by inverses: add –5) | $-3n = -3$ |
| multiply each side by –1 | $3n = 3$ |
| divide each side by 3 | $n = 1$ |

*Check*:

$$5 - 3n = 2$$
$$5 - 3(1) = 2$$
$$5 - 3 = 2$$
$$2 = 2$$

*Example*: Two-thirds a number diminished by 5 is 3. Find the number.

| | |
|---|---|
| write | $\frac{2}{3}n - 5 = 3$ |
| add 5 to each side | $\frac{2}{3}n = 8$ |
| multiply each side by $\frac{3}{2}$ | $n = 12$ |

*Check*:

$$\frac{2}{3}n - 5 = 3$$
$$\frac{2}{3}(12) - 5 = 3$$
$$8 - 5 = 3$$
$$3 = 3$$

*Example*: Ten is diminished by 4 times a number. The result is –2.

*Solution*:

| | |
|---|---|
| write | $10 - 4n = -2$ |
| subtract 10 from each side | $-4n = -12$ |
| multiply each side by –1 | $4n = 12$ |
| divide each side by 4 | $n = 3$ |

*Check*:

$$10 - 4n = -2$$
$$10 - 4(3) = -2$$
$$10 - 12 = -2$$
$$-2 = -2$$

**Exercise Set 8.5**

Write the following as algebraic expressions and solve for n. Check your results.

1. Twelve is decreased by 3 times a number. The result is 6. Find the number.

2. Four diminished by twice a number is zero. What is the number?

3. One-half a number is subtracted from 7. The result is 4. Find the number.

4. Fifty-four is decreased by 4 times a number, resulting in the number 6. Find the number.

5. Twice a number is decreased by 8. If the result is 32, find the original number.

6. Twelve times a number is subtracted from 36. The result is 12. What is the number?

7. If $\frac{3}{4}$ of a number is subtracted from 30 the result is 6. Find the number.

8. Twice a number decreased by 15 is 20. What is the number?

9. What number can be doubled and subtracted from 10 to give zero?

10. Subtract $\frac{3}{5}$ of a number from 50. If the result is 14, find the number.

11. Ten decreased by three times a number is 1. What is the number?

12. Twelve decreased by 3 times a number is 3. Find the number.

13. Two is diminished by a number. If the result is 5, what is the number?

14. Three-fourths of a number decreased by 10 is –4. What is the number?

15. Two-thirds of a number is diminished by 15. If the result is –5, find the number.

## 8.6 *Equations Requiring Simplification before Solving*

Sometimes it is necessary to simplify one or both sides of an equation before solving.

*Example*: In order to solve the equation

$$2x + 4 = 3x - 7,$$

you must have the variable terms together on the same side of the equation, and the constants together on the other side.

Subtracting 2x from each side gives

Now, adding 7 to each side gives

$$\begin{array}{rcl} 4 &=& x - 7 \\ +7 &=& \quad + 7 \\ \hline 11 &=& x \end{array}$$

*Check:*

$$\begin{aligned} 2x + 4 &= 3x - 7 \\ 2(11) + 4 &= 3(11) - 7 \\ 22 + 4 &= 33 - 7 \\ 26 &= 26 \end{aligned}$$

*Example:* Solve the same equation in a different way.

Subtract 3x from each side:

$$\begin{array}{rcl} 2x + 4 &=& 3x - 7 \\ -3x & & -3x \\ \hline -x + 4 &=& -7 \end{array}$$

Subtract 4 from each side:

$$\begin{array}{rcl} -x \; + 4 &=& - \; 7 \\ - 4 & & - \; 4 \\ \hline -x &=& -11 \end{array}$$

Divide each side by $-1$:

$$\frac{-x}{-1} = \frac{-11}{-1}$$

$$x = 11$$

*Example:* Solve for x: $3(x + 5) = 4(x - 2)$.

1. Multiplying through the parenthesis gives: $3x + 15 = 4x - 8$
2. Subtracting 3x from each side gives: $15 = x - 8$
3. Adding 8 to each side gives the result: $23 = x$

*Check*:

$$3(x + 5) = 4(x - 2)$$
$$3(23 + 5) = 4(23 - 2)$$
$$3(28) = 4(21)$$
$$84 = 84$$

*Example*: Solve for n: $\frac{3}{4}(n + 8) = n$

1. Multiplying through the parenthesis gives: $\frac{3}{4}n + 6 = n$
2. Subtracting $\frac{3}{4}n$ from each side gives: $6 = \frac{1}{4}n$
3. Multiplying each side by 4 gives the final result: $24 = n$

*Check*:

$$\frac{3}{4}(n + 8) = n$$
$$\frac{3}{4}(24 + 8) = 24$$
$$\frac{3}{4}(32) = 24$$
$$24 = 24$$

*Example*: Solve for t: $3(t + 4) - 5(t - 2) = 26$.

1. Multiply through the parenthesis: $3t + 12 - 5t + 10 = 26$ (watch the signs!)
2. Collect like terms: $-2t + 22 = 26$
3. Subtract 22 from each side of the equation: $-2t = 4$
4. Divide each side by $-2$ to get the final result: $t = -2$

check to see if: $3(-2 + 4) - 5(-2 - 2) = 26$!

**Exercise Set 8.6**

Solve the following equations.

| | **Answers** |
|---|---|
| 1. $2x + 4x = 30$ | 1. __________ |
| 2. $3x - 5 + 6x = 54$ | 2. __________ |
| 3. $7y + 10 - 3y = -46$ | 3. __________ |
| 4. $8y + 15 = 2y + 45$ | 4. __________ |
| 5. $7z + 11 = 6z$ | 5. __________ |
| 6. $15a - 4 = 11a$ | 6. __________ |
| 7. $21 - 9c = c - 1$ | 7. __________ |
| 8. $2z - 8 + 6z = 18z + 2$ | 8. __________ |
| 9. $2(x + 3) = x + 5$ | 9. __________ |

**Answers**

10. $4w - 2w + 16 = w - 35$ 10. ________

11. $1.7x - 3 = 2x + 6$ 11. ________

12. $-3.7 + 2.5y = .3 - .5y$ 12. ________

13. $\frac{3}{5}x - 4 = 6 - \frac{2}{5}x$ 13. ________

14. $\frac{5}{3}y + 7 - \frac{2}{3}y = -3$ 14. ________

15. $\frac{3}{11}x + 14 = \frac{5}{11}x - 6$ 15. ________

16. $-2(x - 5) = 5x + 3$ 16. ________

17. $4(x - 7) = 2(x + 6)$ 17. ________

18. $9c - 3 + 4c + 7 = 3c + 9$ 18. ________

19. $10w + 7.2 = 2w - 1.6$ 19. ________

20. $35c - 48 = 12c - 2$ 20. ________

## 8.7 *Evaluating Expressions and Formulas*

In many professions and skilled workmanship positions, the ability to use formulas is of great importance. Formulas for areas, perimeters, volumes, interest, heat and energy require algebra skills. In this section, we will focus on evaluating formulas most frequently used. The skills required are precisely those developed in the preceding sections of this chapter.

*Example*: In the formula $A = lw$, find w if $A = 64$ and $l = 10$.

Substitute in $A = lw$

$64 = 10w$

Divide each side by 10 $6.4 = w$

*Example*: In the formula $a^2 + b^2 = c^2$, solve for b if $a = 6$ and $c = 10$.

Substitute in $a^2 + b^2 = c^2$

$6^2 + b^2 = 10^2$

Simplify $36 + b^2 = 100$

Subtracting 36 from each side $b^2 = 100 - 36$

$b^2 = 64$

Taking square root of each side $b = \pm 8$

*Example:* In the formula $y = mx + b$, solve for b if $y = 2$, $m = -2$ and $x = 3$

Substituting the values in the formula:

$$\begin{aligned} 2 &= (-2)(3) + b \\ 2 &= -6 + b \\ +6 &\quad\;\; +6 \\ \hline 8 &= b \end{aligned}$$

*Example:* Solve for b if $a = 8$, $h = 4$ and $A = 35$ if the formula is given:

$$A = \frac{1}{2}[(a + b)]h$$

Substituting $$35 = \frac{1}{2}[(8 + b)](4)$$

Multiply $\frac{1}{2}(4)$ $$35 = (8 + b)(2)$$

Divide each side by 2 $$17.5 = 8 + b$$

Subtract 8 $$9.5 = b$$

**Exercise Set 8.7**

Solve for the indicated variable, given the values of the other variables.

| | | | | |
|---|---|---|---|---|
| 1. $A = Lw$ | solve for a | $L = 70$ | $w = 40$ | |
| 2. $F = MA$ | solve for M | $F = 250$ | $A = 25$ | |
| 3. $r^2 = x^2 + y^2$ | solve for y | $x = 7$ | $r = 9$ | |
| 4. $z = \dfrac{x - m}{s}$ | solve for x | $m = 80$ | $s = 5$ | $z = 1.5$ |
| 5. $y = mx + b$ | solve for b | $y = -1$ | $m = 3$ | $x = -2$ |
| 6. $A = P(1 + r)$ | solve for P | $A = 1100$ | $r = 10\%$ | |
| 7. $V = \pi r^2 h$ | solve for V | $r = 4$ | $h = 6$ | $\pi = 3.14$ |
| 8. $a(c - d) = b$ | solve for a | $c = 5$ | $d = 2$ | $b = 18$ |
| 9. $P = 2L + 2w$ | solve for L | $P = 70$ | $w = 10$ | |
| 10. $A = \frac{1}{2}(bh)$ | solve for h | $A = 21$ | $b = 6$ | |

## 8.8 *Solving Equations Containing No Constants*

If you were solving the equation $x + 5 = 1$, you would subtract 5 from each side:

$$\begin{array}{l} x + 5 = 1 \\ \underline{\quad -5 \quad -5} \\ x \quad = -4 \end{array}$$

to get the result $x = -4$

So, if you have an equation $x + y = z$, to solve for x, you would subtract y:

$$\begin{array}{l} x + y = z \\ \underline{\quad -y \quad -y} \\ x \quad = z - y \end{array}$$

to get the result $x = z - y$

To solve the equation $2x + 6 = 12$, subtract 6

$$\begin{array}{l} 2x + 6 = 12 \\ \underline{\quad -6 \quad -6} \end{array}$$

$$\frac{2x}{2} = \frac{6}{2} \quad \text{then divide by 2}$$

$$x = 3$$

So to solve the equation $ax + b = c$, subtract b

$$\begin{array}{l} ax + b = c \\ \underline{\quad -b \quad -b} \end{array}$$

$$\frac{ax}{a} = \frac{c - b}{a} \quad \text{then divide by a}$$

$$x = \frac{c - b}{a}$$

*Example:* Solve for t: $d = st$

"undo" multiplication by division, so:

$$\frac{d}{s} = \frac{st}{s}$$

$$\frac{d}{s} = t$$

*Example:* Solve for b: $a^2 + b^2 = c^2$

Subtract $a^2$:

$$\begin{array}{l} a^2 + b^2 = c^2 \\ \underline{-a^2 \qquad\quad -a^2} \\ \qquad b^2 = c^2 - a^2 \end{array}$$

"Undo" squaring by taking square root

$$\sqrt{b^2} = \sqrt{c^2 - a^2}$$

$$b = \pm\sqrt{c^2 - a^2}$$

*Example:* Solve for l: $V = lwh$

Divide by wh: $\frac{V}{wh} = \frac{lwh}{wh}$

$$\frac{V}{wh} = l$$

**Exercise Set 8.8**

Solve for the indicated variable in terms of the others.

| | |
|---|---|
| 1. $A = bh$ | for h |
| 2. $P = 2L + 2w$ | for w |
| 3. $P = a + b + c$ | for c |
| 4. $V = Lwh$ | for w |
| 5. $d = st$ | for s |
| 6. $a^2 + b^2 = c^2$ | for c |
| 7. $A = \frac{1}{2}bh$ | for h |
| 8. $PRT = I$ | for P |
| 9. $PRT = I$ | for R |
| 10. $d = 2r$ | for r |
| 11. $A = bh$ | for b |
| 12. $P = 2L + 2w$ | for L |
| 13. $P = a + b + c$ | for a |
| 14. $V = Lwh$ | for h |
| 15. $a^2 + b^2 = c^2$ | for a |
| 16. $A = \frac{1}{2}bh$ | for b |
| 17. $A = s^2$ | for s |
| 18. $PRT = I$ | for T |
| 19. $P + T = C$ | for P |
| 20. $c = \frac{1}{2}(a + b)$ | for b |

## CHAPTER EXERCISES

### Section 8.1

Use n as the variable and write the following as algebraic expressions or equations:

1. A number increased by 5
2. A number increased by 5 is 25.
3. Twice a number increased by 2
4. Twice a number increased by 2 is 12.
5. Three times a number decreased by 5
6. The sum of twice a number and 4 is 6.
7. One-half a number increased by 10
8. Twice the sum of a number and 3
9. A number doubled is 24.
10. One-third of a number is 15.
11. When 4 is added to a number the result is 44.
12. The sum of a number and 10 increased by 6
13. Twice a number diminished by 7 is 5.

### Section 8.2

Write each of the following as an algebraic equation and then solve for the variable n:

14. A number increased by 3 is 15. Find the number.
15. A number decreased by 5 is 25. Find the number.
16. A number increased by 10 is 32. What is the number?
17. A number diminished by 4 is 28. Find the number.
18. 12 more than a number is 32. What is the number?
19. When 15 is added to a number the result is 27. What is the number?
20. When 15 is subtracted from a number the result is 27. What is the number?

Write as algebraic equations and solve for the variable n:

21. What number must be added to 35 so that the sum is 50?

22. What number subtracted from 30 results in 8?

23. What number must be added to 25 to get the result of 32?

24. If 7 is increased by a number the result is 5. Find that number.

25. A number diminished by 5 is 2. Find the number.

26. A number diminished by 5 is 12. Find the number.

27. When 8 is added to a number the result is zero. What is the number?

28. A number increased by .4 is 2.8. Find the number.

29. When 16 is subtracted from a number the result is 20. What is the number?

30. 21 decreased by what number is 37?

31. A number diminished by 1.5 is 3.5. Find the number.

32. A number increased by $1\frac{1}{2}$ is $11\frac{1}{2}$. Find the number.

33. A number diminished by 3 is 17. What is the number?

34. 6 less than a number is 35. What is the number?

35. 26 is 7 less than what number?

36. What number must be added to 7 to give the result −12?

37. 11 is 5 more than what number?

38. What number diminished by 12 results in −17?

39. −9 is 3 more than what number?

40. When −4 is subtracted from a number the result is 10. What is the number?

41. When −13 is added to a number the result is zero. What is the number?

42. −2 less than a number is 10. What is the number?

43. A number increased by −2.5 is zero. Find the number.

**Section 8.3**

Write each of the following as an algebraic equation using n as the variable. Solve.

44. Twice a number is 14. What is the number?

45. One-third of a number is 18. Find the number.

46. Four times a number is 20. What is the number?

47. A number divided by 3 is 21. Find the number.

48. Three-fourths of what number is 18?

49. Two-thirds of what number is 50?

50. A number multiplied by 2.5 is 120. Find the number.

51. Sixty is $\frac{2}{3}$ of what number?

52. $3\frac{1}{2}$ times a number is 90. Find the number.

53. A number divided by 16 is 96. Find the number.

54. What number divided by 1.4 results in the number 7?

55. $\frac{5}{4}$ of what number is 28?

56. A number divided by $\frac{7}{8}$ is 56. Find the number.

57. If a number is multiplied by 2 and then divided by 3, the result is 45. What is the number?

58. If a number is multiplied by 5 and then divided by 4, the result is 8. What is the number?

## Section 8.4

Express each of the following as an algebraic equation using n for the variable. Solve each equation and check your answers.

59. Twice a number increased by 4 is 50. Find the number.

60. Six times a number decreased by 18 is 24. What is the number?

61. One-quarter of what number is 32?

62. If 15 is subtracted from 3 times a number the result is 30. What is the number?

63. 4 times the sum of a number and 7 is 44. Find the number.

64. The sum of two-thirds of a number and 11 is 51. Find the number.

65. If 12 is added to the sum of a number and 6, the result is 55. Find the number.

66. If 5 times a number is decreased by 17 the result is 52. Find the number.

67. The difference between 7 times a number and 22 is 64. Find the number.

68. When 2.75 is subtracted from three times a number, the result is 10.5. Find the number.

69. If the difference between a number and 8 is tripled, the result is 42. What is the number?

70. One-half a number increased by $4\frac{1}{2}$ is $23\frac{1}{2}$. Find the number.

71. Six times a number diminished by 5 is 43. Find the number.

72. Fifteen less than 5 times a number is 100. What is the number?

73. Nine times what number diminished by 10 is –64?

74. When 13 is added to twice a number the result is 39. Find the number.

75. Three times the sum of twice a number and 2.7 is 4.5. Find the number.

76. If 15 is subtracted from 4 times a number the result is –15. What is the number?

77. If twice a number divided by five is increased by seven the result is 43. Find the number.

78. When the sum of three times a number and 11 is divided by 2 the result is 32. What is the number?

**Section 8.5**

Write each of the following as an algebraic equation. Solve the equation and check your answer.

79. If 17 is decreased by twice a number the result is 35. Find the number.

80. When $\frac{1}{3}$ of a number is subtracted from 39, the result is 4. Find the number.

81. 12 diminished by four times a number is 48. Find the number.

82. What number must be subtracted from 18 to produce the result of 11?

83. 7.5 decreased by twice a number is 3.7. Find the number.

84. If $\frac{4}{3}$ of a number is subtracted from –3 the result is 15. What is the number?

85. 10 decreased by the sum of a number and 3 is –16. Find the number.

86. When three times a number is subtracted from 17 the result is 50. Find the number.

87. If 5 is diminished by 5 times a number, the result is 35. What is the number?

88. When –8.8 is decreased by 2.3 times a number, the result is –1.3. What is the number?

89. 174 is diminished by six times a number. The result is 51. What is the number?

90. Five times the difference of 8 and twice a number is 0. What is the number?

### Section 8.6

Solve each of the following equations and check your results.

91. $2n = 38$
92. $3n + 6 = 27$
93. $5n - 16 = 94$
94. $\frac{2}{3}x = 18$
95. $8y + 2 = 18$
96. $2z + 11 = 47$
97. $21 - 3a = 53$
98. $2(x + 1) = 18$
99. $17 - 3y = 29$
100. $3x + 2.1 = -8.7$
101. $36 - a = 22$
102. $c - 12 = -17$
103. $6 - x = 35$
104. $2(4 - y) = -14$
105. $\frac{2}{3}x + \frac{4}{3} = \frac{8}{3}$
106. $.6 - .3z = 1.7$
107. $5x - 2.5 = 0$
108. $9y - 10 = 64$
109. $-2(x + 3) = 16$
110. $17 - 3x = 50$
111. $4(7 + 2x) = -1$
112. $3(x - 8) = 42$
113. $\frac{1}{2}x - 3\frac{1}{2} = 20\frac{1}{2}$
114. $\frac{1}{4} - 3x = 15\frac{1}{4}$
115. $12x - 10 = 86$
116. $\frac{2x}{5} + 7 = 43$
117. $12x - 8 = 0$
118. $2x - 16 = 20$
119. $-3(x + 6) + 10 = -20$
120. $-5(6 - x) - 1 = -7$

### Section 8.7

Solve for the indicated variable given the values for the others:

121. $V = Lwh$ solve for V if $L = 10.5$ $w = 4$ $h = 2$
122. $d = st$ solve for s if $d = 150$ $t = 2.3$

### Section 8.8

Solve for the indicated variable:

123. $p = 2r + 2q$ r
124. $z = xy$ y
125. $r^2 = s$ r
126. $d = \frac{-b}{a}$ b
127. $s = 3rt$ t
128. $g = \frac{1}{2}ef^2$ e
129. $A = \frac{(b + c)h}{2}$ b
130. $b^2 = 4ac$ c

## CHAPTER TEST *Form 8A*

Express each of the following as an algebraic equation. Solve the equation and check your results.

1. Three times a number is 81. What is the number?

2. Four times a number increased by 11 is 47. Find the number.

3. Two-thirds of a number is –24. What is the number?

4. Three times a number diminished by 16 is 7. Find the number.

5. Twice the sum of a number and 10 is 56. Find the number.

6. 18 less than a number is 54. Find the number.

7. What number must be added to 15 to produce the result –39?

8. A number diminished by 4.3 is –6.2. Find the number.

9. Five less than four times a number is –23. What is the number?

10. One-third of a number increased by 11 is –2. Find the number.

11. Four times a number divided by five is –16. What is the number?

12. Twenty-four is three-fifths of what number?

13. What number divided by 1.3 is 2.4?

14. If 12 is subtracted from four times a number the result is 32. Find the number.

15. If –17 is decreased by twice a number the result is 9. What is the number?

16. Solve for n: $2n - 6 = 18$

17. Solve for x: $5 - 2(x + 1) = 3$

18. Solve for a: $25 - \frac{2}{3}a = 0$

19. Solve for z: $10 - (8 - z) = 11$

20. Solve for w: $-8 + 3(7 - w) = -6$

21. Solve for y: $9y = 2y + 21$

22. Solve for x: $3x + 11 = 2x$

23. Solve for z: $18 + 7z = 5z + 12$

24. Solve for w: $-3w + 23 = 5(w - 1)$

25. Solve for x: $1.3x - 3.7 = 8.7x - 2.7$

26. Solve for t: $st - w = x$ if $s = 5$, $w = -2$ and $x = 17$

27. Solve for y: $xyz = 8$ if $x = -2$ and $z = -1$

28. Solve for r: $C = 2\pi r$

29. Solve for d: $\frac{1}{2}d = r$

30. Solve for e: $a + be = 0$

## CUMULATIVE EXERCISES Chapters 1–8 *Set A*

Evaluate the following:

1. $(6)^{-2}$
2. $(-1)^3$
3. $(1.2)^2$
4. $(52^3)^2$
5. $(.5)^2 - 3(2^3)$
6. $-3(-5) + -2(6)$

Combine like terms:

7. $7x^2 + 2x - 8x^2 + x + 9$
8. $(11x^2 + 6x - 1) - (3x^2 + 4)$
9. $-(13y^3 - 8y + 7) + (13y^3)$
10. $(7a^3 - 5a^2 + 3a + 7) + 4a^3$

Evaluate:

11. $(-10)^{-2}$
12. $(-34)^2$
13. $(x^2)(x^3)$
14. $\left(\frac{3}{4}\right)^{-1}$
15. $\left(\frac{-3}{11}\right)^2$

Perform the indicated operations and simplify:

16. $\frac{2}{5} - \frac{1}{3} + \frac{7}{2}$
17. $\frac{12}{15} \cdot \frac{3}{4}$
18. $\frac{-12}{7} \div \frac{1}{14}$
19. $(x - 2)(x + 5)$
20. $(2x + 1)^2$
21. $7ab(2a^2b + 5ab + c)$
22. $(a^2 + 2ab + c^2)(2a)$
23. $(x^2 - 8)(x^2 + 8)$
24. $\frac{3t^2 + 9t^2 - 18t}{3t}$
25. $\frac{2w^2 - 8w + 6}{2w}$

Solve the equations. Check your results.

26. $6y + 4 = -14$
27. $\frac{3}{2}b + 3 = -9$
28. $24 - 4t = 0$

Write the following as algebraic equations and solve.

29. Seven times a number less two is 40. Find the number.

30. 50 less than 5 times a number is 14. Find the number.

31. One-half a number plus 50 is 0. What is the number?

32. Grandpa is four years more than three times Becky's age. How old is Becky if Grandpa is 79? (Hint: Let g represent Grandpa's age and b represent Becky's.)

33. If a number minus 3 is equal to twice itself, what is the number?

34. The sum of two consecutive integers is 113. What are the two integers? (Consecutive integers follow one another: 1, 2, 3, etc. Let n represent the first integer.)

35. The sum of two consecutive EVEN integers is 38. Find the two even integers. (Let n represent the first of the two EVEN integers.)

36. Solve for c: $c + (c + 2) = 96$

37. Solve for x: $4x + 6 = 7x - 15$

38. Solve for x: $3(5 - x) = 8x + 2$

39. Solve: $1.5x + 3 = 2.4x + 2.1$

40. Solve: $.5x + .6 = .7x - 2.8$

41. Solve: $\frac{1}{4}(16 - 8x) = \frac{1}{3}(9 - 12x)$

42. Solve: $-(14 + x) = 2 - (3x - 6)$

43. Solve for d: $\frac{d - f}{2} = g$ if $g = 25$ and $f = 30$

44. Solve for v: $v + xy = w$

45. Solve for a: $\frac{1}{2}(a + b) = c$

## CUMULATIVE TEST Chapters 1-8 *Form A*

**Answers**

1. What is the reciprocal of $\frac{-5}{6}$? 1. ________
2. True of false: $8(a - 3b) = 8a - 24b$ is an example of the associative property of real numbers. 2. ________
3. What is the additive inverse of $-56$? 3. ________
4. Evaluate $(-4.5)^2 - 20$ 4. ________
5. Find $\sqrt{23.76}$ to two decimal places. 5. ________
6. Evaluate $-2|-5|$ 6. ________
7. Simplify $8x^5 \div 4x^2$ 7. ________
8. Evaluate $6(3^2) - 10(2^3)$ 8. ________
9. Multiply $\frac{7}{15} \cdot \frac{3}{21}$ 9. ________
10. Simplify $8\{5 - (3 + 1)^2 - 6\}$ 10. ________
11. Combine like terms: $12x^2y - 6x^2z - 3x^2y + 10x^2$ 11. ________
12. Combine: $-(b^2 + 3ab - c^2) + (6ab - 5c^2)$ 12. ________
13. Multiply $(3x + 2)(x - 5)$ 13. ________
14. Multiply $(7x - 1)(7x + 1)$ 14. ________
15. Divide $3x^2y + 15xy - 18x$ by $3x$ 15. ________
16. Solve: $2z - 14 = -2$ 16. ________
17. Solve: $4(x - 1) = 3(x + 4)$ 17. ________
18. Solve: $\frac{2}{3}x + 6 = \frac{1}{3}x$ 18. ________
19. If 3 times a number less 10 is 5, what is the number? 19. ________
20. If a father is twice his son's age plus 6 years, how old is the son if his Dad is 30? 20. ________
21. Solve for k: $t + m = \frac{1}{2}k$ if $t = 4$ and $m = 7$ 21. ________
22. Solve for r: $2t + 3r = s$ 22. ________

CHAPTER

# 9 Graphing Linear Equations

## 9.0 *Vocabulary Preview*

Define these words without using a dictionary. Formulate their meanings from past experience or knowledge.

| | | |
|---|---|---|
| **axis** | **horizontal** | **plot** |
| **coordinate** | **intersection** | **plane** |
| **counter-clockwise** | **intercept** | **point** |
| **determine** | **linear** | **quadrant** |
| **equation** | **ordered** | **slope** |
| **geometry** | **origin** | **unique** |
| **graph** | **pair** | **vertical** |

## 9.1 *The Coordinate Plane*

In the 17th Century, a mathematician named Des Cartes (Day-cart), attempted to draw **graphs** (visual representations) of algebraic equations. His work led to the development of a branch of mathematics called Analytic Geometry, sometimes referred to as **Coordinate** Geometry.

The basic concepts of this area of mathematics will be treated in this chapter.

The first concept to consider is the relationship between a geometric **point** (location) on a **plane** (flat surface) and its numeric representation.

1. Draw two **intersecting** lines, one **vertical,** the other **horizontal.** The vertical line will be called the y axis, and the horizontal line will be called the x axis.
2. Place arrowheads at both ends of the axes to show that each line contains an infinite number of points and continues in both directions.
3. Mark the point of intersection "o" to indicate that it is the point of **origin.**
4. Moving in a **counter-clockwise** direction, number the **quadrants** with Roman numerals, as shown in figure 9.1.

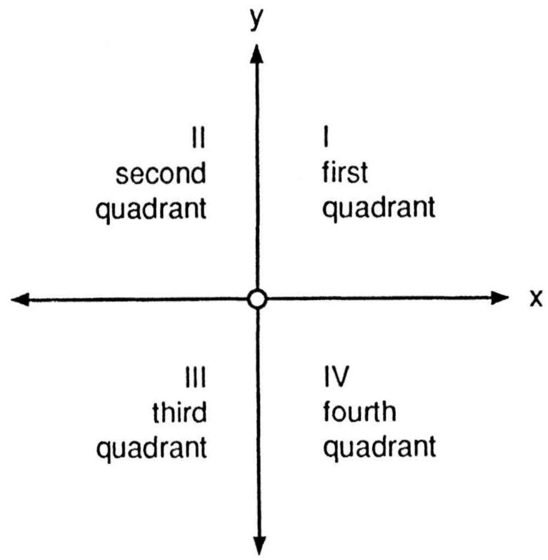


**Figure 9.1**

**Summarizing**

| *Quadrant* | *Relative to Origin* | *Relative to x axis* |
|---|---|---|
| I | right | above |
| II | left | above |
| III | left | below |
| IV | right | below |

## 9.2 *Plotting Points on the Cartesian Plane*

Figure 9.1 is a representation of the Cartesian Plane, so named after the mathematician Des Cartes. The x and y axes are divided into equal units, similar to the real number line. (See figure 9.2 A)

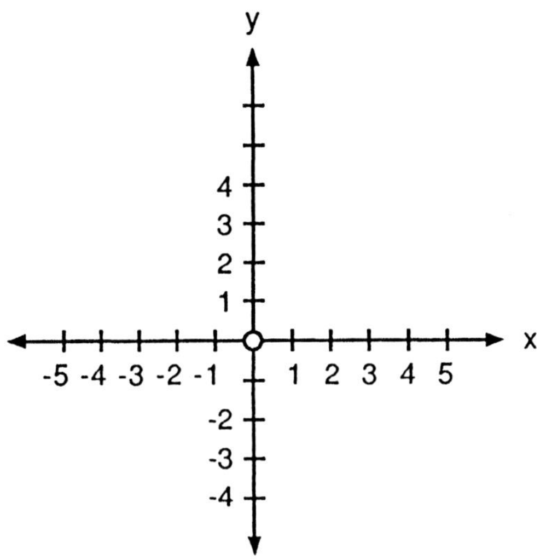


**Figure 9.2A**

Numbers to the **right** of zero on the x axis are **positive.** Numbers to the **left** of zero on the x axis are **negative.**

Numbers which are on the y axis and **above** the x axis are **positive.** Numbers which are on the y axis and **below** the x axis are **negative.**

**Note:** If you move right or left but NOT up or down, the point is on the x axis; if you move up or down but NOT right or left, the point is on the y axis.

**Points on the Cartesian Plane are represented by ordered pairs of numbers, so called because**

**the FIRST number represents the X coordinate**
**and**
**the SECOND number represents the Y coordinate.**

The pairs of numbers are written in parentheses and separated by commas: (x,y) (3,−1) (0,6) etc.

The sign of each number indicates "direction", that is:

if x is positive, the point is to the right of zero
if x is negative, the point is to the left of zero
if y is positive, the point is above zero
if y is negative, the point is below zero.

*Example:* Plot (locate) the following points in the Cartesian Plane:
A = (3,5) B = (−3,5) C = (−3,−5) D = (3,−5)
E = (0,5) F = (0,−5) G = (−3,0) H = (3,0)

(See figure 9.2 B)

For A: start at origin and move 3 units to the right along the x axis, then up 5 units

For B: start at origin; move 3 units left and 5 units up

For C: start at origin; move 3 units left and 5 units down

For D: move 3 units right and 5 units down from zero.

For E: move 5 units up from zero

For F: move 5 units down from zero

For G: move 3 units left from zero

For H: move 3 units right from zero.

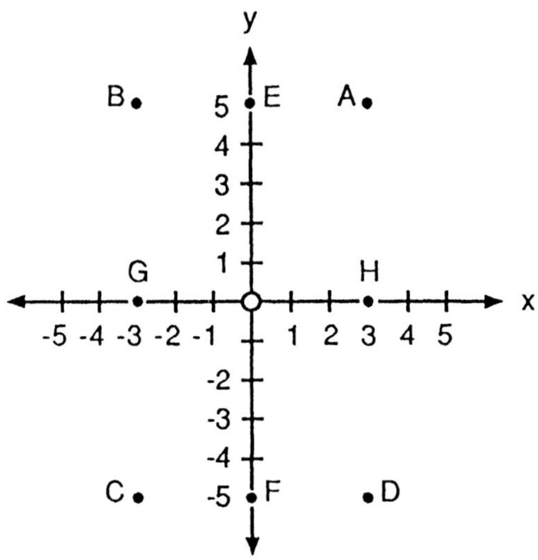


**Figure 9.2B**

***Notice* that each ordered pair has one and only one graphic representation on the plane, and no two points have the same numeric representation.**

We say then, that each point has a UNIQUE ordered pair representing it.

*Example:* State the ordered pair which corresponds to each of the points in figure 9.2 C

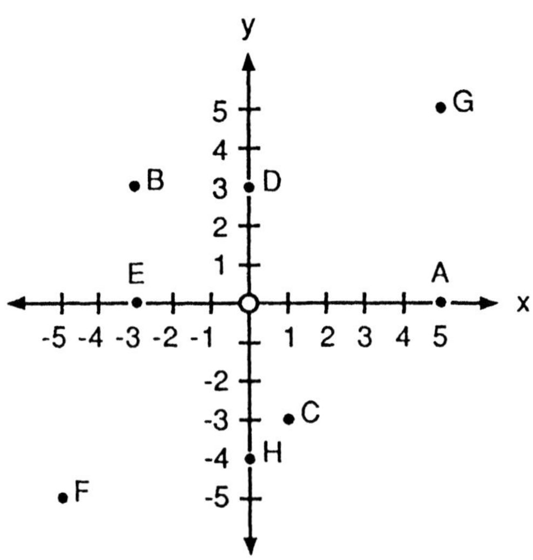


**Figure 9.2C**

Since A is 5 units right of zero and on the x axis A = (5,0)

Since B is 3 units left and 3 units up from zero B = (−3,3)

Since C is 1 unit right and 3 units down from zero C = (1,−3)

Since D is 3 units up from zero and on the y axis D = (0,3)

Since E is 3 units left of zero and on the x axis E = (−3,0)

Since F is 5 units left and 5 units down from zero F = (−5,−5)

Since G is 5 units right and 5 units up from zero G = (5,5)

Since H is 4 units down from zero and on the y axis H = (0,−4)

**Exercise Set 9.2 A**

Graph each ordered pair as a unique point on the Cartesian Plane:

A = (4,1)
B = (1,3)
C = (−1,2)
D = (−2,4)
E = (−2,−3)

F = (−1,−4)
G = (2,−4)
H = (5,−1)
I = (3,0)
J = (−4,0)

K = (0,0)
L = (0,2)
M = (0,−3)

**Exercise Set 9.2 B**

State the unique ordered pair for the points whose locations are given as follows:

1. 3 units right of zero.
2. 6 units right and 4 units down from zero.
3. 2 units left and 7 units down from zero.
4. 5 units up from zero.
5. 12 units down from zero.
6. 9 units right and 1 unit up from zero.

7. 11 units left and 6 units up from zero.

8. 3.5 units left of zero.

9. 1.5 units right and 1.5 units down from zero.

**Exercise Set 9.2 C**

Without graphing, indicate the quadrant in which each of the following points is located:

| | | |
|---|---|---|
| 1. $(-3,1)$ | 2. $(4,-6)$ | 3. $(9,0)$ |
| 4. $(41,-8)$ | 5. $(0,8)$ | 6. $(2,4)$ |
| 7. $(3,-7)$ | 8. $(-1,-1)$ | 9. $(-3,4)$ |
| 10. $(-5,-5)$ | 11. $(8,6)$ | 12. $(12,-3)$ |
| 13. $(7,9)$ | 14. $(0,-5)$ | 15. $(-6,6)$ |

## 9.3 *Graphing Lines on the Cartesian Plane*

In order to draw a line, only two points are required; however, a line contains an infinite number of points.

Look at figure 9.3.

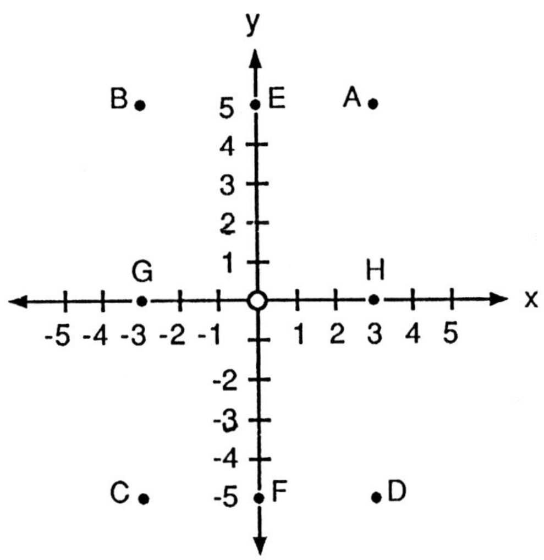


**Figure 9.3**

Connect points B and G. Now connect points G and C. Finally, connect points B and C.

In each case, the result is a vertical line. That's because in each case, the x coordinate is $-3$.

Now connect points B and E; E and A; B and A.

In each case, a horizontal line is the result because the y coordinate for each point is 5.

Connecting points E and F results in the line which is part of the y axis. Note that their x coordinate is zero.

Finally, connecting points G and H results in the line which is part of the x axis. Note that their y coordinates are zero.

In every case, the two end points were not the only points on the line. There were, in fact, an infinite number of points between each two! And, there are an infinite number of points beyond each end-point!

**Just as each point has a unique ordered pair of numbers to represent it, so each line has a unique equation to represent it.**

*Example:* Given that $y = 2x + 5$, name some points which would "satisfy the rule".

Here we must take any value of x, multiply by 2 and add 5.

Let $x = 2$, then $y = 2(2) + 5$
so $y = 4 + 5$ or $y = 9$

Let $x = 0$, then $y = 2(0) + 5$ or $y = 5$

The points (2,9) and (0,5) satisfy the rule.

Other points would be: $(-1,3)$ $(-5,-5)$ $(1,7)$

*Example:* Name several points which will satisfy the equation $y = x$

This is simple! x and y must have the same values so,
$(-5,-5)$ $(-1,-1)$ $(0,0)$ $(5.6,5.6)$ $(-2.3,-2.3)$ are all correct answers.

*Example:* Give ten points which satisfy the equation $y = x + 2$.

Here the rule states: take any value for x and add 2 to get y, so:
(2,4) (0,2) $(-1,1)$ (5,7) (3.6,5.6)
$(-8,-6)$ (4,6) (8.7,10.7) $\left(\frac{1}{2},\frac{5}{2}\right)$ $\left(\frac{-1}{2},\frac{3}{2}\right)$

**Exercise Set 9.3**

List 5 points which would satisfy the rule (equation) given:

1. $y = -x$
2. $y = 2x$
3. $y = x - 1$
4. $y = 2x - 1$
5. $y = x + 3$
6. $y = x - 3$
7. $y = -2x$
8. $y = -2x - 1$
9. $y = -2x + 1$
10. $y = \frac{1}{2}x$

**NOTE:** Answers will vary for this set.

## 9.4 *Graphing Linear Equations*

Some basic concepts were established in Section 9.3:

**a. two points determine a line**

**b. a line consists of an infinite series of points**

**c. all the ordered pairs which represent the points on the line follow the same "rule" which may be written as an algebraic equation.**

The equation for the line is called a **LINEAR** equation and shows the relationship between the x coordinate to the y coordinate of each point on the line.

**Notice** from all the previous examples, that in the linear equation, the exponent for x and y is understood to be 1.

Also notice from previous examples, that the linear equation was most often expressed in the form:

$\mathbf{y = ax + b}$, where a and b are real numbers.

***TO GRAPH A LINEAR EQUATION:***

**1. FIND ORDERED PAIRS THAT SATISFY THE "RULE" (that is, the equation of the line)**

**2. PLOT THE POINTS REPRESENTED BY THOSE ORDERED PAIRS**

**3. CONNECT THE POINTS WITH A STRAIGHT LINE.**

It is usual to "label" the line with its equation.

*Example:* Graph $y = 2x + 1$

*Solution:* 1. Find ordered pairs that make the equation true by substituting any convenient value for x (say 1 or 0) and solving for y.

let $x = 1$

$y = 2x + 1$
$y = 2(1) + 1$
$y = 2 + 1$
$y = 3$

so, (1,3) is a point on the line $y = 2x + 1$.

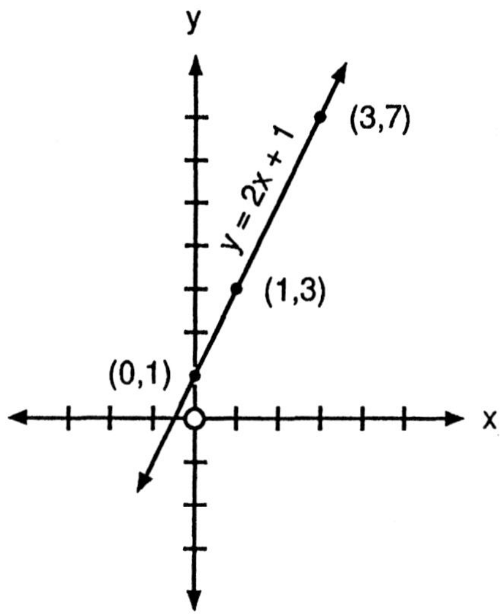


Figure 9.4A

let $x = 0$

$y = 2x + 1$
$y = 2(0) + 1$
$y = 1$

so, (0,1) is another point on the line $y = 2x + 1$.

let $x = 3$

$y = 2x + 1$
$y = 2(3) + 1$
$y = 6 + 1$
$y = 7$

so, (3,7) is a third point on the line.

2. Graph the points (1,3), (0,1) and (3,7).

3. Connect the points with a straight line and "label" the line with its equation.

*Example:* Graph the linear equation $y = x + 1$

*Solution:*

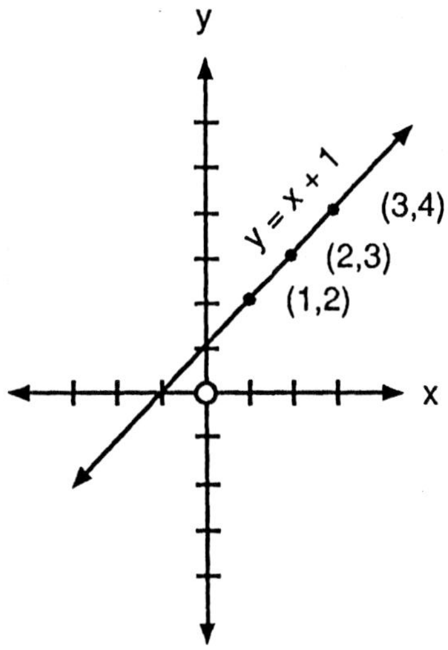


**Figure 9.4B**

1. a. Let $x = 1$ then $y = 1 + 1$
   $y = 2$
   so, (1,2) is on the line.

   b. Let $x = 2$ then $y = 2 + 1$
   $y = 3$
   The point (2,3) is on the line.

   c. Let $x = 3$ then $y = 3 + 1$
   $y = 4$
   (3,4) is also on the line.
2. Graph the points (1,2), (2,3) and (3,4).
3. Connect the points with a straight line.
4. Label the line with its equation.

*Example:* Graph the line $y = 3x - 2$

*Solution:*

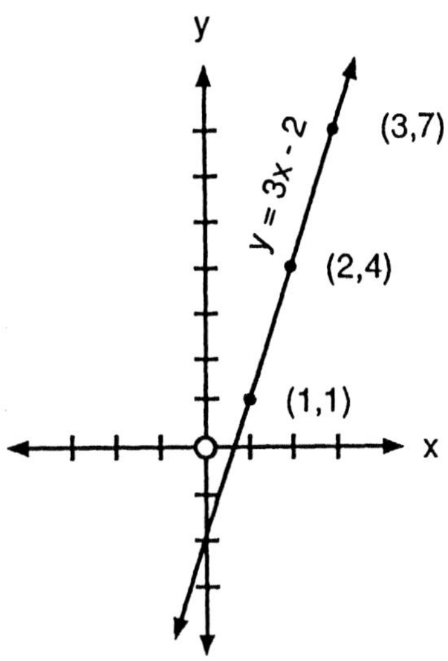


**Figure 9.4C**

1. Let $x = 1$ then $y = 3(1) - 2 = 3 - 2 = 1$
   Let $x = 2$ then $y = 3(2) - 2 = 6 - 2 = 4$
   Let $x = 3$ then $y = 3(3) - 2 = 9 - 2 = 7$
2. Graph the points (1,1), (2,4), and (3,7)
3. Connect the points with a straight line.
4. Label the line with its equation.

*Example:* Graph the line $y = -2x + 1$

*Solution:*

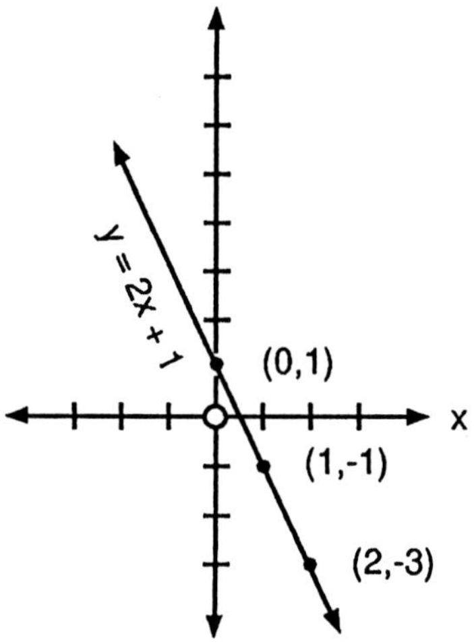


**Figure 9.4D**

1. Let $x = 1$ then $y = -2(1) + 1 = -1$

   Let $x = 2$ then $y = -2(2) + 1 = -3$

   Let $x = 0$ then $y = -2(0) + 1 = 1$

   This information may be summarized in a chart or table:

   | x | y |
   |---|---|
   | 1 | −1 |
   | 2 | −3 |
   | 0 | 1 |

   NOTE: Any convenient value may be selected for x; however, it is "sensible" to pick values that will not result in very high or very low values for y. It's a good idea to choose values like 1, 0, and −1 for x, since these make calculations easy.

2. Graph the points.
3. Connect the points.
4. Label the line.

*Example:* Graph $y = \frac{1}{2}x - 3$

1. Here it is wise to select values for x that are divisible by 2:

| Let x = | then y = |
|---|---|
| −4 | $\frac{1}{2}(-4) - 3 = -2 - 3 = -5$ |
| −2 | $\frac{1}{2}(-2) - 3 = -1 - 3 = -4$ |
| 0 | $\frac{1}{2}(0) - 3 = 0 - 3 = -3$ |
| 2 | $\frac{1}{2}(2) - 3 = 1 - 3 = -2$ |
| 4 | $\frac{1}{2}(4) - 3 = 2 - 3 = -1$ |

2. Graph the points $(-4,-5)$, $(-2,-4)$, $(0,-3)$, $(2,-2)$ and $(4,-1)$.

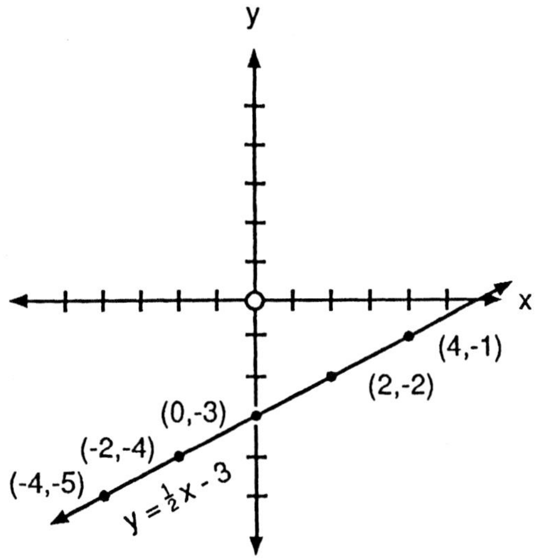


3. Connect the points with a line.

4. Label the line with its equation.

**Figure 9.4E**

**Exercise Set 9.4**

Graph the following lines:

1. $y = x + 4$
2. $y = 3x - 4$
3. $y = 5x - 10$
4. $y = 2x + 4$
5. $y = -3x + 2$
6. $y = -5x + 3$
7. $y = \frac{1}{3}x + 2$
8. $y = \frac{2}{3}x - 4$
9. $y = \frac{-3}{4}x + 5$
10. $y = \frac{-5}{4}x - 1$
11. $y = 2x$
12. $y = -3x$
13. $y = -4x$
14. $y = 3$
15. $y = -1$
16. $x = 3$

## 9.5 *Determining Points on a Line*

In section 9.3 we saw how points fall on a line if their ordered pairs "satisfy the rule" (equation) which defines the relationship between the x coordinate and the y coordinate.

So, to determine if a point lies on a given line:

replace x with the first number in the ordered pair and replace y with the second number of the pair.

**If the statement is true, then the point satisfies the equation and so lies on the given line.**

*Example:* Determine if the point (4,9) is on the line $y = 2x + 1$

Substitute 4 for x and 9 for y in the

equation: $y = 2x + 1$
$9 = 2(4) + 1$

Simplify: $9 = 8 + 1$
$9 = 9$

Since the result is true, (4,9) falls on the line $y = 2x + 1$

*Example:* Determine if the point (0,1) lies on the line $y = 2x + 1$.

Substitute 0 for x and 1 for y in
$y = 2x + 1$:
$1 = 2(0) + 1$

Simplify: $1 = 0 + 1$
$1 = 1$

The result is true. (0,1) is on the line.

*Example:* Is the point (1,0) on the line $y = 2x + 1$?

Substituting: $0 = 2(1) + 1$

Simplifying: $0 = 2 + 1$
$0 = 3$ **NOT TRUE!**

(1,0) does not satisfy the equation, so it is *NOT* on the line $y = 2x + 1$

*Example:* Does the point (0,2) lie on the line $2x + 3y + 6 = 0$?

Substituting: $2(0) + 3(2) + 6 = 0$
$0 + 6 + 6 = 0$
$12 = 0$ **NO!**

*Example:* Does the point (0,2) lie on the line $2x + 3y - 6 = 0$?

Substituting: $2(0) + 3(2) - 6 = 0$
$0 + 6 - 6 = 0$
$0 = 0$ **YES!**

*Example:* Which of the following points are on the line $y = -x + 4$?

a. $(-5,9)$ b. $(0,-4)$ c. $(2,2)$
d. $(3,1)$ e. $(-7,-3)$

a. Substituting $-5$ for x and 9 for y:

$9 = -(-5) + 4$
$9 = 5 + 4$
$9 = 9$ **YES**

b. Substituting 0 for x and $-4$ for y:

$-4 = 0 + 4$
$-4 = 4$ **NO**

c. Substituting 2 for x and 2 for y:

$2 = -2 + 4$
$2 = 2$ **YES**

d. Substituting 3 for x and 1 for y:
$1 = -3 + 4$
$1 = 1$ **YES**

d. Substituting $-7$ for x and $-3$ for y:
$-3 = -(-7) + 4$
$-3 = 7 + 4$
$-3 = 11$ **NO**

*Example:* Does the line $y = 2x - 1$ go through the origin?

To answer the question we must determine if the point (0,0) lies on the line.

Substituting: $0 = 2(0) - 1$
$0 = 0 - 1$
$0 = -1$ **NO!**

The line does not go through the origin.

**Exercise Set 9.5 A**

Determine if the point which follows each equation lies on the line it represents:

| | | | |
|---|---|---|---|
| 1. $y = x + 1$ | (5,6) | 2. $y = 2x$ | (10,20) |
| 3. $y = 5x - 10$ | (1,−5) | 4. $y = -4x$ | (0,−4) |
| 5. $y = -3x + 2$ | (1,−1) | 6. $y = 3x - 4$ | (0,−4) |
| 7. $y = \frac{1}{3}x + 2$ | (6,8) | 8. $y = 2x + 4$ | (−2,0) |
| 9. $y = \frac{-3}{4}x + 5$ | (8,−1) | 10. $y = -5x + 3$ | (−2,13) |
| 11. $y = \frac{1}{2}x - 4$ | (12,−2) | 12. $y = \frac{1}{4}x + 2$ | (8,4) |
| 13. $y = -x - 1$ | (0,1) | 14. $y = 3x - 5$ | (−1,−8) |

**Exercise Set 9.5 B**

Supply the missing coordinate so that the ordered pair will satisfy the given equation:

| | | | |
|---|---|---|---|
| 1. $y = \frac{1}{3}x + 2$ | (6, ) | 2. $y = -4x$ | ( ,−4) |
| 3. $y = 3x + 5$ | (0, ) | 4. $y = \frac{1}{2}x$ | ( ,9) |
| 5. $y = -2x$ | (−1, ) | 6. $y = -5x$ | ( ,0) |
| 7. $y = -x - 1$ | (6, ) | 8. $y = \frac{1}{4}x$ | ( ,4) |
| 9. $y = 5x - 5$ | (−1, ) | 10. $y = 5x - 5$ | ( ,−5) |
| 11. $y = \frac{-1}{2}x$ | (8, ) | 12. $y = \frac{-1}{2}x$ | ( ,−8) |

## 9.6 *The Slope-Intercept Form of the Linear Equation*

In the preceding sections, all of the lines graphed had a certain "form". That form, **$y = mx + b$,** is called the slope-intercept form of the linear equation. It is so called because it clearly shows the **slope** of the line and the **y intercept.** (The y intercept is the point on the y axis through which the line passes.)

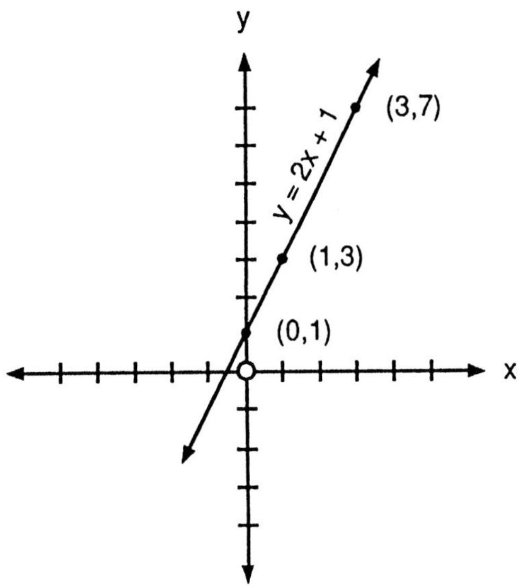


Figure 9.6A

Consider the graph of $y = 2x + 1$ from section 9.4.

Look at points (1,3) and (3,7). y changed 4 units since the y value went from 3 to 7. x changed 2 units since the x value went from 1 to 3.

The change in y compared to the change in x is $\frac{4}{2}$ or 2

Look at the points (0,1) and (1,3). y changed 2 units and x changed 1 unit.

The change in y compared to the change in x is $\frac{2}{1}$ or 2

Suppose we compare the change in y and x for points (0,1) and (3,7). Here, y changed 6 units and x changed 3. Again, the change in y compared to the change in x is 2, since $\frac{6}{3} = 2$.

**We call this relationship of $\frac{\text{the change in y}}{\text{the change in x}}$ the SLOPE of the line.**

Notice two things about the line:

1. its slope is 2
2. it intersects (cuts through) the y axis at $y = 1$.

Now lcok at the equation of the line: $y = 2x + 1$. Do you see the slope and y intercept in the equation?

In the linear form $y = mx + b$, we see that the slope is represented by m, and the place where the line crosses the y axis is represented by b.

*Example:* Find the slope and y intercept for the graph of $y = x + 1$.

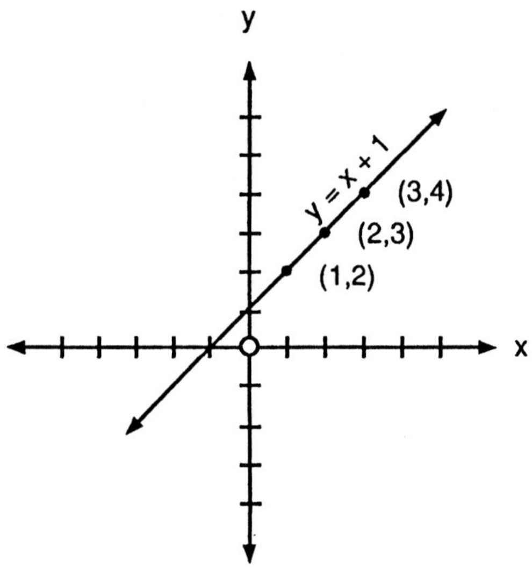


Figure 9.6B

To move from one point to the other, y must change 1 unit and x must change 1 unit.

$$\frac{\text{the change in y}}{\text{the change in x}} = \frac{1}{1} = 1$$

The slope of the line, is 1.

If you note the "rule" that y is always greater than x by 1 you'll see that the line must cross the y axis at $y = 1$.

The y intercept is $y = 1$.

Comparing the equation $y = x + 1$ to the form $y = mx + b$, we see that the slope, m, is 1 and the y intercept, b, is at (0,1).

*Example:* Find the slope and y intercept for the graph of $y = 3x - 2$.

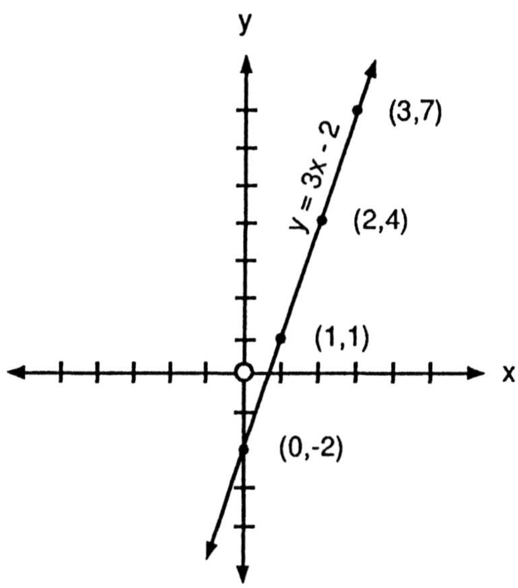


Figure 9.6C

Using the points in the figure at left, note that to move from one point to another y always changes 3 units while x changes 1 unit. So, the slope of the line is 3.

Since the line crosses the y axis at $y = -2$, the y intercept is $y = -2$.

Comparing $y = 3x - 2$ to $y = mx + b$ we see that m (the slope) is 3 and b (the y intercept) is $-2$.

*Example:* Find the slope and y intercept for the graph $y = -2x + 1$.

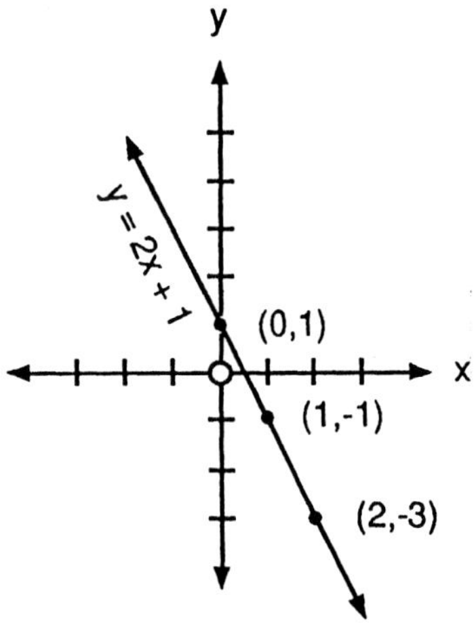


**Figure 9.6D**

If we look at the general form

$y = mx + b$ and the equation $y = -2x + 1$, we see that the slope (m) is $-2$ and the y intercept (b) is 1.

This means that the line should cross the y axis at $y = 1$ and the change in y compared to the change in x should be $-2$.

In this graph, if you move from one point to another, you move 2 units DOWN for every 1 unit right. The slope, then, is $-2$.

*Example:* Graph $y = \frac{1}{2}x - 3$ using the slope and intercept.

Comparing the equation $y = \frac{1}{2}x - 3$ to the form $y = mx + b$

note that m (slope) is $\frac{1}{2}$ and b (y intercept) is $-3$.

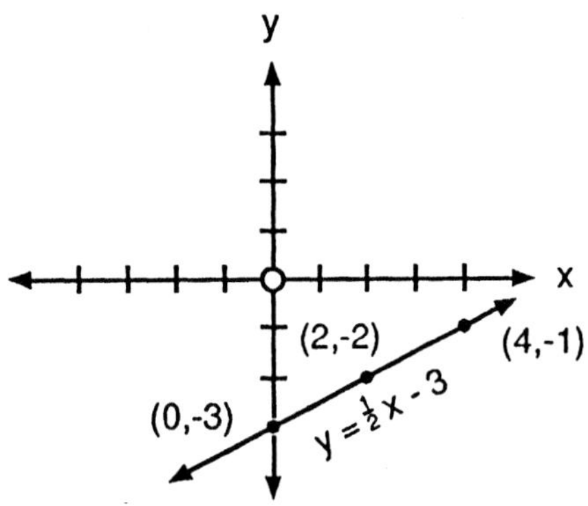


**Figure 9.6E**

The line then crosses the y axis at $(0,-3)$ and starting at that point y changes 1 unit while x changes 2.

Remember that slope is $\frac{\text{change in y}}{\text{change in x}}$

Start at $(0,-3)$: move up 1 and right 2 to $(2,-2)$

Start there and move up 1 and right 2 units to $(4,-1)$

*Example*: Use the slope-intercept method to graph the line $2x + y = 1$.

The equation must be rewritten in the slope-intercept form, so first solve for y:

$y = -2x + 1$.

Rearranging to get the form $y = mx + b$, the equation becomes $y = -2x + 1$.

The slope is $-2$ and b is 1.

this is the same graph as fig. 9.6 D.

*Example*: Use the slope-intercept form to graph the equation $2y - x + 6 = 0$.

Solving for y: $2y = x - 6$

$$y = \frac{1}{2}x - 3$$

The slope is $\frac{1}{2}$ and b is $-3$.

This is the same graph as fig. 9.6 E.

**Exercise Set 9.6**

Find the slope and y intercept of the following lines. Use the information to graph the lines.

1. $y = x + 4$
2. $y = 3x - 4$
3. $y = 2x$
4. $y = 5x - 10$
5. $y = 2x + 4$
6. $y = -3x$
7. $y = -3x + 2$
8. $y = -5x + 3$
9. $2y - 4x = 16$
10. $y = \frac{1}{3}x + 2$
11. $y = \frac{2}{3}x - 4$
12. $y = -4x$
13. $y = \frac{-3}{4}x + 5$
14. $y = \frac{-5}{4}x - 1$
15. $y = 3$
16. $y = -1$
17. $x = 3$
18. $6x + 8y = 16$
19. $y - x + 3 = 0$
20. $x + 2y - 8 = 0$
21. $2y - 4x = 0$

## CHAPTER EXERCISES

### Section 9.2

Plot the following points on the Cartesian plane:

1. (2,5)
2. (−4,1)
3. (9,0)
4. (1,−9)
5. (5,4)
6. (1,−1)
7. (−4,−5)
8. (−5,−5)
9. (3,−2)
10. (−4,5)

### Section 9.2

For each of the following points, tell in which quadrant the point lies.

11. (0,−3)
12. (−4,0)
13. (−1,−2)
14. (−2,−3)
15. (−7,−4)
16. (−1,−6)
17. (6,−4)
18. (6,−3)
19. (−3,0)
20. (0,0)

### Section 9.3

Use the given values for x in the equations to determine the corresponding value of y:

21. $y = x + 5$ — $x = 3$, $x = 2$, $x = 1$
22. $y = 2x - 1$ — $x = 4$, $x = 2$, $x = 0$
23. $y = 4x + 3$ — $x = 1$, $x = 0$, $x = -1$
24. $y = 6x - 2$ — $x = 2$, $x = 0$, $x = -2$
25. $y = -2x + 5$ — $x = 2$, $x = 1$, $x = 0$
26. $y = -3x + 7$ — $x = 3$, $x = 1$, $x = -1$
27. $y = -5x - 2$ — $x = 1$, $x = 0$, $x = -1$
28. $y = 3x - 4$ — $x = 2$, $x = 0$, $x = -2$
29. $y = 5x - 10$ — $x = 3$, $x = 2$, $x = 0$
30. $y = \frac{1}{2}x + 3$ — $x = 6$, $x = 4$, $x = 2$
31. $y = \frac{2}{3}x - 1$ — $x = 6$, $x = 3$, $x = 0$
32. $y = \frac{-3}{2}x + 4$ — $x = 2$, $x = 0$, $x = -2$
33. $y = \frac{5}{6}x - 3$ — $x = 6$, $x = 0$, $x = -6$
34. $y = \frac{-3}{5}x + 3$ — $x = 5$, $x = 0$, $x = -5$

| | | | |
|---|---|---|---|
| 35. $y = \frac{-7}{4}x - 10$ | $x = 4$ | $x = 0$ | $x = -1$ |
| 36. $y = \frac{-3}{8}x$ | $x = 8$ | $x = 0$ | $x = -8$ |
| 37. $y = \frac{5}{3}x$ | $x = 3$ | $x = 0$ | $x = -3$ |
| 38. $y = 2(x + 1)$ | $x = 1$ | $x = 0$ | $x = -1$ |
| 39. $y = -3(x - 4)$ | $x = 1$ | $x = 2$ | $x = 3$ |
| 40. $y + 2x = 7$ | $x = 1$ | $x = 0$ | $x = -1$ |

**Section 9.4**

For each of your answers in problems 21 through 40, graph the three points and connect them with a straight line.

**Section 9.5**

For each of the following, determine if the point lies on the given line:

| | | | |
|---|---|---|---|
| 41. $y = 2x + 1$ | $(1,3)$ | 42. $y = x - 4$ | $(5,1)$ |
| 43. $y = -3x + 1$ | $(3,-8)$ | 44. $y = -3x + 7$ | $(-1,4)$ |
| 45. $y = x + 5$ | $(3,2)$ | 46. $y = -2x + 3$ | $(-1,1)$ |
| 47. $y = -3x + 7$ | $(-2,1)$ | 48. $y = 6x - 2$ | $(0,0)$ |
| 49. $y = -5x - 2$ | $(0,-2)$ | 50. $y = 3x - 4$ | $(-3,-13)$ |
| 51. $y = 2x + 1$ | $(-3,7)$ | 52. $y = x - 4$ | $(-3,-7)$ |
| 53. $y = -3x + 1$ | $(-4,-11)$ | 54. $y = -3x + 7$ | $(0,7)$ |
| 55. $y = x + 5$ | $(-5,5)$ | 56. $y = -2x + 3$ | $(-4,11)$ |
| 57. $y = -3x + 7$ | $(0,7)$ | 58. $y = 6x - 2$ | $(-1,4)$ |
| 59. $y = -5x - 2$ | $(5,8)$ | 60. $y = 3x - 4$ | $(-0,-4)$ |
| 61. $y + 4x = 3$ | $(2,-5)$ | 62. $3x + 4y = 9$ | $(3,0)$ |
| 63. $5x - 4y = 27$ | $(3,-3)$ | 64. $4x - 3y = 0$ | $\left(\frac{1}{2},\frac{2}{3}\right)$ |
| 65. $5x - 4y = 1$ | $\left(\frac{2}{5},\frac{3}{4}\right)$ | 66. $x = 4$ | $(4,6)$ |
| 67. $y = -x$ | $(-3,3)$ | 68. $y = -4$ | $(5,-4)$ |
| 69. $2x = 4y$ | $(8,0)$ | 70. $x = -1$ | $(-1,4)$ |

## Section 9.5

For each of the following linear equations, find the missing coordinate so that the ordered pair will satisfy the given equation:

71. $y = 5x - 1$ (2, )
72. $y = -x + 3$ (7, )
73. $y = \frac{2}{5}x - 7$ (10, )
74. $y = \frac{-3}{4}x + 1$ (20, )
75. $y = -6x + 3$ (−2, )
76. $y = -7x + 4$ (0, )
77. $y = -2x + 10$ (−3, )
78. $y = 12 - 4x$ (−3, )
79. $y = 5x$ ( ,10)
80. $y = -2x + 4$ ( ,12)
81. $y = 7x - 3$ ( ,18)
82. $y = -2x + 2$ ( ,−8)
83. $y = \frac{2}{3}x$ ( ,14)
84. $y = 3x - 1$ ( ,8)
85. $y = -5x + 3$ ( ,18)
86. $y = \frac{4}{5}x + 1$ ( ,17)
87. $y = 7$ (3, )
88. $y = -6$ (−3, )
89. $x = 2$ ( ,−4)
90. $x = -4$ ( ,+3)

## Section 9.6

Find the slope and y intercept for the following equations. Graph the lines they represent.

91. $y = x + 5$
92. $y = 2x - 1$
93. $y = 4x + 3$
94. $y = 6x - 2$
95. $y = -2x + 5$
96. $y = -3x + 7$
97. $y = -5x - 2$
98. $y = 3x - 4$
99. $y = 5x - 10$
100. $y = \frac{1}{2}x + 3$
101. $y = \frac{2}{3}x - 1$
102. $y = \frac{-3}{2}x + 4$
103. $y = \frac{5}{6}x - 3$
104. $y = \frac{-3}{5}x + 3$
105. $y = \frac{-7}{4}x - 10$
106. $y = \frac{-3}{8}x$
107. $y = \frac{5}{3}x$
108. $y = 2(x + 1)$
109. $y = -3(x - 4)$
110. $y + 2x = 7$
111. $3x + 2y = 8$
112. $4x - y = 1$
113. $y = 7$

## CHAPTER TEST *Form 9A*

1. Plot the points:

   a. $(-3,1)$ b. $(4,2)$ c. $(0,-6)$ d. $(-5,-5)$

   e. $(2,4)$ f. $(4,6)$ g. $(0,8)$ h. $(-1,0)$

   i. $(3,7)$ j. $(-3,4)$

2. State the quadrant in which each of the following points is located. If the point is on an axis, state if x or y.

   a. $(9,0)$ b. $(-4,-3)$ c. $(-2,2)$ d. $(-0.5,-6)$

   e. $\left(\frac{1}{2},0\right)$ f. $(41,-8)$ g. $(1.5,9.2)$ h. $(-11,-31)$

   i. $(7,6)$ j. $(-10,-5)$

3. Plot any two points which satisfy the following equations, and graph the line represented by each:

   a. $y = -2$ b. $y = 3x - 4$

   c. $y = \frac{2}{3}x$ d. $y = 6 - x$

4. Does the point $(3,-12)$ lie on the line $y = 5x - 3$?

5. Find the y value of the point which lies on the line $y = -2x - 4$ if the x value is $x = 5$.

Match the following equations with the letter of the line which best represents its graph:

6. $y = \frac{2}{3}x + 4$ 7. $y = -3x + 5$ 8. $x = 7$ 9. $y = -2$

a. 

b. 

c. 

d. 

Find the slope and y intercept of the lines represented by the following equations. Graph the lines.

10. $y = 2x - 1$ 11. $y = -3x + 2$

**CUMULATIVE EXERCISES Chapter 1-9** *Set A*

Evaluate the following:

1. $(-2)^3$
2. $(7)^{-2}$
3. $(6^3)^2$
4. $3(-.6) - 4(2.1)^2$

Combine like terms to simplify:

5. $6y^2 - 3y + 4y^2 - 6y + 2$
6. $4x - \{6 - 3x\}$
7. $-(8x^3 + 4x - 6) - (2x - 3x^2 - 10x^3)$
8. $2(4a^2 - 3a + 5) + \frac{1}{2}(4a^3 - 6a^2 + 4a - 6)$

Evaluate:

9. $(a^2)(a^3)$
10. $\frac{w^3}{w^{-2}}$
11. $\frac{7}{3} - \frac{2}{5} + \frac{11}{6}$
12. $\frac{11}{3} \div \frac{-33}{15}$
13. $3 + \frac{1}{5} - \frac{5}{2}$
14. $4y(3y^2 - 9xy + 2x^2)$
15. $(-30)^{-2}$
16. $\left(\frac{-3}{11}\right)^{-2}$
17. $(a + 4)(a - 3)$
18. $(m^2 - 3)(m^2 + 3)$
19. $(4a + 1)^2$
20. $5xy(3x^2 + 4xy + 7y^2)$
21. $\frac{14z^3 - 21z^2 + 49z}{-7z}$
22. $\frac{x^2}{4} \cdot \frac{-5}{x^3}$

Solve and check the results.

23. $11d - 4 = 29$
24. $8a + 6 = 16$
25. $9y + 4 = -5$
26. $4x - 2 = 10 - 2x$
27. $\frac{4}{5}x - 3 = -27$
28. $45 - t = 3$

For numbers 29–34, write the equation and solve:

29. Three less than 4 times a number is 25. Find the number.

30. Joe's father is twice his age plus five years. If Joe is 21 years old, how old is his father?

31. Four times a number plus five is euivalent to three times the number. What is the number?

32. The sum of two consecutive odd integers is 204. What are the integers?

33. One-third of a number increased by three is equal to twenty-five. What is the number?

34. the sum of three consecutive even integers is 102. Find the integers.

Solve for the indicated variable:

35. Solve for x: $3(2 + x) = 18$

36. Solve for y: $4y + 2.4 = -(2.1 + y)$

37. Solve for k: $z = 2k + y$

38. Solve for s: $z = \frac{x - m}{s}$

39. Solve for a: $F = ma$ if $F = 45$ and $m = 90$

40. Solve for x: $\frac{1}{3}(x + y) = z$

Without plotting the points, determine the quadrant in which each is located:

41. (7,–3)  42. (–1,4)  43. (–2,–1)  44. $\left(2, \frac{1}{2}\right)$

In the equation $y = 6x - 5$, find the values for y if:

45. $x = -2$  46. $x = 9$  47. $x = \frac{-1}{3}$  48. $x = \frac{5}{6}$

Find the slope of the lines represented by each of the following:

49. $y = 7x - 2$  50. $y = 8 - 3x$

51. $y = \frac{2}{3}x + 3$  52. $y = \frac{4}{3}x + 2$

Graph the lines represented by the following equations:

53. $y = 2x + 5$  54. $y = 5 - x$  55. $2x + y - 6 = 0$

**CUMULATIVE TEST Chapters 1–9** *Form A*

**Answers**

1. What property of real numbers is illustrated by: $(a + 4)(y) = a(y) + 4(y)$ — 1. ________
2. What is the additive inverse of –72? — 2. ________

Evaluate:

**Answers**

3. $6x^3 \cdot 3x^2$ — 3. ________
4. $-2(5^2) + |-15|$ — 4. ________
5. $\frac{15}{7} \div \frac{45}{3}$ — 5. ________

Simplify:

**Answers**

6. $3 + 4\{2 - (9 - 12)^2\}$ — 6. ________
7. $-2(x^2 - 4x) + 3(4 - x^2)$ — 7. ________

Multiply:

**Answers**

8. $(2x + 7)(x - 4)$ — 8. ________
9. $(x^2 + 4)(x^2 - 3)$ — 9. ________
10. Divide $44x^4 - 11x^2 + 22x$ by $11x$ — 10. ________

Solve:

**Answers**

11. $4x - 2 = 10$ — 11. ________
12. $5(x + 2) = 2(x - 3)$ — 12. ________
13. The sum of three consecutive integers is 54. What is the smallest of the integers? — 13. ________

14 Solve for g: $d = \frac{1}{2}(gt^2)$ — 14. ________

15. Solve for W: $2L + 2W = P$ — 15. ________
16. What is the slope of the line $y = -3x + 2$? — 16. ________
17. What is the y intercept of $y = -3x + 2$? — 17. ________
18. Is the point (–2,7) on the line $y = 7x + 4$. — 18. ________

Graph the lines represented by the following equations:

**Answers:**

19. $y = 5 - 2x$
20. $y = 2x - 5$

# CHAPTER 10

# Factoring Polynomials

## 10.0 *Vocabulary Preview*

Define these words without using a dictionary. Formulate their meanings from past experience or knowledge.

| | | |
|---|---|---|
| **binomial** | **monomial** | **roots** |
| **common** | **polynomial** | **square** |
| **difference** | **prime** | **trinomial** |
| **factor** | | |

## 10.1 *Monomials Containing Common Factors*

| | | | |
|---|---|---|---|
| The monomial $8x^2$ can be factored as or $2^3 \cdot x^2$. | [ $2 \cdot 2$ ] | $\cdot\ 2\ \cdot$ | [ $x \cdot x$ ] |
| The monomial $4x^3$ can be factored as or $2^2 \cdot x^3$. | [ $2 \cdot 2$ ] | $\cdot\ x\ \cdot$ | [ $x \cdot x$ ] |

Their **greatest common factor** is $2 \cdot 2 \cdot x \cdot x$ or $2^2 \cdot x^2$ or $4x^2$.

| | | | | | | | | |
|---|---|---|---|---|---|---|---|---|
| Likewise, $12xy^2$ factors into | $2 \cdot 2\ \cdot$ | [ 3 ] | $\cdot$ | [ x ] | | $\cdot$ | [ y ] | $\cdot\ y$ |
| $3x^2y$ factors into | | [ 3 ] | $\cdot$ | [ x ] | $\cdot\ x\ \cdot$ | | [ y ] | |
| and $6xy^3$ factors into | $2\ \cdot$ | [ 3 ] | $\cdot$ | [ x ] | | $\cdot$ | [ y ] | $\cdot\ y \cdot y$ |

Their **greatest common factor** is $3 \cdot x \cdot y$

*Example:* The monomials $7a^2b^3c$, $21ab^2c$ and $70ac$ have a greatest common factor of $7ac$.

Note that:

1. each COEFFICIENT contains the common factor 7;
2. each term contains the common factor $a^1$
3. each term contains the common factor $c^1$.
4. no power of b is common to *all* three terms.

*Example*:

$$9x^3 = \boxed{3 \cdot 3} \cdot \boxed{x \cdot x} \cdot x$$
$$54x^2y = \boxed{3 \cdot 3} \cdot 3 \cdot 2 \cdot \boxed{x \cdot x} \cdot y$$

Their **greatest common factor** is $3^2 \cdot x^2$ or $9x^2$

*Example*:

$$5x^2y^2 = \boxed{5} \cdot \boxed{x \cdot x} \cdot \boxed{y \cdot y}$$
$$15x^3y^5 = 3 \cdot \boxed{5} \cdot \boxed{x \cdot x} \cdot x \cdot \boxed{y \cdot y} \cdot y \cdot y \cdot y$$

Their **greatest common factor** is $5x^2y^2$

*Example*:

$$\frac{1}{4}a^2 = \boxed{\frac{1}{2} \cdot \frac{1}{2}} \cdot \boxed{a \cdot a}$$
$$\frac{1}{8}a^4 = \boxed{\frac{1}{2} \cdot \frac{1}{2}} \cdot \frac{1}{2} \cdot \boxed{a \cdot a} \cdot a \cdot a$$

Their **greatest common factor** is $\frac{1}{2} \cdot \frac{1}{2} \cdot a \cdot a$ or $\frac{1}{4}a^2$

**Exercise Set 10.1**

Find the greatest common factors of the following monomials:

**Answers**

1. $5x$ and $15x^3$ — 1. ________
2. $7xy^2$, $14x^2y$ and $28xy$ — 2. ________
3. $13x^2$ and $26xy$ — 3. ________
4. $2c^3$, $10c^5$ and $50c^2$ — 4. ________
5. $24x^2y^3z$, $12xz^3$, $15x^5z^2$ — 5. ________
6. $2t^2$ and $t^4$ — 6. ________
7. $3r^4s$ and $15rs$ — 7. ________
8. $4t^5$, $8t^2$ and $10t$ — 8. ________
9. $12x^2$, $24xy^2$, $21x^2y^2$ — 9. ________
10. $\frac{3}{4}r^2$, $\frac{1}{2}r$ and $\frac{1}{4}r^3$ — 10. ________

## 10.2 *Polynomials Containing Common Factors*

Each term of the polynomial $4y^2 + 2xy$ contains the common factors of 2 and y since

$$\begin{aligned} 4y^2 &= 2 \cdot 2 \cdot y \cdot y \\ \text{and} \quad 2xy &= 2 \cdot x \cdot y \end{aligned}$$

The common factor, 2y, is the greatest common factor of both terms, and divides each term of the expression as follows:

$$\frac{4y^2 + 2xy}{2y} = \frac{4y^2}{2y} + \frac{2y}{2y} = 2y + 1$$

The polynomial may be expressed in factored form, that is:

$$4y^2 + 2xy = 2y(2y + 1).$$

*Example:* Factor the trinomial $3xy + 6x^2y^2 - 9x^3y$.

$$\begin{array}{rcl|c|ccc|c|c} 3xy & = & & 3 \cdot x & & & \cdot & y & \\ 6x^2y^2 & = & 2 \cdot & 3 \cdot x & \cdot x & & \cdot & y & \cdot y \\ 9x^3y & = & 3 \cdot & 3 \cdot x & \cdot x & \cdot x & \cdot & y & \end{array}$$

Since the greatest common factor is 3xy, divide each term of the trinomial by 3xy

$$\begin{aligned} \frac{3xy + 6x^2y^2 - 9x^3y}{3xy} &= \frac{3xy}{3xy} + \frac{6x^2y^2}{3xy} - \frac{9x^3y}{3xy} \\ &= 1 + 2xy - 3x^2 \end{aligned}$$

In factored form:

$$3xy + 6x^2y^2 - 9x^3y = 3xy(1 + 2xy - 3x^2)$$

*Example:* $2a^2b - 2ab^2 + 6ab^5 = 2ab(a - b + 3b^4)$

since the common factor 2ab divides each term as follows:

$$\frac{2a^2b}{2ab} = a, \quad \frac{-2ab^2}{2ab} = -b \text{ and } \frac{6ab^5}{2ab} = 3b^4$$

To check your answer, multiply through the parenthesis:

$$2ab(a - b + 3b^4) = 2a^2b - 2ab^2 + 6ab^5$$

which is the original expression.

*Example*: Factor: $7r^2s^2 + 14r^3s - 28r^5s^3$

The expression contains the common factor $7r^2s$.

The first term divided by $7r^2s$ is s.

The second term divided by $7r^2s$ is 2r.

The third term divided by $7r^2s$ is $-4r^3s^2$.

So: $7r^2s^2 + 14r^3s - 28r^5s^3 =$

$7r^2s\,(s + 2r - 4r^3s^2)$

**Exercise Set 10.2**

Write the following expressions in factored form:

1. $3x^2 + 12xy + 9x^5$
2. $y^3 - 16y^2$
3. $5t^5 - 10t^7$
4. $3x^2 - 27xy^2 + 9y$
5. $8r^4 + 16rs + 24r^2s^3$
6. $4a^3 - 8a$
7. $6t^2 + 3t + 24t^5$
8. $6c^2 - 3cd^2 + 12cd^3$
9. $3xy + 6x^2y^2 + 12x^4y^3$
10. $7st^2 + 14s^2t - 28s^2t^3$
11. $6a^2 + 24ab + 18a^5$
12. $8v^3 - 4v$
13. $12st^3 - 3st^2 + 6s^2$
14. $6g^2h^2 + 12g^4h^3 + 3gh$
15. $r^2s^2 + 2rs^2 - 8rt + st$
16. $4y^2z - 84y^2 + 4yz$

## 10.3 *Factoring Binomials—Differences of Squares*

The expressions:

$x^2 - 9 \qquad 4x^2 - 25 \qquad 9a^2 - b^2 \qquad$ and $\qquad z^6 - 1$

share similar characteristics:

**the two terms which make up the expressions are "SQUARES"**

$x^2$ is a square since it is the product of $x \cdot x$;

9 is a square since it is the product of $3 \cdot 3$.

$4x^2$ is the result of squaring $2x$, so $4x^2$ is a square.

$25 = 5^2$

$9a^2 = (3a)^2$; $b^2$ is a square since it is $b \cdot b$;

$z^6 = z^3 \cdot z^3$ or $(z^3)^2$ and $1 = 1^2$

**the expressions all indicate subtraction**

An expression of this type is called *THE DIFFERENCE OF TWO SQUARES.*

If you review Section 7.7, you will recall that:

$$(x + 3)(x - 3) = x^2 - 9 \qquad (2x + 5)(2x - 5) = 4x^2 - 25$$
$$(3a + b)(3a - b) = 9a^2 - b^2 \qquad (z^3 + 1)(z^3 - 1) = z^6 - 1$$

Notice that these products yield the same expressions as those above, which represent the differences of two squares.

In reverse:

$x^2 - 9$ in factored form is $(x + 3)(x - 3)$

$4x^2 - 25$ in factored form is $(2x + 5)(2x - 5)$

$9a^2 - b^2$ in factored form is $(3a + b)(3a - b)$

and $z^6 - 1 = (z^3 + 1)(z^3 - 1)$

**TO FACTOR THE DIFFERENCES OF TWO SQUARES, then:**

1. **take the square root of the first term**
2. **take the square root of the second term**
3. **write the expression as the product of a sum and a difference of those roots.**

*Example:* Factor $x^2 - y^2$

since this expression is the difference of two squares

$x^2 = x \cdot x$ and $y^2 = y \cdot y$

$x^2 - y^2$ factors into $(x + y)(x - y)$

*Example:* Factor $c^2 - 16d^6$.

$c^2 = c \cdot c$ and $16d^6 = 4d^3 \cdot 4d^3$

$c^2 - 16d^6 = (c + 4d^3)(c - 4d^3)$

*Example:* Factor $x^4 - y^2$

$x^4 = x^2 \cdot x^2$ and $y^2 = y \cdot y$

$x^4 - y^2 = (x^2 + y)(x^2 - y)$.

*Example:* Factor $4x^2 - 16$

In this case, you must first factor out the common factor 4; then factor the difference of two squares.

$$4x^2 - 16 = 4(x^2 - 4)$$
$$= 4(x + 2)(x - 2)$$

*Example:* Factor $3x^4 - 27$.

First factor out the common factor 3:

$3x^4 - 27 = 3(x^4 - 9)$

Then, since $x^4 = x^2 \cdot x^2$ and $9 = 3 \cdot 3$

$3x^4 - 27 = 3(x^4 - 9) = 3(x^2 + 3)(x^2 - 3)$

*Example:* Factor $x^4 - 81$.

$x^4 = x^2 \cdot x^2$ and $81 = 9 \cdot 9$ so

$x^4 - 81$ factors into $(x^2 + 9)(x^2 - 9)$.

BUT $x^2 - 9$ is itself the difference of two squares, so it can be factored into $(x + 3)(x - 3)$.

COMPLETELY FACTORED, then:

$$x^2 - 81 = (x^2 + 9)(x + 3)(x - 3).$$

**Exercise Set 10.3**

Factor completely:

1. $m^2 - n^2$
2. $9a^4 - 16b^2$
3. $2c^2 - 2d^2$
4. $25r^2 - 4s^2$
5. $7a^2 - 28b^2$
6. $c^2d^2 - c^2d^4$
7. $36s^4 - 9t^2$
8. $a^3b^2 - ab^4$
9. $x^2 - 121$
10. $x^6 - y^4$

## 10.4 *Factoring Trinomials with Leading Coefficient 1*

To factor the trinomial $x^2 + 5x + 6$, you must consider

1. the factors of the first term
2. the factors of the last term
3. the sign of the last term, and
4. the sign and coefficient of the middle term.

In the above expression,

1. the factors of the first term are $x \cdot x$
2. the factors of the last term are either $6 \cdot 1$ or $3 \cdot 2$
3. the sign of the last term is positive, so $(-6)(-1)$ and $(-2)(-3)$ are also possible factors
4. the sign of the middle term is positive, and it has coefficient 5.
5. the combination $+2$ and $+3$ will add to $+5$, and at the same time will multiply to $+6$.

The result of factoring $x^2 + 5x + 6$ is $(x + 2)(x + 3)$.

To check the answer, multiply:

$$(x + 2)(x + 3) = x^2 + 3x + 2x + 6 = x^2 + 5x + 6.$$

The result is the original expression.

*Example:* Factor $x^2 - 9x + 20$.

1. The factors of the first term are $x \cdot x$
2. The factors of the last term are $20 \cdot 1$ or $10 \cdot 2$ or $5 \cdot 4$
3. The sign of the last term is positive so $(-20)(-1)$, $(-10)(-2)$ or $(-5)(-4)$ are also possible factors.
4. The sign of the middle term is negative and has coefficient 9, so we want the combination of $-5$ and $-4$ for our factors, since they add to $-9$ while multiplying to $+20$.

   Therefore in factored form: $x^2 - 9x + 20 = (x - 5)(x - 4)$

*Check:* $(x - 5)(x - 4) = x^2 - 4x - 5x + 20 = x^2 - 9x + 20$

*Example:* Factor $x^2 + 5x - 50$.

1. The factors of $x^2$ are $x \cdot x$.
2. The factors of 50 are $50 \cdot 1$ or $25 \cdot 2$ or $10 \cdot 5$.
3. The sign of the last term is negative, so one of the factors must be negative. The following are possible combinations: $(-50)(1)$, $(50)(-1)$, $(25)(-2)$, $(-25)(2)$, $(10)(-5)$ or $(-10)(5)$.
4. The sign of the middle term is positive and the coefficient is 5, so
5. The combination of 10 and $-5$ add to 5 while multiplying to $-50$

   Therefore, in factored form: $x^2 + 5x - 50 = (x + 10)(x - 5)$.

*Check:* $(x + 10)(x - 5) = x^2 - 5x + 10x - 50 = x^2 + 5x - 50$.

*Example*: Factor $x^2 - 2x - 15$

1. The factors of $x^2$ are $x \cdot x$
2. The factors of –15 are (15)(–1) or (–15)(1) or (5)(–3) or (3)(–5)
3. The coefficient of the middle term is 2 and it is negative, so
4. The combination of –5 and +3 gives us the desired result: they add to –2 while multiplying to –15.

   Therefore, $x^2 - 2x - 15 = (x - 5)(x + 3)$.

*Check*: $(x - 5)(x + 3) = x^2 + 3x - 5x - 15 = x^2 - 2x - 15$.

*Example*: Factor completely: $5x^2 - 25x + 30$.

This situation is a little different! FIRST factor out the common factor 5: this gives $5(x^2 - 5x + 6)$. Now factor the trinomial as before.

1. $x^2 = x \cdot x$
2. +6 has sets of factors 6 and 1, 3 and 2, –6 and –1 or –3 and –2
3. we want a combination that will add to –5 (the coefficient of the middle term) while multiplying to +6 (the last term)
4. that combination is –3 and –2

   So, factored completely

   $5x^2 - 25x + 30 = 5(x^2 - 5x + 6) = 5(x - 3)(x - 2)$

**Exercise Set 10.4**

Factor completely:

1. $y^2 + 4y + 3$
2. $y^2 - 4y + 3$
3. $y^2 - 2y - 3$
4. $y^2 + 2y - 3$
5. $z^2 + 5z + 4$
6. $z^2 - 5z + 4$
7. $c^2 + 2c + 1$
8. $c^2 - 2c + 1$
9. $c^2 + 2c - 3$
10. $c^2 - 2c - 3$
11. $a^2 + 2ab + b^2$
12. $a^2 - 2ab + b^2$

## 10.5 *General Rules for Factoring Algebraic Expressions*

***Look for the greatest common factor.*** **Divide the entire expression by that factor, in order to obtain the second factor.**

*Example*: $3xy + 12x^2y - 9xy^2 = 3xy(1 + 4x - 3y)$

***Look for a "difference of two squares".*** **That will factor into a sum and difference of the square roots of the two terms.**

*Example*: $9x^2 - 16y^2 = (3x + 4y)(3x - 4y)$

*Example:* $8d^2 - 50 = 2(4d^2 - 25) = 2(2d + 5)(2d - 5)$

***When factoring trinomials:***
**If the leading coefficient is 1, find the factors of the last term whose sum will equal the coefficient of the middle term.**

*Example*: $a^2 + 4b + 3 = (a + 3)(a + 1)$ since $(+3)(+1) = +3$; and $(+3) + (+1) = +4$

*Example*: $a^2 - 4b + 3 = (a - 3)(a - 1)$ since $(-3)(-1) = +3$; and $(-3) + (-1) = -4$

*Example:* $a^2 - 3a - 4 = (a - 4)(a + 1)$ since $(-4)(+1) = -4$; and $(-4) + (+1) = -3$

*Example:* $a^2 + 3a - 4 = (a + 4)(a - 1)$ since $(+4)(-1) = -4$; and $(+4) + (-1) = +3$

*Example:* $6t^3 + 24t^2 + 12t = 6t(t^2 + 4t + 4) = 6t(t + 2)(t + 2)$

**Exercise Set 10.5**

Factor completely:

1. $4x^2 - 16y^2$
2. $t^3 - 4t^2 - 5t$
3. $5a - 10a^2$
4. $10y^2 - 15y$
5. $100 - 4x^2$
6. $y^6 - 4y^4$
7. $9x^3 - 36x$
8. $a^4 - b^4$
9. $2w^2 - 8w - 24$
10. $x^3 - 8x^2 + 7x$

## CHAPTER EXERCISES

### Section 10.1

Find the greatest common factor for the following:

1. $5a$ and $10a^2$
2. $2xy$ and $4xy^2$
3. $6x^2$ and $3xy$
4. $2ab$, $12ab$ and $6ab$
5. $6x$, $10x$ and $15x$
6. $10y^2$, $15y$ and $21y^2$
7. $22c^2$ and $55c^3$
8. $20xy^2$, $28x^2y^2$, $21x^2y^2$
9. $9ab^2c$, $12a^2b^2c$, $15ab^2c^2$
10. $0.2x$ and $1.0x^2$
11. $0.3x^2$ and $1.2x^2$
12. $1.2w^2$, $0.6w^2$, $1.5w^2$
13. $25yz$, $50y^2z$, $35yz$
14. $-8a$, $32a$, $-6a$
15. $33b^2$, $-9b$, $-15b^2$
16. $\frac{2s}{5}, \frac{3s}{5}$
17. $\frac{4a}{7}, \frac{-3a^2}{7}, \frac{2a}{7}$
18. $\frac{7z^2}{3}, \frac{5z^2}{3}$
19. $3r^2t^3$, $15r^3t^2$
20. $-2a^3b^2$, $8a^2b^3$
21. $5a^2bc^2$, $35a^3b^2c^2$
22. $8ab^2c$, $16bc^2d$, $6c^2a$
23. $-27ax^2$, $18a^2x^3$, $-2x^3$
24. $8t^3$, $12t^2$, $16t^4$

### Section 10.2

Factor the following polynomials:

25. $3x + 6x^2$
26. $2x^2 - 8x$
27. $6x^3 - 9x^2$
28. $18y^2 + 27y$
29. $3x^2y - 6x^2$
30. $72xy^2 + 18x^2y^2 - 27xy$
31. $4a^2b - 10a^2b^2 - 16a^2b^3$
32. $25abc^2 - 35abc$
33. $2.3a^2 + 4.6a^3$
34. $0.7a^2 - 4.9a$
35. $\frac{2}{3}x^5 - \frac{5}{3}x^3$
36. $ab^2c^2 + 2ab^2c^2$
37. $r^2 - 4r$
38. $2pr + pr^2$
39. $\frac{1}{2}pr^2 - \frac{3}{2}pr$
40. $mx^2 + mx$
41. $\frac{1}{2}hb + \frac{1}{2}hx$
42. $a^3x^2 - a^2x + a$

**Section 10.3**

Factor completely:

43. $a^2 - b^2$
44. $y^2 - w^2$
45. $z^2 - 16$
46. $c^2 - 144$
47. $36 - x^2$
48. $100 - d^2$
49. $9x^2 - 36$
50. $121c^2 - 64d^2$
51. $25k^2 - 169$
52. $196x^2 - 4y^2$
53. $x^4 - a^2$
54. $y^6 - x^4$
55. $16a^8 - y^6$
56. $49y^4 - 64y^{12}$
57. $a^2y^2 - b^2z^2$
58. $b^2c^4 - d^2f^6$
59. $.04 - w^4$
60. $\frac{9}{4} - x^2$
61. $\frac{121}{25} - y^2$
62. $\frac{64}{25} - x^4$

**Section 10.4**

Factor the trinomials:

63. $x^2 - 5x + 6$
64. $x^2 - 5x - 6$
65. $y^2 + 7y + 12$
66. $y^2 - 8y + 15$
67. $w^2 - 4w - 12$
68. $z^2 + 6z - 40$
69. $a^2 - 13a + 22$
70. $b^2 + 23b - 50$
71. $x^2 + 12x + 32$
72. $y^2 + 7y - 18$
73. $w^2 + 18w + 81$
74. $x^2 + 4x - 32$
75. $y^2 - y - 72$
76. $x^2 + 2x - 3$
77. $w^2 + 14w + 33$
78. $x^2 - 8x + 7$
79. $y^2 + 9y - 22$
80. $z^2 + 12z + 36$
81. $b^2 + 16b + 64$
82. $d^2 + 26d + 169$
83. $a^2 - 11a - 60$
84. $k^2 + 2k - 48$
85. $d^2 + 9d + 20$
86. $x^2 - 9x + 18$
87. $x^2 - 38x + 72$
88. $y^2 + 8y - 33$
89. $y^2 - 14y - 15$
90. $t^2 - 13t + 40$

**CHAPTER TEST** *Form 10A*

Find the greatest common factor of the following:

1. $7x^2$ and $21y^2$

2. $15x^2$ and $6x$

3. $10x^2$, $25x^3$ and $35x$

4. $12x^3y$ and $18x^2y^2$

5. $ay^3x^2$ and $by^2$

Factor completely:

6. $20x^2 - 12y^2$

7. $9x^3 - 15x^2 + 3x$

8. $by^3 + bz^3$

9. $50z^4 + 32z^2 - 8z^6$

10. $w^2 - 16$

11. $81 - x^2$

12. $9y^2 - 25w^2$

13. $a^2y^2 - c^2x^2$

14. $x^6 - y^8$

15. $x^2 - 12x + 35$

16. $y^2 + 5y - 24$

17. $w^2 + 15w + 36$

18. $d^2 - 6d - 27$

19. $26 - 15y + y^2$ (Hint: rearrange, putting the $y^2$ term first)

20. $x^2 + x - 12$

## CUMULATIVE EXERCISES Chapters 1–10 *Set A*

1. Find the reciprocal of $\frac{-7}{3}$.

2. Is the following an example of the commutative property of real numbers?
$a^2 + (b + 4) = (a^2 + b) + 4$

Evaluate the following:

3. $6.5(3.1) + 4.2$

4. $(1.1)^2$

5. $\frac{5}{2} - \frac{3}{5} + 4.2$

6. $\frac{5}{13} \cdot \frac{39}{7}$

7. $4(-3)^2 - 2\,|-4|$

8. $(8)^{-2}$

9. $(2^3)^2$

10. $y^3(y^4)^2$

Perform the indicated operations and simplify:

11. $12a^2 + 4ab - 6ab + 3a^2 + 2ab$

12. $14(2x - 1) - (8x + 7)$

13. $-(c^2 + 2c + 5) + 2(3 - 4c - c^2)$

14. $(2x - 1)^2$

15. $5a(4a^2 - 3a + 5)$

16. $2ab(2a^2 - 3ab + b^2)$

17. $\left(\frac{-7}{2}\right)^2$

18. $\frac{4y^3}{3} \div \frac{16y^5}{9}$

19. $\frac{9x^2 - 18x}{3x}$

20. $(x - 4)(3x + 2)$

21. $(x + y)(2x + y)$

Solve and check results:

22. $10a - 4 = 16$

23. $16 - 8y = -42$

24. $3(x + 2) = -15$

25. Solve for c: $b^2 + 4ac = 5$

26. Solve for m: $y = mx + b$

27. Solve for a: $v = \frac{1}{2}at^2$ if $t = 2$ and $v = 12$

28. Solve for x: $7.1x = 2.1x - 4.8$

29. Seven times a number increased by 4 gives the same result as multiplying the number by 6. What is the number?

30. A nine foot board is cut into three pieces. The second piece is one foot longer than the first, and the third piece is twice as long as the first. Find the length of each piece.

31. Four times Andy's age increased by nine is equal to his 81 year old Grandfather's age. How old is Andy?

32. The sum of three consecutive odd integers is 201. What are the numbers?

State the quadrant in which the following points are located:

33. $(17,18)$

34. $\left(\frac{-1}{4}, \frac{-1}{6}\right)$

35. $\left(-12, \frac{7}{8}\right)$

36. $(-1,7)$

37. $(-1,3)$

38. $\left(\frac{1}{2}, \frac{3}{4}\right)$

Give the slope of the lines represented by the equations:

39. $y = 8 - 7x$

40. $y = 9x + 2$

41. $y = \frac{5}{7}x + 6$

42. $y = \frac{-8}{3}x$

Find the value for x or y which will satisfy the equations:

43. $y = 3x - 8 \quad (0,?)$

44. $y = 7x - 2 \quad (?,5)$

45. $y = 8x + 5 \quad \left(\frac{1}{2},?\right)$

46. $y = 2x + 7 \quad (?,1)$

Graph the lines represented by the following equations:

47. $y = 7 - 2x$

48. $y = x + 7$

Factor completely:

49. $9x^2 + 27$

50. $a^2b^2 - 4ab^2 + 3ab$

51. $x^2 - 16$

52. $25a^2 - 9$

53. $x^2 - 3x + 2$

54. $5x^2 + 15x - 50$

55. $x^4 - 16$

56. $a^2 - 2ab + b^2$

## CUMULATIVE TEST Chapters 1–10 *Form A*

**Answers**

1. What is the reciprocal of –41.5? 1. ________
2. What is the additive inverse of 13.86? 2. ________
3. Evaluate: $7(-2)^2 + 3 - 21$ 3. ________
4. Evaluate: $\frac{45}{3} \div \frac{-135}{7}$ 4. ________
5. Multiply: $(y^2 - 3)(y + 2)$ 5. ________
6. Combine and simplify: $\frac{3}{4} - \frac{7}{2} + \frac{8}{3}$ 6. ________
7. Divide $a^2 - 4ab + 2ab^2$ by a 7. ________
8. Solve: $3x - 4 = 7$ 8. ________
9. Solve: $5(x + 3) = 2(4x + 5)$ 9. ________
10. Add 24 to a number and the result is the same as multiplying it by 5. What is the number? 10. ________
11. Solve for h: $A = \frac{1}{2}bh$ 11. ________
12. What is the slope of the line whose equation is $y = 8 - 14x$? 12. ________

For questions 13–16, write true or false. If false, change the statement to make it true.

**Answers**

13. The point (–7,–3) is in the fourth quadrant. 13. ________
14. (5,–2) is a point on the line $y = 4x - 22$. 14. ________
15. The slope of the line $y = 2 - \frac{5}{11}x$ is $\frac{5}{11}$. 15. ________
16. The point (0,–8) is on the positive x axis. 16. ________
17. Graph the line $y = \frac{1}{3}x + 2$ 17. ________

Factor the following:

**Answers**

18. $7a^2 + 14a^3$ 18. ________
19. $36y^2 - 1$ 19. ________
20. $x^2 - 2x - 99$ 20. ________
21. $y^4 - 1$ 21. ________
22. $x^2 - 11x + 30$ 22. ________

CHAPTER

# 11 Algebraic Fractions

## 11.0 *Vocabulary Preview*

Define these words without using a dictionary. Formulate their meanings from past knowledge and experience.

| | | | |
|---|---|---|---|
| **common** | **denominator** | **factor** | **fraction** |
| **identity** | **multiple** | **numerator** | **square** |

## 11.1 *Reducing Algebraic Fractions*

Before attempting this section on algebraic fractions, it would be helpful to review:

Section 5.2: Common Factors and Reducing Fractions
Section 6.5: Multiplication of Terms Involving Exponents
Section 6.6: Division of Terms Involving Exponents
Section 10.1: Polynomials Containing Common Factors

To reduce a fraction which has monomials in numerator and denominator, simply "divide off" the factors which are common to each. To reduce the fraction to LOWEST terms, divide off the GREATEST common factor.

*Example:* $$\frac{8x^2}{4x^3} = \frac{2 \cdot 2 \cdot 2 \cdot x \cdot x}{2 \cdot 2 \cdot x \cdot x \cdot x} = \frac{2}{x}$$

The greatest common factor was $4x^2$.

*Example:* $$\frac{12xy^2}{3x^2y} = \frac{2 \cdot 2 \cdot 3 \cdot x \cdot y \cdot y}{3 \cdot x \cdot x \cdot y} = \frac{4y}{x}$$

The greatest common factor was $3xy$.

*Example:* $$\frac{18r^3s^2t^2}{6rs^2t^2} = \frac{2 \cdot 3 \cdot 3 \cdot r \cdot r \cdot r \cdot s \cdot s \cdot t \cdot t}{2 \cdot 3 \cdot r \cdot s \cdot s \cdot t \cdot t} = 3r^2$$

The greatest common factor was $6rs^2t^2$.

It is obvious that we are dividing the terms with like bases after dividing the coefficients. Recall the rule for division from Section 6.6

*Example*: Reduce the fractions by dividing off the greatest common factor of numerator and denominator:

$$\frac{5x}{15x^3} = \frac{5x\,(1)}{5x\,(3x^2)} = \frac{1}{3x^2}$$

$$\frac{7xy^2}{14x^2y} = \frac{7xy\,(y)}{7xy\,(2x)} = \frac{y}{2x}$$

$$\frac{24x^2y^3z}{15x^5z^2} = \frac{3x^2z\,(8y^3)}{3x^2z\,(5x^3z)} = \frac{8y^3}{5x^3z}$$

**Exercise Set 11.1**

Reduce the fractions by dividing off the greatest common factor from numerator and denominator:

1. $\frac{32xy}{48y^2}$
2. $\frac{7r^2s^2t}{14r^5st^2}$
3. $\frac{7x^2y^2z}{14xy^5z}$
4. $\frac{32p^2qr^5}{24pqr^3}$
5. $\frac{15x^2}{25x^3}$
6. $\frac{xy^7z^5}{x^6yz^3}$
7. $\frac{3a^2b^3c}{9ab^2c^5}$
8. $\frac{32x^2y^2}{64xy}$
9. $\frac{2cd}{20c^2}$
10. $\frac{15x^2y^3}{5y^2}$
11. $\frac{7r^2s^2}{28r^5}$
12. $\frac{14r^3s}{7r^3s^2}$
13. $\frac{24x^4y}{8xy^3}$
14. $\frac{-27xy^2}{-9y}$
15. $\frac{a^2bc^3}{abc}$
16. $\frac{g^6hi^7}{gh^2i^4}$

## 11.2 *Multiplication and Division of Algebraic Fractions*

Review the following before studying this section:

Section 3.3: Multiplication of Real Numbers
Section 3.4: Division of Real Numbers
Section 5.3: Multiplication of Fractions
Section 5.4: Division of Fractions

**To multiply algebraic fractions:**

1. **express the products of the numerators and denominators as a single term**
2. **find the greatest common factor of the numerator and denominator**
3. **divide off the greatest common factors**
4. **multiply the remaining factors in numerator and denominator.**

*Example:* Multiply: $\frac{3x^2y^3}{8xy} \cdot \frac{2x^5y^3}{y^2}$

$$\frac{3x^2y^3 \cdot 2x^5y^3}{8xy \cdot y^2} = \frac{2xy^3(3x^6y^3)}{2xy^3(4)}$$

Reduce the fraction $= \frac{3x^6y^3}{4}$

*Example:* Multiply: $\frac{3}{2x^2} \cdot \frac{6x}{xy^2} \cdot \frac{4x^2y^2}{x^3y^3}$

Reducing the second and third fractions to lowest terms and then multiplying:

$$\frac{3}{2x^2} \cdot \frac{6}{y^2} \cdot \frac{4}{xy} = \frac{3 \cdot 6 \cdot 4}{2x^2 \cdot y^2 \cdot xy}$$

$$= \frac{3 \cdot 6 \cdot 2}{x^2 \cdot y^2 \cdot xy}$$

$$= \frac{36}{x^3y^3}$$

*Example:* Multiply: $\frac{2x}{5} \cdot \frac{10y^2}{4xy} \cdot \frac{3z}{2y}$

Multiplying first, gives $\frac{60xy^2z}{40xy^2}$

which reduces to $\frac{3z}{2}$

An alternate solution to the previous expression could be obtained by reducing the middle fraction first:

$$\frac{2x}{5} \cdot \frac{10y^2}{4xy} \cdot \frac{3z}{2y} = \frac{2x}{5} \cdot \frac{5y}{2x} \cdot \frac{3z}{2y}$$

then dividing off common factors from numerators and denominators results in $\frac{3z}{2}$

Remember that division is the inverse operation of multiplication.

To divide fractions, use the reciprocal of the divisor and multiply.

*Example:* Divide: $\frac{6x^2}{y^3} \div \frac{3xy^2}{4xz^2}$

Use the reciprocal of the divisor and multiply

$$\frac{6x^2}{y^3} \cdot \frac{4xz^2}{3xy^2} =$$

$$\frac{6x^2}{y^3} \cdot \frac{4z^2}{3y^2} =$$

$$\frac{6 \cdot 4 \cdot x^2 \cdot z^2}{3 \cdot y^3 \cdot y^2} = \frac{2 \cdot 4x^2z^2}{y^5} = \frac{8x^2z^2}{y^5}$$

*Example:* Divide: $\frac{18rst}{7r^2}$ by $\frac{6r}{14}$

Simplify the left fraction first:

$$\frac{18st}{7r} \div \frac{6r}{14}$$

Use the reciprocal of the divisor and multiply:

$$\frac{18st}{7r} \cdot \frac{14}{6r}$$

Dividing off the common factor of 7 and 14 and the common factor of 6 and 18:

$$\frac{3st}{r} \cdot \frac{2}{r} = \frac{6st}{r^2}$$

**Exercise Set 11.2**

Multiply or divide as indicated. Reduce fractions when possible.

1. $\frac{2t}{5} \div \frac{10t^2}{3}$

2. $\frac{2x}{5} \cdot \frac{40}{7x}$

3. $\frac{4y^2z}{3xy} \div \frac{5y}{6}$

4. $\frac{7}{x^2} \cdot \frac{4xy}{21}$

5. $\frac{3a^2b}{4d} \div \frac{ab}{6d^2}$

6. $\frac{4xy}{7} \div \frac{2x^2y^2}{3}$

7. $\frac{y^3}{5} \cdot \frac{10x^2}{y}$

8. $\frac{x^2y^2z^5}{7} \div \frac{xz^3}{14}$

9. $\frac{1}{5y} \cdot \frac{7}{4y^3} \div \frac{7}{10y^5}$

10. $\frac{abc^5}{14d} \cdot \frac{2d^2}{ac} \div \frac{4ab}{ad^3}$

## 11.3 *Addition and Subtraction of Algebraic Fractions*

Before studying this section, review:

Section 5.6: Addition and Subtraction of Fractions
Section 10.1: Polynomials Containing Common Factors

**Fractions may be added or subtracted ONLY IF they have the same denominators.**

Remember two important facts:

1. multiplying a number by 1 does not change its "identity"
2. any number divided by itself is 1.

**To add or subtract two fractions:**

1. **find the LCM (least common multiple) of the denominators**
2. **write each fraction with the same denominator (the LCM)**
3. **add or subtract the numerators as directed**
4. **write the sum or difference over the common denominator.**

*Example:* Add the fractions $\frac{5}{x} + \frac{7}{x}$

Since the denominators are alike, simply combine numerators. The result is $\frac{12}{x}$

*Example:* Add the fractions $\frac{3x}{10} + \frac{x}{5}$

Since 10 is the least common multiple of the denominators (which we'll now call the LEAST COMMON DENOMINATOR), multiply the second fraction by $\frac{2}{2}$

Then $\frac{3x}{10} + \frac{x\,(2)}{5\,(2)} = \frac{3x}{10} + \frac{2x}{10} = \frac{5x}{10} = \frac{x}{2}$

*Example:* Subtract $\frac{6}{a + b} - \frac{14}{a + b}$

Since the denominators are alike, simply combine the numerators.

$$\frac{6 - 14}{a + b} = \frac{-8}{a + b}$$

*Example:* Add $\frac{4xy}{7z} + \frac{2x}{z}$

The common denominator is 7z, so multiply the second fraction by $\frac{7}{7}$

$$\frac{4xy}{7z} + \frac{2x(7)}{z(7)} = \frac{4xy}{7z} + \frac{14x}{7z}$$

Since the denominators are now alike, combine numerators. The result is $\frac{4xy + 14x}{7z}$

*Example:* Subtract the fractions $\frac{11s}{4} - \frac{7s}{12}$

The least common denominator is 12, so multiply the first fraction by $\frac{3}{3}$.

$$\frac{11s(3)}{4(3)} - \frac{7s}{12} = \frac{33s}{12} - \frac{7s}{12} = \frac{26s}{12}$$

Reducing the fraction gives the result $\frac{13s}{4}$

*Example:* Combine $\frac{3xy}{7} + \frac{2xy}{3} - \frac{xy}{2}$

The common denominator is 42, so

multiply the first fraction by $\frac{6}{6}$

multiply the second fraction by $\frac{14}{14}$

multiply the third fraction by $\frac{21}{21}$:

$$\frac{3xy(6)}{7(6)} + \frac{2xy(14)}{3(14)} - \frac{xy(21)}{2(21)} =$$

$$\frac{18xy}{42} + \frac{28xy}{42} - \frac{21xy}{42} = \frac{25xy}{42}$$

**Exercise Set 11.3**

Add or subtract as indicated:

1. $\frac{3}{x} - \frac{1}{2x}$
2. $\frac{x^2}{4} - \frac{4}{3}$
3. $\frac{x^2y}{2} + \frac{3x^2y}{4} - \frac{1}{8}$
4. $\frac{3}{4x} + \frac{2}{x}$
5. $\frac{1}{5} + \frac{x}{10} - \frac{y}{5}$
6. $\frac{2}{5t} - \frac{1}{5}$
7. $\frac{2y}{3} + \frac{6y}{7}$
8. $\frac{1}{x} - \frac{3}{x} + \frac{1}{2x}$
9. $\frac{t}{3} - \frac{4t}{5}$
10. $\frac{t+1}{4} - \frac{1}{2}$
11. $\frac{6}{y} - \frac{2}{y} + \frac{1}{2y}$
12. $\frac{3}{y^2} + \frac{7}{y^2}$

13. $\frac{5y}{6} + \frac{1}{3}$

14. $\frac{5}{8x} - \frac{2}{4x}$

15. $\frac{m^2}{3} + \frac{2m^2}{9}$

16. $\frac{5a}{6} - \frac{2a}{3} + \frac{5}{3}$

17. $\frac{15x}{2} + \frac{5}{4} - 2x$

18. $\frac{4}{y} - \frac{9}{y}$

19. $\frac{2}{a + 3} - \frac{5}{a + 3}$

20. $\frac{a + 6}{6} + \frac{2}{3}$

## 11.4 *Solving Equations Containing Fractions*

The same procedures that are used to solve ordinary equations, are used to solve equations containing fractions; except that the fractions must often be dealt with first.

*Example:* Solve for x: $\frac{4x}{7} = 5$

Multiplying each side by 7 gives

$$(7)\frac{4x}{7} = 5(7)$$

and $4x = 35$

Dividing each side by 4 gives the result

$$x = \frac{35}{4}$$

*Example:* Solve for x: $\frac{x + 2}{3} = 7$

Multiplying each side by 3 gives

$(3)\frac{x + 2}{3} = 7(3)$ which simplifies to

$x + 2 = 21$ Now subtracting 2 from each side gives the result

$x = 19.$

*Example:* Solve for x: $\frac{x}{6} - \frac{x}{2} = 5$

The common denominator is 6 so we have

$$\frac{x}{6} - \frac{x(3)}{2(3)} = 5$$

$$\frac{x}{6} - \frac{3x}{6} = 5$$

Combine: $$\frac{-2x}{6} = 5$$

Multiply by 6: $$-2x = 30$$

Divide by $-2$: $$x = -15$$

An alternate method for solving this equation is to multiply through the equation by 6:

$$(6)\frac{x}{6} - (6)\frac{x}{2} = (6)5$$

$$x - 3x = 30$$

$$-2x = 30$$

dividing by $-2$ $\quad x = -15$

*Example:* Solve for y: $\frac{2y}{3} = 5 + \frac{3y}{2}$

Subtract $\frac{3y}{2}$ from each side of the equation:

$$\frac{2y}{3} - \frac{3y}{2} = 5$$

The common denominator is 6 so

$$\frac{(2)2y}{(2)3} - \frac{(3)3y}{(3)2} = 5$$

$\frac{4y}{6} - \frac{9y}{6} = 5$ says that $\frac{-5y}{6} = 5$

Multiplying each side by 6 gives:

$$-5y = 30.$$

Now dividing each side by –5 gives y = –6

An alternate method of solving this equation would be to multiply through the terms by 6:

$$(6)\frac{2y}{3} = (6)5 + (6)\frac{3y}{2}$$

$$4y = 30 + 9y$$

subtracting 9y $\qquad -5y = 30$

dividing by –5 $\qquad y = -6$

**Exercise Set 11.4**

Solve the equations; check results.

1. $\frac{x}{2} = -6$
2. $\frac{2x}{7} = 4$
3. $\frac{3y}{5} = -3$
4. $\frac{-x}{5} = 5$
5. $\frac{x}{4} - \frac{x}{5} = 2$
6. $\frac{4x-3}{5} = 3$
7. $\frac{y-2}{3} = 5$
8. $\frac{2z+4}{7} = 2$
9. $\frac{4z+1}{3} = 3$
10. $\frac{3x}{2} = 5 - \frac{x}{2}$
11. $\frac{4z}{3} = 7 + \frac{z}{3}$
12. $\frac{y}{6} - \frac{2y}{3} = 9$
13. $\frac{7z+2}{3} = -4$
14. $\frac{y}{2} - 3 = \frac{y}{4} + 1$

## CHAPTER EXERCISES

### Section 11.1

Reduce the following fractions to lowest terms:

1. $\frac{6x}{5x}$
2. $\frac{8y^2}{2y^2}$
3. $\frac{6y^3}{7y}$
4. $\frac{21x^3}{7x}$
5. $\frac{24x^2y}{3xy}$
6. $\frac{7x^4y^2}{28x^2y^2}$
7. $\frac{6a^2bc}{15ab^2c}$
8. $\frac{-15x^2yz^2}{20x^3yz}$
9. $\frac{24ab^2c^3}{-40a^2bc^2}$
10. $\frac{-12a^5yx^3}{27a^4y^3z}$

### Section 11.2

Perform the indicated operations. Reduce answers when possible.

11. $\frac{2}{5} \cdot \frac{15x^2}{1}$
12. $\frac{14c^2}{3} \cdot \frac{15}{2}$
13. $\frac{7x^2}{3} \cdot \frac{21}{14x}$
14. $\frac{ab^2}{3} \cdot \frac{28a}{b^2}$
15. $\frac{12c^2}{3} \div \frac{1}{3}$
16. $\frac{c^2}{6} \div \frac{4}{3}$
17. $32x \div \frac{8x}{5}$
18. $\frac{14xy^2}{5} \div \frac{7y^2}{30x}$
19. $\frac{12ab^2c}{7a} \cdot \frac{c}{4ab^2}$
20. $\frac{28xy}{15a} \div \frac{14x^2}{5ax}$
21. $\frac{-22t^2}{3v} \cdot \frac{6v^2s}{55ts}$
22. $\frac{9vs^3}{-4t^2} \div \frac{-s^2}{2tv}$
23. $\frac{xy^3}{33ax} \cdot \frac{27ay}{x^3} \cdot \frac{44ay}{y^2}$
24. $\frac{2t}{12v^2} \cdot \frac{17t^2}{32} \div \frac{t^3}{64}$
25. $\frac{-3x}{40} \cdot \frac{-18xy}{5x}$
26. $\frac{-9cd^2}{x^2} \div \frac{72cd}{-8x}$
27. $\frac{-75}{t^5x^6} \cdot \frac{-t^2x^3}{12}$
28. $\frac{a^2by}{a} \cdot \frac{20y}{6dy^2} \div \frac{-4ab}{18d}$

29. $\frac{2k^2m}{w} \div \frac{4m}{2w^2k} \div \frac{6}{k}$

30. $\frac{-131t^3}{s^3} \div \frac{21t}{2s} \div \frac{-22}{s^3}$

Perform the indicated operations. Simplify the results.

31. $\frac{3x}{2} + \frac{x}{2}$

32. $\frac{8}{x} - \frac{5}{x}$

33. $\frac{5}{y^2} + \frac{6}{y^2}$

34. $\frac{10}{b} + \frac{6}{b}$

35. $\frac{2}{w} - \frac{7x}{w}$

36. $\frac{7c}{11} + \frac{2}{11}$

37. $\frac{4}{5x} + \frac{3x}{5x}$

38. $\frac{9y}{4} + \frac{2y}{4} - \frac{y}{4}$

39. $\frac{1}{x} + \frac{3y}{x} - \frac{4}{x}$

40. $\frac{6c}{y} + \frac{4}{y} - \frac{2c}{y}$

41. $\frac{2}{x} + \frac{1}{2x}$

42. $\frac{1}{4y} - \frac{5}{2y}$

43. $\frac{2}{x^2} + \frac{3}{x}$

44. $\frac{11}{c} - \frac{15}{c^2}$

45. $\frac{1}{2x^2} - \frac{3}{4x^2}$

46. $\frac{3d}{8t} + \frac{3d}{2t}$

47. $\frac{4}{y^2} + \frac{3}{2y} - \frac{4}{2y}$

48. $\frac{-4}{6z} + \frac{2}{3z} + \frac{2}{6z}$

49. $\frac{5x^2}{10} - \frac{2x}{5} + \frac{x}{10}$

50. $\frac{12}{x+3} - \frac{7}{x+3}$

51. $\frac{-3}{2-y} + \frac{4}{2-y}$

52. $\frac{2}{a+3} - \frac{2a}{a+3}$

53. $\frac{7y}{6t} - \frac{2}{3t}$

54. $\frac{7}{6} + \frac{x+2}{12}$

55. $\frac{y+2}{15} - \frac{1}{5}$

56. $\frac{vt}{25} - \frac{2vt}{5}$

**Section 11.4**

Solve the following equations and check your answers.

57. $\frac{x}{3} = 7$

58. $\frac{y}{3} = 6$

59. $\frac{2x}{5} = 14$

60. $\frac{3x}{8} = 24$

61. $\frac{-3a}{4} = 15$

62. $\frac{-5c}{7} = -5$

63. $\frac{x-2}{3} = 7$

64. $\frac{y+3}{6} = 2$

65. $\frac{5-y}{4} = 1$

66. $\frac{6+3y}{3} = 1$

67. $\frac{11-15z}{4} = -1$

68. $\frac{2x+3}{7} = 1$

69. $\frac{y}{6} + \frac{y}{3} = 1$

70. $\frac{w}{4} - \frac{3w}{8} = 7$

71. $\frac{z}{3} + \frac{z}{3} = 20$

72. $\frac{3x}{11} - \frac{2x}{22} = 6$

73. $\frac{-3}{10} + \frac{x}{5} = 7$

74. $\frac{2y}{6} - \frac{2y}{3} = 8$

75. $\frac{y}{3} - 2 = 4$

76. $3 + \frac{x}{5} = 7$

## CHAPTER TEST *Form 11A*

Reduce the fractions to lowest terms:

1. $\frac{26x^3}{13x^3}$

2. $\frac{3x^2y}{15xy^2}$

Perform the indicated operations and simplify your answers:

**Answers**

3. $\frac{2x}{5} \cdot \frac{10}{x}$ — 3. ________

4. $\frac{15x}{c} \div \frac{x}{c}$ — 4. ________

5. $\frac{3}{7} \cdot \frac{35y^2}{3}$ — 5. ________

6. $\frac{4x^2y^2}{x} \cdot \frac{3y}{6x}$ — 6. ________

7. $\frac{a^2b}{16} \div \frac{b}{4}$ — 7. ________

8. $\frac{-7y^2z}{2c} \div \frac{14}{c^2}$ — 8. ________

9. $\frac{7y}{4} + \frac{2y}{4}$ — 9. ________

10. $\frac{y}{15} - \frac{2y}{3}$ — 10. ________

11. $\frac{3}{2x} - \frac{5}{x}$ — 11. ________

12. $t - \frac{3t}{5}$ — 12. ________

13. $\frac{x}{x+6} - \frac{2}{x+6}$ — 13. ________

14. $\frac{y^2}{9} + \frac{2y^2}{3} + 1$ — 14. ________

Solve the equations and check your answers:

**Answers**

15. $\frac{y}{5} + 2 = 4$ — 15. ________

16. $\frac{x}{2} - 3x = 10$ — 16. ________

17. $\frac{x}{3} - 5 = x + 7$ — 17. ________

18. $\frac{a}{5} - 4 = \frac{a}{10} + 1$ — 18. ________

19. $\frac{y-3}{4} = 10$ — 19. ________

20. $\frac{y}{4} + 1 = \frac{y}{6} - 1$ — 20. ________

**CUMULATIVE EXERCISES Chapters 1–11** *Set A*

Evaluate the following:

1. $\sqrt{257.4}$
2. $3|-2.8|$
3. $(-7)^2$
4. $(1.2)^{-2}$
5. $\left(\frac{3}{10}\right)^2$
6. $-4(1.3) + 2(-2.4)$

Combine like terms:

7. $4a^2b^2 + 3a^2b - 6a^2b^2 - 2a^2b$
8. $3x^2 + 5x^3 - 6x^2 + 3x^3 - x - 2x^2$

Perform the indicated operation and simplify:

9. $\frac{2}{7} + \frac{1}{6} - \frac{3}{7}$
10. $(-21)^2$
11. $(3x - 2)^2$
12. $\frac{x^3}{5} \div \frac{5^5}{20}$
13. $(-4x)(8x^3 - 7x^2 + 2x - 7)$
14. $(y - 9)(3y + 5)$
15. $\frac{18z^3 + 4z^2 - 8z}{2z}$

Write the equation and solve:

16. One-third a number is equal to the number increased by 18. What is the number?
17. The sum of two consecutive odd integers is 176. Find the integers.
18. The ages of the three Baxter children sum to 59 years. Jim, the middle child is 7 years older than Janet, the youngest child. Jeff, the oldest, is twice Janet's age. How old are the Baxter children?
19. In a Math 117 class at Nassau Community College, an average of 80 will give a student a B grade. If a student has taken four exams and received the grades: 71, 88, 72 and 81, what grade must he get on the last exam to get a B?

20. Solve for a: $4(a + 2) = -12$

21. Solve for x: $4(x + 2) = x - a$

22. Solve for t: $d = \frac{1}{2}gt$

State the slope and y intercept for the following:

23. $y = 7x - 5$

24. $y = \frac{-3}{7}x + 2$

25. $y = \frac{1}{7}x - \frac{2}{3}$

26. $y = \frac{7}{4} - 2x$

27. $y = 8x + 4$

28. $y = 2x - \frac{1}{7}$

Graph the following:

29. $y = x + 4$

30. $y = 2x - 1$

31. $y = \frac{2}{3}x + 1$

32. $y = \frac{4}{3}x - 2$

Factor completely:

33. $8x + 24$

34. $3x^2y - xy + 4x^2y^2$

35. $3x^2 - 27$

36. $a^4 - 3a^3$

37. $16a^2 - 25$

38. $y^2 - 0.01$

39. $k^2 - m^2$

40. $x^2 + 8x + 15$

41. $x^2 + 6x - 7$

42. $x^2 + 2x - 15$

Reduce the following fractions:

43. $\frac{24a^2b^3}{18a^5b^2}$

44. $\frac{7d^5z^2}{21dz^5}$

45. $\frac{28x^2y^3z^4}{12x^3y^7z^3}$

46. $\frac{39a^5b^2c^3}{13a^3b^4c^5}$

Perform the indicated operations. Simplify answers.

**Answers**

47. $\frac{3x^2}{8} \cdot \frac{14}{9x^3}$ 47. ______

48. $\frac{21y^3}{z^2} \div \frac{18y^2}{3z}$ 48. ______

49. $\frac{32a^2b}{45} \div \frac{8ab^2}{15}$ 49. ______

50. $\frac{42x^3}{y^2} \cdot \frac{2y^3}{6x^5}$ 50. ______

51. $\frac{27}{a^3} + \frac{13}{a^3}$ 51. ______

52. $\frac{11}{5b} - \frac{3}{b}$ 52. ______

53. $\frac{7}{a+2} + \frac{3a}{2(a+2)}$ 53. ______

54. $\frac{17}{x+y} + \frac{3}{2(x+y)} - \frac{4}{3(x+y)}$ 54. ______

Solve for the variable:

**Answers**

55. $\frac{x}{5} + 3 = 9$ 55. ______

56. $\frac{a+5}{2} = 8$ 56. ______

57. $\frac{x-8}{6} = \frac{x}{2}$ 57. ______

58. $\frac{x}{2} + 3 = \frac{x}{3} - 1$ 4. ______

## CUMULATIVE TEST Chapters 1–11 *Form A*

**Answers**

1. Find $\sqrt{148}$ to two decimal places. 1. ________
2. What is the additive inverse of –23.4? 2. ________
3. Evaluate $-(-7)^2$ 3. ________
4. Combine: $-(6x^2 + 5x - 3) + (9x^2 - 3x + 1)$ 4. ________
5. Multiply: $(6x + 3)(x - 1)$ 5. ________
6. Divide $8x^2y + 28xy^2$ by $4xy$ 6. ________
7. Factor completely: $6x^3 - 3x^2$ 7. ________
8. Solve for x: $4x - 3 = 2x + 1$ 8. ________
9. Evaluate: $\frac{7}{15} \div \left(\frac{-30}{21}\right)$ 9. ________
10. Evaluate: $7\{5 - [3 + 92 - 4^2]\}$ 10. ________
11. Multiply: $(7x - 3)^2$ 11. ________
12. The sum of a number and the number increased by 5 is equal to 55. Find the number. 12. ________
13. Simplify: $\frac{64a^3b^4c^2}{12ab^5c^2}$ 13. ________
14. Add: $\frac{3x^2 + 4}{5} + \frac{6x^2 - 1}{2}$ 14. ________
15. Multiply: $\frac{6x^2y}{11z} \cdot \frac{22z^2}{xy^2}$ 15. ________
16. Solve for x: $4ax + 3 = 19$ 16. ________
17. Factor: $a^2 - 1$ 17. ________
18. Factor: $a^2 + 18a + 77$ 18. ________
19. Divide: $\frac{23x^2y^2}{4} \div \frac{xy^2}{20}$ 19. ________
20. Graph the line represented by $y = 3x - 11$ 20. ________

CHAPTER

# 12 Second Degree Equations

## 12.0 *Vocabulary Preview*

Define these words without using a dictionary. Formulate their meaning from past experience or knowledge.

**degree** **factored** **formula** **quadratic**

## 12.1 *Solving Second Degree Equations in Factored Form*

Consider the following conditions:

a. Twelve times a number is zero.
b. The product of two numbers is zero. One of the numbers is 5.

Clearly, the missing number in each case is zero, since **the product of two numbers is zero only if one of the numbers is zero.**

Now consider:

c. A number plus 16 is zero.
d. A number minus 16 is zero.

Here the answers are different.

The first number is $-16$, since $(-16) + (16) = 0$.
The second number is 16, since $16 + (-16) = 0$.

*Example:* Find the missing number in order to make the statements true:

a. $(?)(-23) = 0$ answer: 0

b. $(100)(?) = 0$ answer: 0

c. $(? - 5) = 0$ answer: 5

d. $(? + 5) = 0$ answer: $-5$

e. $(? - 5)(? + 5) = 0$ answers: 5 and $-5$

You know that if the product of two numbers is zero, one of them must be zero.

If $x \cdot y = 0$, then $x = 0$ or $y = 0$ or **both** are zero.

*Example:* If $x(x - 3) = 0$, then one of the factors must be zero,

so, $x = 0$ or $(x - 3)$ is zero.

In order for $(x - 3)$ to be zero, x must be 3.

**This problem, then, has TWO solutions:**
The statement is true for $x = 0$ and $x = 3$.

*Check:* for $x = 0$ $\quad 0(0 - 3) = 0(-3) = 0$
for $x = 3$ $\quad 3(3 - 3) = 3(0) = 0$

*Example:* If $x(2x + 4) = 0$, then either $x = 0$ or $(2x + 4) = 0$.

In order for $2x + 4$ to be zero, 2x must equal $-4$ and so x must be $-2$.

Again, this equation has **two** solutions.

The statement is true for $x = 0$ and $x = -2$

*Check* for $x = 0$
$0[2(0) + 4] = 0(0 + 4) = 0(4) = 0$

for $x = -2$
$-2[2(-2) + 4] = -2[-4 + 4] = -2(0) = 0$

*Example:* Find the values for t that would make the statement true: $(t - 3)(t + 5) = 0$

*Solution:* Since the product is zero, one of the factors must be zero.

If $t - 3 = 0 \quad t = 3$
If $t + 5 = 0 \quad t = -5$

The TWO values, $t = 3$ and $t = -5$ satisfy the equation.

*Check* for $t = 3$ $\quad (3 - 3)(3 + 5) = (0)(8) = 0$
for $t = -5$ $\quad [3 - (-5)](-5 + 5) = 8(0) = 0$

*Example:* Find the values for g which will make the statement true: $(2g + 4)(g - 1) = 0$

*Solution:* Since the product is zero, either $2g + 4 = 0$ or $g - 1 = 0$.

Let $2g + 4 = 0$ then
$2g = -4$ and
$g = -2$

*Check:* $2(-2) + 4 = 0$
$-4 + 4 = 0$
$0 = 0$ Okay!

Let $g - 1 = 0$ then
$g = 1$

*Check:* $1 - 1 = 0$
$0 = 0$ Okay!

The **two** values which satisfy the equation $(2g + 4)(g - 1) = 0$ are $g = -2$ and $g = 1$

*Example:* What values of w satisfy the equation: $(2w - 1)(2w + 1) = 0$?

*Solution:* Since the product is zero, one factor must be zero, so

| | | | |
|---|---|---|---|
| | $2w - 1 = 0$ | or | $2w + 1 = 0$ |
| solving: | $2w = 1$ | or | $2w = -1$ |
| and so | $w = \frac{1}{2}$ | or | $w = \frac{-1}{2}$ |

*Check:* for $w = \frac{1}{2}$:

$$\left[2\left(\frac{1}{2}\right) - 1\right]\left[2\left(\frac{1}{2}\right) + 1\right] =$$
$$[1 - 1][1 + 1] =$$
$$[0][2] = 0$$

for $w = -\frac{1}{2}$

$$\left[2\left(-\frac{1}{2}\right) - 1\right]\left[2\left(-\frac{1}{2}\right) + 1\right] =$$
$$[-1 - 1][-1 + 1] =$$
$$[-2][0] = 0$$

**Exercise Set 12.1**

Solve for two values which will satisfy each equation:

1. $(x - 5)(x + 5) = 0$
2. $(x - 3)(2x + 6) = 0$
3. $x(3x - 9) = 0$
4. $(5x - 10)(x + 2) = 0$
5. $(x - 1)(x + 1) = 0$
6. $x(3x - 18) = 0$
7. $(2x - 3)(x - 4) = 0$
8. $(x + 7)(x - 7) = 0$

## 12.2 *Solving Second Degree Equations in Non-Factored Form*

In previous chapters, we dealt with algebraic EXPRESSIONS of degree higher than 1, but we have only solved EQUATIONS of degree 1 (which we called **linear** equations).

An expression or equation whose highest exponent is 2, is called a second degree or **QUADRATIC** expression or equations.

*Examples:* The following are examples of quadratic (second degree) equations:

| | | |
|---|---|---|
| $x^2 - 9 = 0$ | $4x^2 - 25 = 5$ | $9a^2 - 2 = 0$ |
| $x^2 + 5x - 50 = 0$ | $x^2 + 5x = 6$ | $y^2 + 4y = -3$ |

In each case, the highest exponent of the variable was 2.

In any case from above, the equation could be set equal to zero if it were not already so:

$4x^2 - 25 = 5$ is the same as $4x^2 - 30 = 0$
$x^2 + 5x = 6$ is the same as $x^2 + 5x - 6 = 0$
$y^2 + 4y = -3$ is the same as $y^2 + 4y + 3 = 0$.

**To solve quadratic equations in non-factored form:**

1. **Set the equation equal to zero.**
2. **Factor**

**After factoring, follow the same procedure as in section 12.1:**

3. **Set each factor equal to zero.**
4. **Solve the resulting linear equations**
5. **Check the solutions in the *original* equation.**

*Example:* What values of x make the statement true: $x^2 + 3x - 4 = 0$?

*Solution:* Factor the quadratic equation into $(x + 4)(x - 1) = 0$.

Since the product is zero, one factor must be zero.

If $x + 4 = 0$, $x = -4$ and if $x - 1 = 0$, $x = 1$.

The two solutions are $x = -4$ and $x = 1$.

*Check:*

| for $x = -4$ | for $x = 1$ |
|---|---|
| $(-4 + 4)(-4 - 1) = 0$ | $(1 + 4)(1 - 1) = 0$ |
| $(0)(-5) = 0$ | $(5)(0) = 0$ |
| $0 = 0$ | $0 = 0$ |

*Example:* Solve for y: $y^2 + 5y = 50$

*Solution:* Set the quadratic equation to equal 0 by subtracting 50 from each side: $y^2 + 5y - 50 = 0$

Factor the quadratic equation into: $(y + 10)(y - 5) = 0$

Since the product is zero, one of the factors must be zero:

If $y + 10 = 0$, $y = -10$.

If $y - 5 = 0$, $y = 5$.

The two values for y which satisfy the equation are $y = -10$ and $y = 5$.

*Check:* $[(-10) + 10](10 - 5) = 0(5) = 0$

$(5 + 10)(5 - 5) = 15(0) = 0$.

*Example:* What values for t make the statement true?

$$4t^2 - 16 = 0$$

*Solution:* Factor the quadratic equation completely: $4(t + 2)(t - 2) = 0$

The product is zero, so one of the factors is zero. Since 4 cannot be zero, one of the other factors must be.

Let $t + 2 = 0$. Then $t = -2$

Let $t - 2 = 0$. Then $t = 2$.

The solutions to the equation are $t = -2$ and $t = 2$.

*Check:* for $t = -2$

$$\begin{aligned} 4(-2)^2 - 16 &= 0 \\ 4(4) - 16 &= 0 \\ 16 - 16 &= 0 \\ 0 &= 0 \end{aligned}$$

for $t = 2$

$$\begin{aligned} 4(2)^2 - 16 &= 0 \\ 4(4) - 16 &= 0 \\ 16 - 16 &= 0 \\ 0 &= 0 \end{aligned}$$

The solutions $t = -2$ and $t = 2$ are correct.

**IMPORTANT NOTE:** Always check your solutions in the ORIGINAL equation.

**Exercise Set 12.2**

Find all values which satisfy the following equations:

1. $x^2 - 121 = 0$
2. $x^2 + 5x + 6 = 0$
3. $x^2 - 9x + 20 = 0$
4. $x^2 - 2x - 15 = 0$
5. $5x^2 - 25x + 30 = 0$
6. $y^2 + 4y + 3 = 0$
7. $y^2 - 4y + 3 = 0$
8. $z^2 - 2z + 1 = 0$
9. $y^2 - 4y - 5 = 0$
10. $z^2 - 2z - 3 = 0$
11. $y^2 - 2y - 3 = 0$
12. $y^2 + 2y - 3 = 0$
13. $z^2 + 5z + 4 = 0$
14. $z^2 - 5z + 4 = 0$
15. $2w^2 - 8w - 24 = 0$
16. $10y^2 - 15y = 0$

**In #15 and #16, remember to factor completely first.**

17. $z^2 - 36 = 0$
18. $z^2 + 12z = -36$
19. $k^2 - 11k = 60$
20. $t^2 + 23t = 50$

**In #18, 19 and 20, set the equation equal to zero before solving.**

## 12.3 *The General Form of the Quadratic Equation*

All of the equations which you solved in section 12.2 may be written in the same "form":

$$y = ax^2 + bx + c$$

This is called the general form of the quadratic equation.

In each case:

y is 0
a is the coefficient of the $x^2$ term
b is the coefficient of the x term
c is the constant.

NOTE: If the x term is missing, b is zero.

If the constant term is missing, c is zero.

*Example:* Using the even numbered problems from Exercise Set 12.2, write the equations in the general form and state the values of a, b and c.

| | | | |
|---|---|---|---|
| 2. $0 = x^2 + 5x + 6$ | $a = 1$ | $b = 5$ | $c = 6$ |
| 4. $0 = x^2 - 2x - 15$ | $a = 1$ | $b = -2$ | $c = -15$ |
| 6. $0 = y^2 + 4y + 3$ | $a = 1$ | $b = 4$ | $c = 3$ |
| 8. $0 = z^2 - 2z + 1$ | $a = 1$ | $b = -2$ | $c = 1$ |
| 10. $0 = z^2 - 2z - 3$ | $a = 1$ | $b = -2$ | $c = -3$ |
| 12. $0 = y^2 + 2y - 3$ | $a = 1$ | $b = 2$ | $c = -3$ |
| 14. $0 = z^2 - 5z + 4$ | $a = 1$ | $b = -5$ | $c = 4$ |
| 16. $0 = 10y^2 - 15y$ | $a = 10$ | $b = -15$ | $c = 0$ |
| 18. $0 = z^2 + 12z + 36$ | $a = 1$ | $b = 12$ | $c = 36$ |
| 20. $0 = t^2 + 23t - 50$ | $a = 1$ | $b = 23$ | $c = -50$ |

**Exercise Set 12.3**

Use the odd-numbered problems from Exercise Set 12.2 and

a. Write each equation in the form $y = ax^2 + bx + c$

b. state the values for a, b and c.

## 12.4 *The Quadratic Formula*

**The quadratic formula may be used to solve second degree equations. However, the formula requires several steps, which must be worked very carefully.**

To solve for the **VARIABLE** by formula, use

$$\textbf{the variable} = \frac{-b \pm \sqrt{b^2 - 4ac}}{2a}$$

**LET'S TAKE IT A STEP AT A TIME!**

Again using the even numbered problems from Exercise Set 12.2:

| Equation | a | b | c | −b | $b^2$ | 4ac | $b^2 - 4ac$ | 2a |
|---|---|---|---|---|---|---|---|---|
| 2. $x^2 + 5x + 6 = 0$ | 1 | 5 | 6 | −5 | 25 | 24 | 1 | 2 |
| 4. $x^2 - 2x - 15 = 0$ | 1 | −2 | −15 | 2 | 4 | −60 | 64 | 2 |
| 6. $y^2 + 4y + 3 = 0$ | 1 | 4 | 3 | −4 | 16 | 12 | 4 | 2 |
| 8. $z^2 - 2z + 1 = 0$ | 1 | −2 | 1 | 2 | 4 | 4 | 0 | 2 |
| 10. $z^2 - 2z - 3 = 0$ | 1 | −2 | −3 | 2 | 4 | −12 | 16 | 2 |

Once you know what values to substitute, you are ready to solve the equation using the quadratic formula.

#2 $\quad x^2 + 5x + 6 = 0$

$$x = \frac{-b \pm \sqrt{b^2 - 4ac}}{2a}$$

$$x = \frac{-5 \pm \sqrt{25 - 4(1)(6)}}{2(1)}$$

$$x = \frac{-5 \pm \sqrt{25 - 24}}{2}$$

$$x = \frac{-5 \pm \sqrt{1}}{2}$$

$$x = \frac{-5 + 1}{2} \quad \text{and} \quad x = \frac{-5 - 1}{2}$$

$$x = \frac{-4}{2} \qquad x = \frac{-6}{2}$$

$$x = -2 \qquad x = -3$$

The solutions are $x = -2$ and $x = -3$

**NOTE:** The ± in the formula means that you must add AND subtract the value of the root.

#4 $x^2 - 2x - 15 = 0$

$$x = \frac{-b \pm \sqrt{b^2 - 4ac}}{2a}$$

$$x = \frac{2 \pm \sqrt{4 - 4(1)(-15)}}{2(1)}$$

$$x = \frac{2 \pm \sqrt{4 - (-60)}}{2}$$

$$x = \frac{2 \pm \sqrt{64}}{2}$$

$x = \frac{2 + 8}{2}$ and $x = \frac{2 - 8}{2}$

$x = \frac{10}{2}$ $\quad$ $x = \frac{-6}{2}$

$x = 5$ $\quad$ $x = -3$

The solutions are $x = 5$ and $x = -3$.

#6 $y^2 + 4y + 3 = 0$

$$y = \frac{-b \pm \sqrt{b^2 - 4ac}}{2a}$$

$$y = \frac{-4 \pm \sqrt{16 - 4(1)(3)}}{2(1)}$$

$$y = \frac{-4 \pm \sqrt{16 - 12}}{2}$$

$$y = \frac{-4 \pm \sqrt{4}}{2}$$

$y = \frac{-4 + 2}{2}$ and $y = \frac{-4 - 2}{2}$

$y = \frac{-2}{2}$ $\quad$ $y = \frac{-6}{2}$

$y = -1$ $\quad$ $y = -3$

The solutions are: $y = -1$ and $y = -3$.

#8 $z^2 - 2z + 1 = 0$

$$z = \frac{-b \pm \sqrt{b^2 - 4ac}}{2a}$$

$$z = \frac{2 \pm \sqrt{4 - 4(1)(1)}}{2(1)}$$

$$z = \frac{2 \pm \sqrt{4 - (4)}}{2}$$

$$z = \frac{2 \pm \sqrt{0}}{2}$$

$z = \frac{2 + 0}{2}$ and $z = \frac{2 - 0}{2}$

$z = \frac{2}{2}$ $\qquad z = \frac{2}{2}$

$z = 1$ $\qquad z = 1$

In this case, there is only one solution and it is $z = 1$.

#10 $z^2 - 2z - 3 = 0$

$$z = \frac{-b \pm \sqrt{b^2 - 4ac}}{2a}$$

$$z = \frac{2 \pm \sqrt{4 - 4(1)(-3)}}{2(1)}$$

$$z = \frac{2 \pm \sqrt{4 - (-12)}}{2}$$

$$z = \frac{2 \pm \sqrt{16}}{2}$$

$z = \frac{2 + 4}{2}$ and $z = \frac{2 - 4}{2}$

$z = \frac{6}{2}$ $\qquad z = \frac{-2}{2}$

The solutions are $x = 3$ and $x = -1$

**Exercise Set 12.4 A**

Use the quadratic formula to solve the following:

1. $x^2 - 121 = 0$
2. $x^2 - 9x + 20 = 0$
3. $5x^2 - 25x + 30 = 0$
4. $y^2 - 4y + 3 = 0$
5. $y^2 - 4y - 5 = 0$
6. $y^2 - 2y - 3 = 0$
7. $z^2 + 5z + 4 = 0$
8. $2w^2 - 8w - 24 = 0$
9. $z^2 - 36 = 0$
10. $k^2 - 11k = 60$

The above exercises are the odd numbered problems from Exercise Set 12.2 on page 212. You must get the same answers using the quadratic formula as you did by factoring.

**Exercise Set 12.4 B**

Solve by the quadratic formula:

1. $y^2 + 2y - 3 = 0$
2. $z^2 - 5z + 4 = 0$
3. $10y^2 - 15y = 0$
4. $z^2 + 12z = -36$
5. $t^2 + 23t = 50$

The above exercises are problems #12, 14, 16, 18 and 20 from Exercise Set 12.2. You must get the same answers using the quadratic formula as you did by factoring.

## CHAPTER EXERCISES

### Section 12.1

Solve the following equations:

1. $(x - 3)(x + 1) = 0$
2. $(x - 2)(x - 1) = 0$
3. $(x - 6)(x + 2) = 0$
4. $(x - 4)(x - 5) = 0$
5. $(x + 3)(x - 2) = 0$
6. $(x - 7)(x + 4) = 0$
7. $(y - 4)(y + 4) = 0$
8. $(z - 3)(z - 3) = 0$
9. $(w + 6)(w - 3) = 0$
10. $(a - 1)(a + 1) = 0$
11. $(r + 10)(r - 9) = 0$
12. $(s + 3)(s + 2) = 0$
13. $(t - 12)(t - 1) = 0$
14. $(y - 50)(y + 8) = 0$
15. $(y - 3)(2y - 1) = 0$
16. $(x + 4)(3x + 6) = 0$

### Section 12.2

Solve the following equations. Check answers.

17. $x^2 - 5x + 6 = 0$
18. $x^2 - 5x + 4 = 0$
19. $x^2 - 25 = 0$
20. $y^2 - 64 = 0$
21. $x^2 - 7x + 10 = 0$
22. $y^2 - 9y + 20 = 0$
23. $z^2 + 10z + 21 = 0$
24. $w^2 + 2w - 3 = 0$
25. $y^2 + 6y - 7 = 0$
26. $z^2 + 3z - 28 = 0$
27. $x^2 - 2x - 24 = 0$
28. $0 = -42 + x + x^2$
29. $w^2 + 6 - 7w = 0$
30. $169 - y^2 = 0$
31. $12 + 7x + x^2 = 0$
32. $0 = y^2 + 55 - 16y$
33. $y^2 - 11y + 30 = 0$
34. $x^2 + 9x + 20 = 0$
35. $x^2 - 9x - 90 = 0$
36. $y^2 - 14y + 48 = 0$
37. $x^2 - 7x = 0$
38. $y^2 + 5y = 0$
39. $15x - x^2 = 0$
40. $8x^2 + 2x = 0$
41. $3x^2 - 15x = 0$
42. $7x = x^2$
43. $12y = 6y^2$
44. $10x = x^2$

**Section 12.2**

Solve and check your answers.

45. $x^2 = 25$
46. $y^2 = 49$
47. $16 = z^2$
48. $121 = w^2$
49. $-42 + 13y = y^2$
50. $3b + 18 = b^2$
51. $16c - 39 = c^2$
52. $x^2 = 28x - 75$

**Section 12.3**

In each of the equations #17 through 52, write the equation in general form, $y = ax^2 + bx + c$, and state the values of a, b and c.

**Section 12.4**

Use the quadratic formula to solve each of the following equations.

53. $x^2 + 5x + 4 = 0$
54. $x^2 - 2x + 1 = 0$
55. $x^2 - 7x + 10 = 0$
56. $y^2 + 6y - 7 = 0$
57. $w^2 + 2w - 3 = 0$
58. $x^2 - 7x + 6 = 0$
59. $y^2 + 10y + 21 = 0$
60. $y^2 - 2y - 8 = 0$
61. $x^2 - 9x - 10 = 0$
62. $y^2 + 6y + 9 = 0$
63. $y^2 + 5y = 0$
64. $x^2 - 3x = 0$
65. $x^2 - 4 = 0$
66. $y^2 = 36$
67. $x^2 + 3x = 18$
68. $x^2 = 3x - 2$
69. $z^2 = 28 - 3z$
70. $y^2 = 9$

**CHAPTER TEST** *Form 12A*

Solve the following equations:

**Answers**

1. $(y-5)(y-6)=0$ — 1. ________
2. $(x+12)(x-3)=0$ — 2. ________
3. $(w-4)(w+4)=0$ — 3. ________
4. $x^2+5x-6=0$ — 4. ________
5. $y^2+2y-15=0$ — 5. ________
6. $x^2+8x+7=0$ — 6. ________
7. $w^2-7w+10=0$ — 7. ________
8. $x^2-49=0$ — 8. ________
9. $y^2+12y=0$ — 9. ________
10. $x^2+2x=24$ — 10. ________
11. $0=x^2-10x+25$ — 11. ________

Write the following equations in general form and state the values of a, b, and c:

**Answers**

12. $x^2-6x+8=0$ — 12. ________________________________
13. $y^2+2y=0$ — 13________________________________
14. $y^2=121$ — 14. ________________________________
15. $x^2=16-3x$ — 15. ________________________________

Solve the following by using the quadratic formula:

**Answers**

16. $x^2+x-12=0$ — 16. ________

## CUMULATIVE EXERCISES Chapters 1–12 *Set A*

1. What is the additive inverse of $\frac{9}{17}$?

2. What is the reciprocal of 1.9?

Evaluate:

3. $\left(\frac{1}{2}\right)^3$ 4. $(-1)^9$ 5. $-2^3$ 6. $|-22.9|$

7. Combine like terms: $9x^2y - 4xy^2 - 2x^2y + xy^2$

Perform the indicated operation and simplify:

8. $-2(3x - 27) - 4(5x - 3y) + 2y$

9. $\frac{3}{5} - \frac{2}{3} + \frac{1}{2}$

10. $(4x - 1)^2$

11. $8xy(7xy^2 + 4x^2y + 3)$

12. $(7x - 2)(9x - 1)$

13. $-6[-(8t - rs) + (-11t - 12rs)]$

14. $|(-3)^2| - |-6|$

Determine if the given ordered pair is a solution of the equation:

15. $y = 8 - x$ (6,2)

16. $y = 11 - 2x$ (5,1)

17. $y = 4x - 13$ (−1,2)

18. $y = \frac{1}{3}x$ (12,4)

19. $y = \frac{5}{2}x - 3$ (0,3)

20. $y = 8 - 7x$ $\left(\frac{8}{7},0\right)$

State the slope of the line represented by each equation:

21. $y = 13x - 4$

22. $y = \frac{1}{5}x - \frac{7}{3}$

23. $y = \frac{2}{3}x + 3$

24. $y = 36 - \frac{5}{4}x$

Graph the following linear equations:

25. $y = 8 - x$

26. $y = x - 6$

27. $y = 2x + 3$

28. $y = 3x - 2$

Factor completely:

29. $8m^2n + 24m^5n^2 - 32m^3n^4$

30. $15a^2b^2 + 25a^2b - 6$

31. $6c^3d^2 - 14c^3d + 2c^2d$

32. $x^2 - 144$

33. $a^2 - 0.25$

34. $25x^2y^2 - 4z^2$

35. $x^2 + 8x + 15$

36. $a^2 - 9a + 8$

37. $a^2 - 3a - 4$

38. $x^4 - 8x^3 - 33x^2$

39. Three times a number, increased by 4, is the same as 32 diminished by the number. Find the number.

40. When 8 is subtracted from 3 times a number, the result is the same as twice the number plus 13. What is the number?

41. The sum of two consecutive integers is 113. Find the integers.

Perform the indicated operations.

42. $\frac{2y^2}{3z^4} \cdot \frac{18z^3}{5}$

43. $\frac{38a^2b^3}{7} \div \frac{19ab}{-21}$

44. $\frac{17a}{b+2} + \frac{13a}{b+2}$

45. $\frac{6}{x-2} - \frac{3}{2(x-2)}$

Solve the equations:

46. $\frac{x-2}{5} = 11$

47. $\frac{x-5}{2} = 9$

48. $\frac{2x}{3} - \frac{1}{2} = \frac{2}{5}$

49. $\frac{x-3}{6} = 9$

Solve:

50. $(x + 2)(x - 2) = 0$
51. $(3x - 1)(2x + 5) = 0$
52. $(6x - 1)(9x + 18) = 0$
53. $(x + 8)(x - 5) = 0$

Solve by factoring:

54. $x^2 - 9 = 0$
55. $25x^2 - 4 = 0$
56. $3x^2 - 6x = 0$
57. $8x^2 - 5x = 0$
58. $x^2 - 2x + 1 = 0$
59. $x^2 + 12x + 35 = 0$

Solve by quadratic formula:

60. $x^2 + 7x + 10 = 0$
61. $x^2 - 9x + 18 = 0$
62. $x^2 + 2x - 5 = 0$
63. $x^2 - 6x - 7 = 0$
64. $x^2 - 9x - 10 = 0$
65. $x^2 - x - 110 = 0$

## CUMULATIVE TEST Chapters 1–12 *Form A*

1. What is the reciprocal of $\left(\frac{-27}{3}\right)$

Perform the indicated operations:

2. $\frac{1}{7} + \frac{3}{5} - 6$

3. $\frac{-1}{3}(-2)^2$

4. $6[8 - (5 - 3)^2]$

5. $\frac{2x}{x + 3} - \frac{6}{x}$

6. $(4x - 3)(5x + 2)$

7. $(1 - 2x)^2$

Solve for the variable:

8. $4(a + 6) = 5a + 1$

9. $\frac{x + 7}{5} - 4 = 3$

Factor:

10. $a^3b^2 + 6a^2b^5$

11. $a^2b^2 - z^2$

12. $x^2 - 100$

13. The sum of three consecutive even integers is 42. What are the integers?

14. One-half a number equals the number plus 8. What is the number?

Solve:

15. $(3x + 7)(5x - 8) = 0$

16. $x(2x + 5) = 0$

17. $x^2 + 9x = 0$

18. $x^2 - x - 2 = 0$

19. $x^2 - 16 = 0$

20. Solve by using the quadratic formula: $x^2 - 3x - 4 = 0$

CHAPTER

# 13 An Introduction to Trigonometry

## 13.0 *Vocabulary Preview*

Define these words without using a dictionary. Formulate their meanings from past knowledge or experience:

| | | |
|---|---|---|
| **acute** | **degree** | **ray** |
| **adjacent** | **hypotenuse** | **right angle** |
| **angle** | **opposite** | **triangle** |

## 13.1 *Basic Trigonometric Definitions*

***Ray:*** A half-line which originates at a fixed point and extends indefinitely in some direction.

***Angle:*** Two rays having a common end point (vertex) form an angle.

***Right angle:*** An angle whose measure is 90 degrees.

***Acute angle:*** An angle which measures more than zero degrees but fewer than 90 degrees.

***Triangle:*** A closed figure formed by three line segments.

***Right triangle:*** A triangle containing one right angle.

## 13.2 *Basic Properties of Triangles*

The sum of the angles in a triangle is 180 degrees.

The sum of any two sides of a triangle is greater than the third side.

The smallest side of a triangle is the side opposite the smallest angle.

The largest side of a triangle is the side opposite the largest angle.

## 13.3 *The Right Triangle*

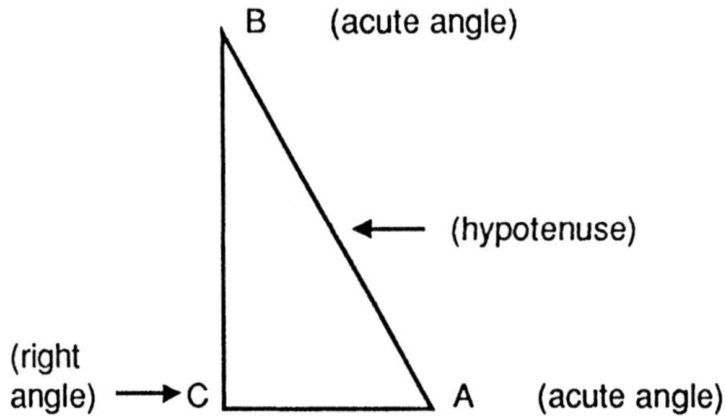


**Property 1: The sum of the two acute angles of a right triangle is 90 degrees.**

**Property 2: The largest side of the right triangle (the hypotenuse) is opposite the right angle.**

*Example:* In a right triangle ABC, angle C = 90° and angle B = 20°. Find the measure of angle A.

From property 1 we know that the sum of angles A and B is 90°, therefore we have:

$$A + 30° = 90°$$

subtracting 30° from each side gives:

$$A = 60°$$

*Example:* In right triangle ABC, angle A is 22° more than angle B. Find the measure of each angle. (Angle C = 90°)

Let angle B measure x degrees, then angle A measures x + 22°.

A + B = 90°, therefore:
(x + 22°) + x = 90°, and combining
2x + 22° = 90°.

Subtracting 22° from each side:

$$2x = 68°$$
$$x = 34° \text{ (measure of angle B)}$$

To find the measure of angle A:

$$(x + 22°) + 34° = 90°$$
$$x + 56° = 90°$$
$$x = 90° - 56°$$
$$x = 34° \text{ (measure of angle A)}$$

**Exercise Set 13.3 A**

Given right triangle ABC, with angle C the right angle. Find the measure of the other angle(s) if:

1. angle A is 50 degrees
2. angle B is 30 degrees
3. angle B is 77 degrees
4. angle A is 14 degrees
5. angle A is 12 degrees less than angle C
6. angle A equals angle B
7. angle A is half of angle B
8. angle B is 10 degrees more than angle A
9. the sum of angle C and angle B is 130 degrees
10. the difference between angle C and angle A is 70 degrees
11. angle A is $\frac{1}{6}$ of angle C
12. angle B is 15 degrees more than twice angle A
13. angle A is 4 times angle B
14. angle C is 10° greater than twice angle B

**Property 3: The square of the hypotenuse is equal to the sum of the squares of the other two sides.**

**This property of right triangles is called the Pythagorean Theorem and is expressed algebraically as:**

$$c^2 = a^2 + b^2$$

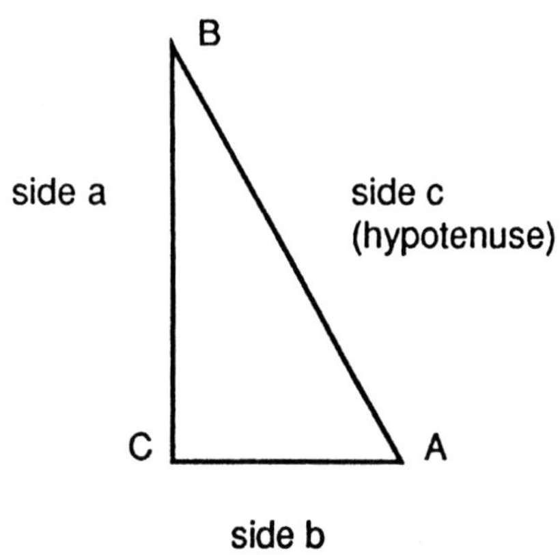

*Example:* In right triangle ABC, side a = 4, and side b = 3. Find the measure of side c.

Using the Pythagorean theorem:

$$c^2 = a^2 + b^2$$
$$c^2 = 4^2 + 3^2$$
$$c^2 = 16 + 9$$
$$c^2 = 25$$
$$c = \sqrt{25} = 5$$

*Example:* In right triangle ABC, side c measures 38 units, and side b measures 10 units. Find the measure of side a.

Using the Pythagorean theorem:

$$c^2 = a^2 + b^2$$
$$(38)^2 = a^2 + (10)^2$$
$$1444 = a^2 + 100$$
$$1344 = a^2$$
$$\sqrt{1344} = a$$
$$36.\overline{6} = a$$

**Exercise Set 13.3 B**

For right triangle ABC, angle C is the right angle. Find the measure of the indicated side:

1. a = 9 b = 3 c = ?
2. a = ? b = 2 c = 5
3. a = 6 b = ? c = 10
4. a = ? b = 15 c = $5\sqrt{13}$

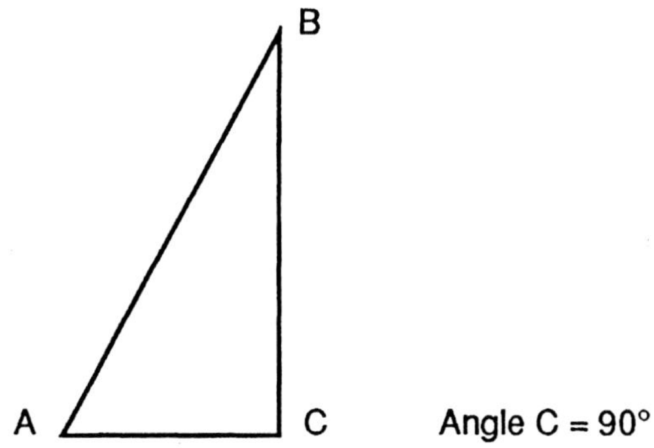

5. $a = 2$ $b = ?$ $c = 2\sqrt{2}$

6. $a = 11$ $b = 15$ $c = ?$

7. $a = ?$ $b = 6$ $c = 12$

8. $a = 9$ $b = ?$ $c = \sqrt{140}$

9. $a = 12$ $b = \sqrt{3}$ $c = ?$

10. $a = 5$ $b = 3$ $c = ?$

## 13.4 *The Trigonometric Relationships*

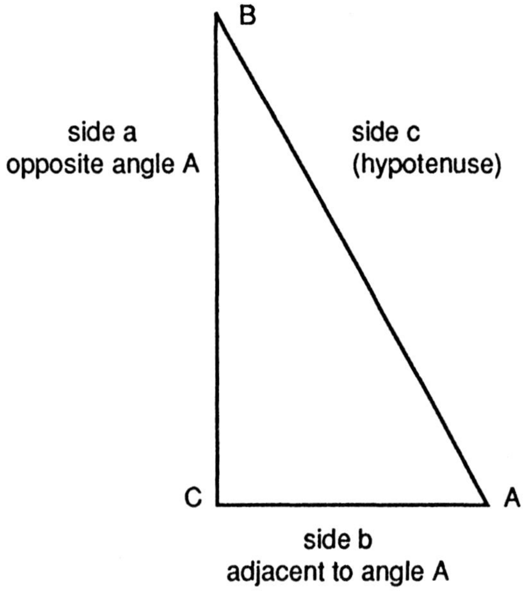


Given the right triangle ABC above, we define the following relationships:

$$\textbf{sine A} = \frac{\textbf{side opposite A}}{\textbf{hypotenuse}} = \frac{a}{c}$$

$$\textbf{cosine A} = \frac{\textbf{side adjacent to A}}{\textbf{hypotenuse}} = \frac{b}{c}$$

$$\textbf{tangent A} = \frac{\textbf{side opposite A}}{\textbf{side adjacent to A}} = \frac{a}{b}$$

*Note:* Hereafter we will use abbreviations as follows:

**sine A** $\rightarrow$ **sin A**
**cosine A** $\rightarrow$ **cos A**
**tangent A** $\rightarrow$ **tan A**

**opposite side** $\rightarrow$ **opp**
**adjacent side** $\rightarrow$ **adj**
**hypotenuse** $\rightarrow$ **hyp**

*Example:* Right triangle ABC has sides which measure as follows:

a = 3, b = 4 and c = 5. Find all the trig relationships for angles A and B. (Angle C measures 90°.)

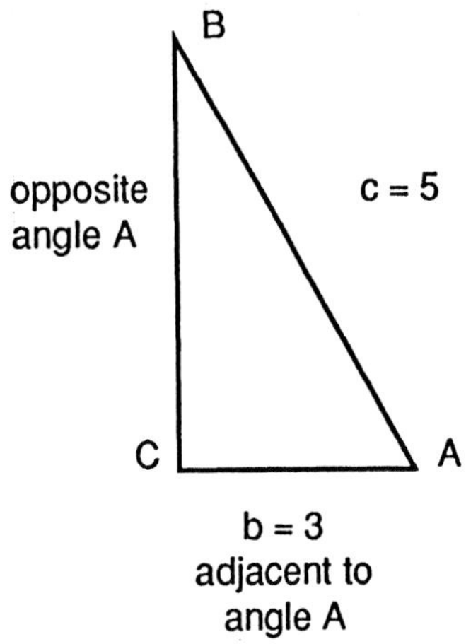


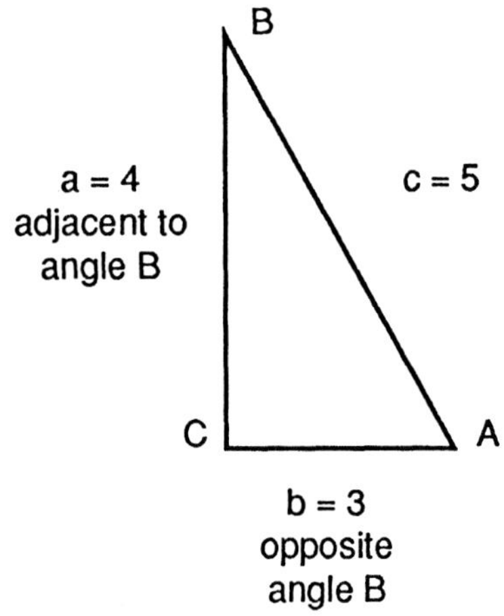


$\sin A = \frac{a}{c} = \frac{4}{5} = .8000$ $\qquad$ $\sin B = \frac{b}{c} = \frac{3}{5} = .6000$

$\cos A = \frac{b}{c} = \frac{3}{5} = .6000$ $\qquad$ $\cos B = \frac{a}{c} = \frac{4}{5} = .8000$

$\tan A = \frac{a}{b} = \frac{4}{3} = 1.333$ $\qquad$ $\tan B = \frac{a}{b} = \frac{3}{4} = .7500$

**Exercise Set 13.4 A**

Find the sine, cosine and tangent of the indicated angles given the following information: (In each case, angle C = 90°.)

| | | |
|---|---|---|
| 1. side a = 9 | side b = 3 | Angle A |
| 2. side b = 2 | side c = 5 | Angle B |
| 3. side a = 6 | side c = 10 | Angle B |
| 4. side b = 15 | side c = $5\sqrt{13}$ | Angle A |
| 5. side a = 2 | side c = $2\sqrt{2}$ | Angle A |
| 6. side a = 11 | side b = 15 | Angle B |
| 7. side b = 6 | side c = 12 | Angle B |
| 8. side a = 9 | side c = $\sqrt{140}$ | Angle A |
| 9. side a = 12 | side b = $\sqrt{3}$ | Angle A |
| 10. side a = 5 | side b = 3 | Angle B |

**Note:** It is time consuming to calculate these ratios each time one is needed. Therefore, tables have been developed which enable us to find the ratios at a glance. The following is a partial trig table:

| angle | sine | cosine | tangent |
|---|---|---|---|
| 5° | 0.0872 | 0.9962 | 0.0875 |
| 10° | 0.1736 | 0.9848 | 0.1763 |
| 15° | 0.2588 | 0.9659 | 0.2679 |
| 20° | 0.3420 | 0.9397 | 0.3640 |
| 25° | 0.4226 | 0.9063 | 0.4663 |
| 30° | 0.5000 | 0.8660 | 0.5774 |
| 35° | 0.5736 | 0.8196 | 0.7002 |
| 40° | 0.6428 | 0.7660 | 0.8391 |
| 45° | 0.7071 | 0.7071 | 1.0000 |
| 50° | 0.7660 | 0.6428 | 1.1918 |
| 55° | 0.8192 | 0.5736 | 1.4281 |
| 60° | 0.8660 | 0.5000 | 1.7321 |
| 65° | 0.9063 | 0.4226 | 2.1445 |
| 70° | 0.9397 | 0.3420 | 2.7475 |
| 75° | 0.9659 | 0.2588 | 3.7321 |
| 80° | 0.9848 | 0.1736 | 5.6713 |
| 85° | 0.9962 | 0.0872 | 11.430 |
| 90° | 1.0000 | 0.0000 | — |

*Example:* If you want to know the cosine of 70°, move down the angle column to 70° and then across to the cosine column.

The cosine of 70° is 0.3420

**Note:** Many scientific calculators are programmed to produce the same ratios.

**Exercise Set 13.4 B**

Using the above table, find the following:

1. sin 30°
2. tan 15°
3. cos 60°
4. sin 5°
5. tan 10°
6. cos 50°
7. tan 45°
8. tan 75°
9. cos 90°
10. sin 45°
11. cos 45°
12. cos 30°
13. sin 60°
14. tan 85°
15. cos 20°

*Example:* In right triangle ABC, angle A measures 30° and side a measures 10 inches. Find the other two sides and the other acute angle.

Since ABC is a right triangle, angle C = 90°. Therefore, angle B = 180° − (90° + 30°) = 60°.

Since $\sin A = \frac{a}{c}$

$$\sin 30° = \frac{10}{c}$$

$$0.5000 = \frac{10}{c}$$

$$0.5000\ c = 10$$

$$c = \frac{10}{0.5000}$$

$$c = 20$$

To find the third side, we may use any appropriate trig relationship.

$$\tan A = \frac{a}{b}$$

$$\tan 30° = \frac{10}{b}$$

$$0.5774 = \frac{10}{b}$$

$$0.5774\ b = 10$$

$$b = \frac{10}{0.5774}$$

$$b = 17.32$$

**Exercise Set 13.4 C**

Find the missing side(s) and angle(s) for right triangle ABC, given the following data. (angle C is 90°)

Express sides to the nearest tenth and angle measure to the nearest five degrees.

| | sides | | | angles | |
|---|---|---|---|---|---|
| | a | b | c | A | B |
| 1. | | | 24 | 30° | |
| 2. | 12 | 15 | | | |
| 3. | | 14 | | | 55° |
| 4. | 5 | 12 | | | |
| 5. | | | 46 | | 70° |
| 6. | 8 | | 32 | | |
| 7. | | 11 | | 45° | |
| 8. | | | 100 | | 60° |
| 9. | | 34 | | 30° | |

## CHAPTER EXERCISES

### Section 13.3

In triangle ABC, find the measure of the remaining angle(s), given the following information:

1. Angle C is a right angle and the measure of angle A is 74 degrees.

2. Angle B measures 38 degrees and angle A is twice angle B.

3. Angle C measures 90 degrees, and the measure of angle A is equal to the measure of angle B.

4. Angle A measures 84 degrees, and the measure of angle A is half the measure of angle B.

5. Angle C measures 100 degrees. Angle B measures 38°.

6. The measure of angle A is 3 times that of angle B. The measure of angle C is twice the sum of angles A and B.

7. Angle C is a right angle. Angle B measures one-half that of angle C.

8. Angles A, B and C are equal in measure.

### Section 13.3

For right triangle ABC, with angle C the right angle, use the Pythagorean Theorem to find the side not given:

9. $a = 11$ $\quad b = 9$ $\quad c = ?$

10. $a = ?$ $\quad b = 8$ $\quad c = 15$

11. $a = 1.2$ $\quad b = 2.4$ $\quad c = ?$

12. $a = 15$ $\quad b = ?$ $\quad c = 225$

13. $a = ?$ $\quad b = 34$ $\quad c = 1597$

14. $a = 125$ $\quad b = ?$ $\quad c = 514$

15. $a = 10$ $\quad b = 16$ $\quad c = ?$

**Section 13.4**

Use the right triangle below to find the trig relationships indicated:

16. sin X = ?
17. cos Y = ?
18. tan Y = ?
19. cos X = ?
20. sin Y = ?
21. tan X = ?

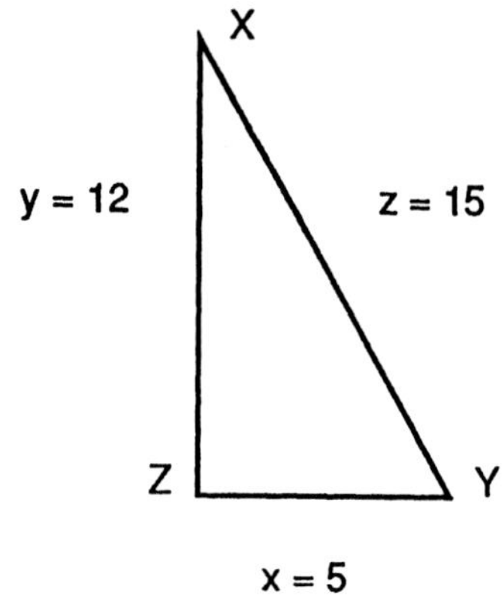


**Section 13.4**

Using the partial table for trig relationships from page 233, find the following:

22. tan 65°
23. cos 10°
24. sin 80°
25. tan 20°
26. sin 90°
27. cos 5°

**Section 13.4**

In right triangle ABC, angle C measures 90°, angle A measures 50° and side c is 200 units long. Find:

28. the measure of angle B
29. the length of side a
30. the length of side b

**CHAPTER TEST** *Form 13A*

For all questions involving right triangle ABC, angle C is the right angle.

1. In right triangle ABC, the measure of angle B is 16°. What is the measure of angle A?

2. The hypotenuse of the right triangle ABC is 20 units long. The shortest side of the triangle is 9 units long. What is the length of the third side?

3. In triangle XYZ, angle Z measures 105° and angle Y measures twice that of angle X. Find the measures of angles Y and X.

4. In right triangle ABC, side a = 9 units and side b = 16 units. Find the length of side c.

5. Using the triangle of question 4, find the tangent of angle B.

6. Use the table on page 233 to find the cosine of 60°.

7. Use the table on page 233 to find the sine of 30°.

For questions 8, 9 and 10, draw and label the triangle in order to see which trig relationship to use for the solution to the problem. Use angle C = 90°.

8. In right triangle ABC, the hypotenuse is 50 units and the measure of one angle is 30°. What is the length of the side opposite the 30° angle?

9. In right triangle ABC, side c = 110 units and side a = 55 units. What is the measure of angle A?

10. In right triangle ABC, side a = 7 units and side b = 4 units. Find the measure of angle A to the nearest degree.

In questions 11 through 15, use the Pythagorean Theorem to find the third side of right triangle ABC when two sides are given. Answers should be rounded to the nearest tenth.

11. a = 3 and c = 5

12. b = 8 and c = 10

13. a = 6 and b = 10

14. a = 0.5 and b = 0.1

15. b = 4 and c = 9

**CUMULATIVE EXERCISES Chapters 1–13** *Set A*

Evaluate:

1. $\sqrt{0.125}$

2. $2^2(|-4|) - 6(2^3)$

3. $\left(\frac{1}{3}\right)^3$

4. $(-4)^3$

5. $-4(2.3) + 6(1.2) + 3(3.6)$

6. $\left(\frac{3}{5}\right)^2$

Combine like terms to simplify:

7. $9m^2n - 3m^2n^2 - 13m^2n + 6m^2n^2$

8. $18x^2y^2 + 9x^2y - 12x^2y^2 - 6x^2y - 3$

Perform the indicated operation and simplify:

9. $3(x - 4) - 5(5x - 2) + 2x$

10. $(2x + 3)^3$

11. $5ab(3b - 6ab + 5a)$

12. $(6x - 2)(5x - 1)$

13. $-2[3(a - 4b) + 2(5a + 6b)]$

14. The sum of three consecutive odd integers is 51. Find the numbers.

15. A piece of ribbon is 90 inches long. It is cut into three pieces as follows: the second piece is twice as long as the first, the third piece is 10 inches longer than the first. Find the length of each piece. (Hing: Let x represent the length of the first piece.)

Solve the following equations:

16. $7x + 2 = 6$

17. $y + 4 = 2y + 7$

18. $\frac{1}{3}x - 5 = 2$

19. $a + 6 = 2a - 6$

20. $\frac{x}{3} - 5 = 2$

21. $\frac{x + 4}{2} = 11$

22. $\frac{x}{2} - 1 = \frac{x}{3} + 4$

23. $\frac{x + 1}{4} = 7$

24. $5(x + 2) = 3x + 2$

25. $2(x - 1) = 5$

State the slope and y intercept of the following linear equations:

26. $y = 4x - 1$

27. $y = 1 - 4x$

Graph the lines represented by:

28. $y = \frac{1}{2}x - 1$

29. $y = x - 4$

Factor completely:

30. $7a^2b - 14ab^2 + 21a^2b^2$

31. $25xy - 15x - 10y$

32. $169x^2y^2 - 1$

33. $9y^2 - 1$

34. $16x^3 - x^2$

35. $y^4 - 64$

36. $a^2b^2 - 144c^2$

37. $x^2 - 12x - 28$

38. $x^2 - 11x + 30$

39. $x^3 - 4x^2 + 4x$

Solve by factoring:

40. $9y^2 + 18y = 0$

41. $y^2 - 4 = 0$

42. $x^2 + 2x - 8 = 0$

43. $x^2 + 12x + 27 = 0$

Solve by using the quadratic formula:

44. $x^2 - 2x - 3 = 0$

45. $x^2 + 5x + 4 = 0$

Perform the indicated operations and simplify:

46. $\frac{a^2b^2c}{4ab^3c} \cdot \frac{6ab}{3a}$

47. $\frac{7c^2d^2}{4} \div \frac{cd}{2d}$

48. $\frac{20}{3a} + \frac{6}{a}$

49. $\frac{b}{2} - \frac{3b}{8} + \frac{5b}{4}$

50. $\frac{x - 3}{6} + \frac{x}{2} - 3$

In right triangle ABC, with angle C the right angle, find the indicated side. Answers may be left in radical form.

| | | |
|---|---|---|
| 51. a = 6 | b = 8 | c = ? |
| 52. a = 5 | b = ? | c = 13 |
| 53. a = ? | b = 9 | c = 20 |
| 54. a = 8 | b = 3 | c = ? |
| 55. a = 3 | b = 3 | c = ? |

From the table on page 231, find the indicated values:

56. sin 30°

57. tan 45°

58. cos 40°

59. sin 60°

Given the right triangle below, with angle C the right angle, find:

60. side a

61. angle B

62. side b

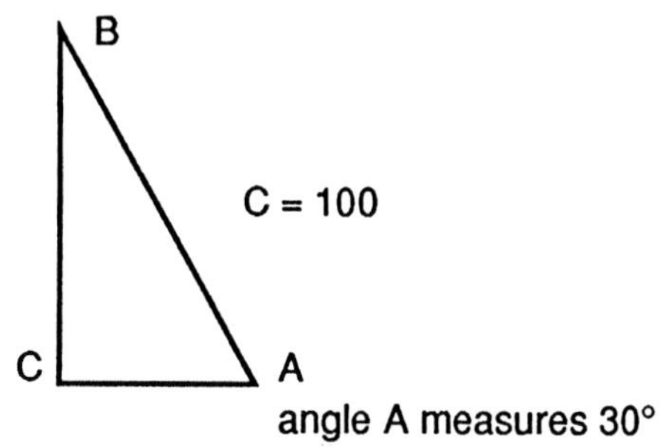

## CUMULATIVE TEST Chapters 1–13 *Form A*

1. What property of real numbers is illustrated by:
   $x + (y + z) = (x + y) + z$
2. Evaluate $\sqrt{491}$ to two decimal places

Perform the indicated operations and simplify:

3. $(11y + 2)(4y - 1)$
4. $\dfrac{48x^2y^3 + 24x^2y^2 - 16x^2y^4}{-8xy^2}$
5. $(4^2)(|-21|) - 10^2$

Factor completely:

6. $x^3 - 4x$
7. $a^2 - 121$
8. $c^2x^2 + cx^2 - 56x^2$
9. What is the slope and y intercept of $y = 4x - 2$?
10. Graph the linear equation: $y = x - 2$
11. Simplify: $\dfrac{114a^2b^4c}{4a^5b^3c^9}$
12. Multiply: $\dfrac{4m^2n}{9} \cdot \dfrac{-27}{n^2}$
13. Divide: $\dfrac{16p^2q^4}{m^2} \div \dfrac{8p^4q}{3m^4}$

Solve for x:

14. $4x + 3 = 7x - 9$
15. $8x^2 - 16x = 0$
16. $\dfrac{x + 4}{5} - 3 = \dfrac{x}{3} + 1$
17. The sum of 4 consecutive integers is $-10$. Find the integers.
18. For right triangle ABC, with $<C = 90°$, $a = 8$, and $b = 3$. Find the hypotenuse, side c.
19. In triangle ABC, angle $B = 37°$ and angle $C = 90°$. Find the measure of angle A.
20. A right triangle has a hypotenuse of 100 units and one side of 80 units. Find the third side.

From the table on page 233, find:

21. sin 65°
22. cos 80°
23. cos 10°
24. tan 70°
25. sin 10°

CHAPTER

# 14 Applications to Geometry

## 14.0 *Vocabulary Preview*

Define these words without using a dictionary. Formulate their meanings from past experience or knowledge.

| | | | |
|---|---|---|---|
| **area** | **equation** | **perspective** | **square** |
| **constant** | **height** | **plane** | **line** |
| **cubic** | **length** | **property** | **volume** |
| **diameter** | **perimeter** | **radius** | **width** |

## 14.1 *Perimeters of Plane Figures*

One application of mathematics to the world around us, is in the field of **Geometry,** which helps us to understand and use the objects and materials which fill our lives. In earlier chapters, you placed numbers on the real number line, discussing direction and position. In chapter nine, the discussion was expanded to include the real plane and graphs of linear equations. In this chapter, we will investigate some properties of closed figures, specifically perimeter, area, volume and surface area.

A closed figure in a plane, whose sides are straight line segments is called a **polygon.** The simplest of these polygons is the triangle, which has three sides. A four sided figure is a quadrilateral.

**The distance around a polygon is called the perimeter which is found by adding the lengths of all sides.**

*Example:* A triangle has sides measuring 8, 17 and 20 units. Its perimeter may be expressed algebraically as:

$$P = a + b + c$$
$$P = 8 + 17 + 20$$
$$P = 45 \text{ units.}$$

Four sided figures, quadrilaterals, are very common in the world in which we live.

Some special quadrilaterals are:

square: 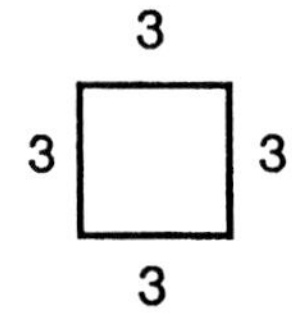

all sides are equal

rectangle: 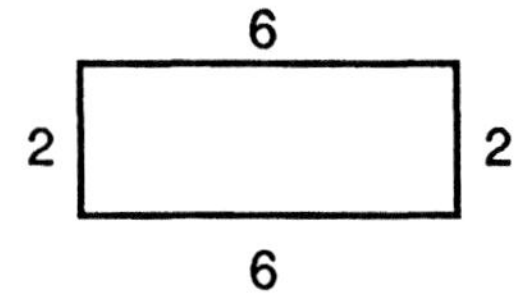

opposite sides are equal

parallelogram: 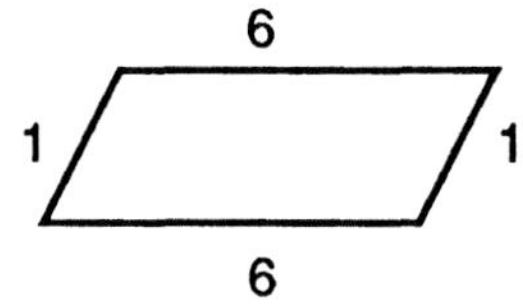

opposite sides are equal

rhombus: 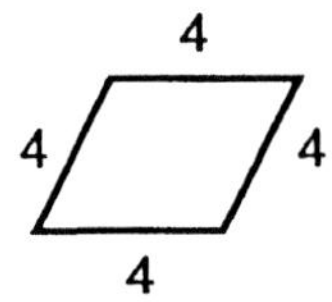

all sides are equal

trapezoid: 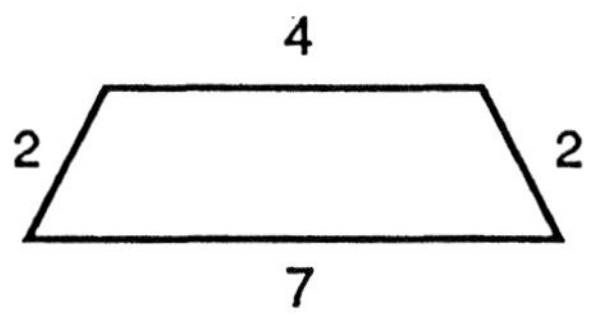

slant sides are sometimes equal

*Example:* Find the perimeter of the square above.

Since perimeter is the sum of the length of the sides,

$P = 3 + 3 + 3 + 3 = 12$
or $P = 4(3) = 12$

*Example:* Find the perimeter of the rectangle above.

$P = 6 + 2 + 6 + 2 = 16$
or $P = 2(6) + 2(2) = 12 + 4 = 16$
or $P = 2(6 + 2) = 2(8) = 16$

*Example:* Find the perimeter of the parallelogram above.

$P = 6 + 1 + 6 + 1 = 14$
or $P = 2(6) + 2(1) = 12 + 2 = 14$
or $P = 2(6 + 1) = 12 + 2 = 14$

*Example:* Find the perimeter of the rhombus.

$P = 4 + 4 + 4 + 4 = 16$
or $P = 4(4) = 16$

*Example:* The perimeter of a square is 20 feet. Find the length of a side.

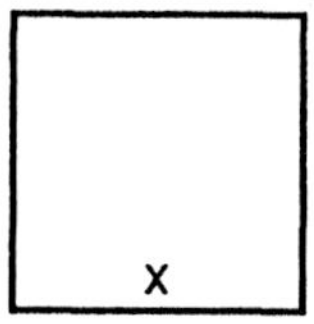


Sketch the figure. Represent the length of the side by x

Since the sides are equal:

$$P = x + x + x + x$$

Then: $P = 4x$

Substituting: $20 = 4x$

Dividing: $5 = x$

*Check:* If the side of the square is 5, its perimeter is

$P = 5 + 5 + 5 + 5 = 20$ (as given)

*Example:* The perimeter of a rectangle is 26. Its length is 8. Find its width.

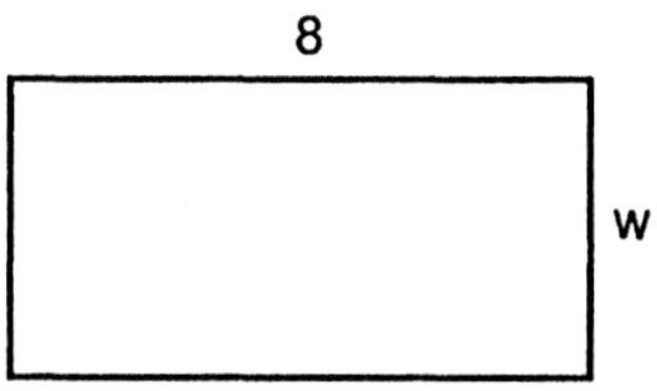


Sketch the figure. Represent the width by w.

$$P = 8 + w + 8 + w$$
$$26 = 16 + 2w$$
$$10 = 2w$$
$$5 = w$$

*Check:* If the width is 5 and the length is 8, the perimeter of the rectangle is

$P = 8 + 5 + 8 + 5 = 26$ (as given)

Another very common plane figure is the **circle.** It is not a polygon, since it is not made up of straight line segments. Its perimeter, the distance around the circle, is called the **circumference.**

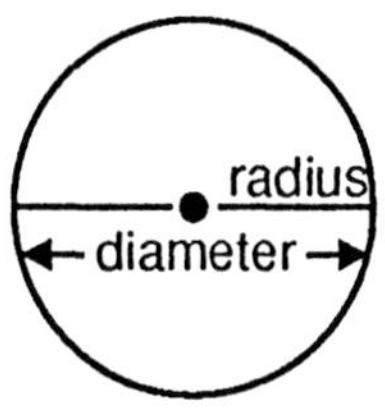


The diameter of the circle is twice the radius. ($d = 2r$)

The radius of a circle is one-half its diameter. ($r = \frac{1}{2}d$)

The circumference of the circle is slightly more than 3 times the diameter. Since Pi (the Greek letter $\pi$) is approximately equal to 3.14, the formula for the circumference of the circle is given as:

$$\mathbf{C = \pi d}$$
**or** $$\mathbf{C = \pi(2r) = 2\pi r}$$

*Example:* Find the circumference of a circle whose diameter is 20 inches.

Using 3.14 for $\pi$ and 20 for d:

$C = \pi d$
$C = 3.14(20)$
$C = 62.8$ inches

*Example:* Find the circumference of a circle whose radius is 10 inches.

Using 3.14 for $\pi$ and 10 for r:

$C = 2\pi r$
$C = 2(3.14)(10)$
$C = 62.8$ inches

*Example:* Find the circumference of a circle if its diameter is 14 ft. Use $\pi = 3.14$.

$C = \pi d$
$C = 3.14(14)$
$C = 43.96$ ft.

*Example:* The circumference of a circle is 22 yards. Find the diameter of the circle to the nearest yard.

(Use $\pi = 3.14$ and let d = diameter.)

$$C = \pi d$$
$$22 = 3.14(d)$$
$$\frac{22}{3.14} = d$$
$$7 \text{ yds.} = d$$

*Example:* If the circumference of a circle is 44 meters, find the radius of the circle to the nearest meter. Use $\pi = 3.14$.

(Hint: recall that the diameter of a circle is twice its radius; that is, $d = 2r$)

$$C = \pi d$$
$$C = \pi(2r)$$
$$44 = 3.14(2r)$$
$$44 = 6.28r$$
$$\frac{44}{6.28} = r$$
$$7 \text{ ft.} = r$$

## Exercise Set 14.1

Find the perimeters of the following figures:

1. 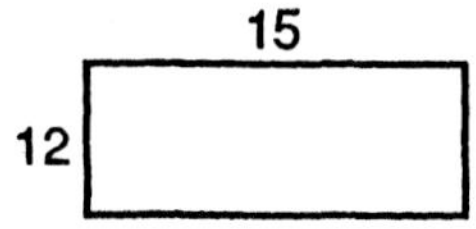


2. 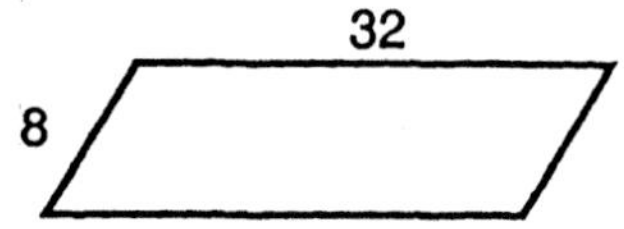


3. 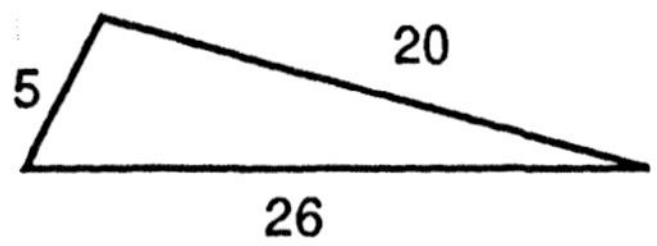


4. 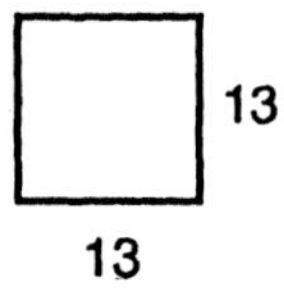


5. 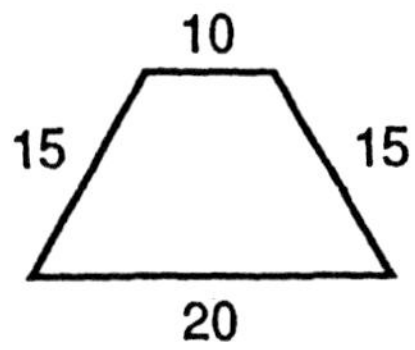


6. 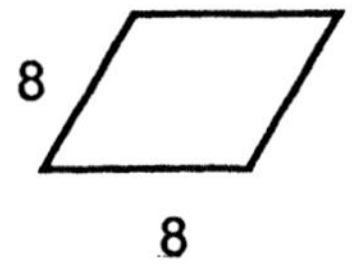


7. A rectangular picture frame is 8 inches long. Find its width if the perimeter is 38 inches.
8. A triangle has sides 2 feet and 5 feet. Find the third side if the perimeter is 13 feet.
9. A gardener has a lot which is 50 feet long. If he has 150 feet of fence to enclose it, how wide can he make the garden?
10. Find the length of a parallelogram if its width is 15 and its perimeter is 50.
11. A square has a perimeter of 28 yards. Find the length of its side.
12. A triangle has sides of 22 and 13. Find the third side if the perimeter is 55.
13. A rhombus has a perimeter of 100 feet. Find the length of its side.
14. One side of a triangle is 2 feet more than a second side. The third side is 17. Find the length of its sides if the perimeter is 33.
15. A carpenter is to make a window frame from 10 feet of lumber. The length of the window must be 3 feet more than the width. Find the dimensions of the window.

16. A pentagon has 5 equal sides. Find the length of the sides if the perimeter is 125 meters.
17. A triangle has all sides equal. Its perimeter is 46.5 units. What is the measure of each side?
18. The length of a rectangle is three times its width. If the perimeter is 80 units, find the length and width.
19. The measurements of the sides of a triangle happen to be three consecutive integers. If the perimeter of the triangle is 33, find the measurements of the sides.

For the following problems, use $\pi = 3.14$.

20. Find the circumference of a circle when the diameter is 3 units.
21. Find the circumference of a circle if the radius is 10 units.
22. If the circumference of a circle is 22 units, find its diameter.
23. Find the circumference of a circle if its diameter is 15 ft.
24. The circumference of a circle is 100 ft. Find its diameter.
25. Find the circumference of a circle if the radius is 35 units.

## 14.2 *Areas of Plane Figures*

In section 14.1, we discussed the concept of perimeter or "the distance around" plane figures. Now we will investigate the concept of area or the space enclosed within a plane figure.

For example, if we consider a rectangle which is 2 inches wide by 3 inches long, its area is given by the formula:

| | |
|---|---|
| | Area = length · width |
| therefore | A = 3 inches · 2 inches |
| or | A = 6 $\text{inches}^2$ or $\text{in}^2$ |
| which we will call | A = 6 square inches |
| or abbreviate | A = 6 sq. in. |

Note several important ideas about area:

1. It is the total space contained within the figure.

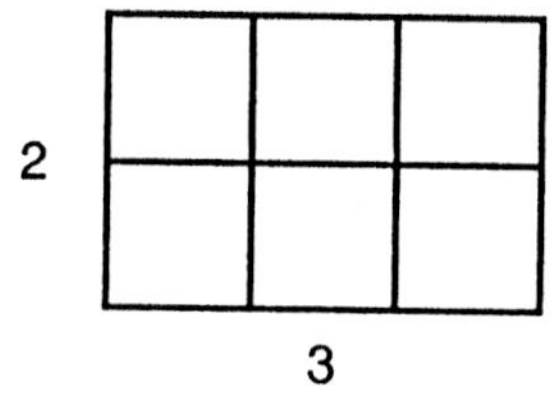


2. The "space" is found by multiplying the dimensions of the figure.
3. The units representing the dimensions are also multiplied. This always results in square units.

*Area Formulas for Plane Figures*

| | |
|---|---|
| **Triangle** | $A = \frac{1}{2}(\text{base} \cdot \text{height})$ <br> $A = \frac{1}{2}bh$ |
| **Rectangle** | $A = \text{length} \cdot \text{width}$ <br> $A = lw$ |
| **Square** | $A = \text{side squared}$ <br> $A = s^2$ |
| **Parallelogram** | $A = \text{base} \cdot \text{height}$ <br> $A = bh$ |
| **Trapezoid** | $A = \frac{1}{2}\text{height} \cdot (\text{sum of the bases})$ <br> $A = \frac{1}{2}h(b_1 + b_2)$ |
| **Circle** | $A = \pi \cdot \text{radius squared}$ <br> $A = \pi r^2$ |

**Note** the similarities in some of the formulas. It will help you remember them.

*Example:* Find the area of a triangle whose base is 20 ft. and whose height is 6 ft.

1. Sketch the figure and represent the data given about it.

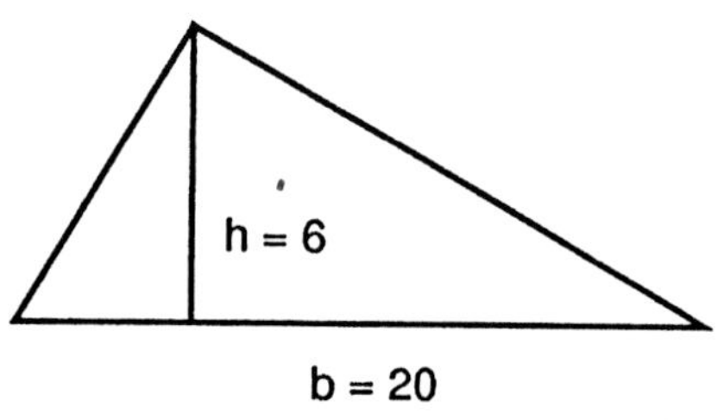

2. Since the figure is a triangle use:

$$A = \frac{1}{2}bh$$

3. Substitute the given data in the formula.

$$A = \frac{1}{2}(20)(6)$$

4. Solve the resulting equation.

$$A = 10(6)$$
$$A = 60$$

5. Label the answer in square units.

$$A = 60 \text{ sq. ft.}$$

*Example:* Find the area of a parallelogram with a base of 12 yds. and height of 10 yds.

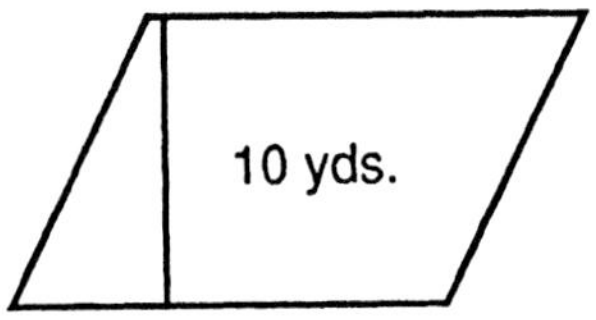


Since the area of a parallelogram is $A = bh$

1. Substitute the given base and height
$$A = 12(10)$$
2. Solve $A = 120$
3. Label the answer in square units.

*Example:* Find the area of a trapezoid if it has bases of 16 ft. and 20 ft. and a height of 8 ft.

1. The formula to use is $A = \frac{1}{2}h(b_1 + b_2)$
2. Substitute the given bases and height
$$A = \frac{1}{2}(8)[16 + 20]$$
3. Solve $A = 4(36)$
$$A = 144 \text{ square units.}$$

*Example:* The area of a rectangle is 437 sq. ft. If its length is 23 ft, find its width.

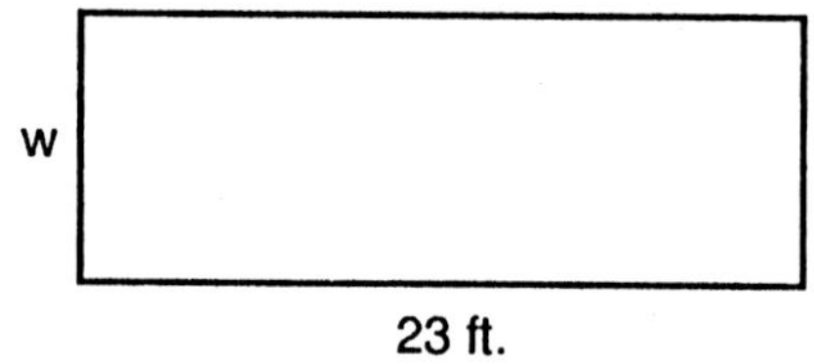


Since the area of a rectangle is

$$\begin{aligned} A &= lw \\ 437 &= 23w \\ 19 &= w \\ \text{width} &= 19 \text{ ft} \end{aligned}$$

***Rules for Solving Area Problems***

**1. Draw the geometric figure and represent the given data on the figure.**

**2. Select the appropriate formula.**

**3. Substitute the given data into the formula.**

**4. Solve the resulting equation.**

**5. Label the answer in square units.**

**Exercise Set 14.2**

1. A kitchen floor is a rectangle 16 ft long and 10 ft wide. Find the area of the floor.
2. The football field at a local school is rectangular in shape. If the dimensions of the field are 120 yards in length by 52 yards in width, find the area of the field.
3. The Chin family has a square patio attached to their home. If the length of one side of the patio is 5 meters, find the area.
4. Find the area of a parallelogram if its base is 34 inches and its height is 7 inches.
5. Find the area of a trapezoid if it has a height of 20 yds. and bases of 36 yds. and 18 yds.
6. The roof face of a garage is a triangle with a base of 40 ft. and a height of 21 ft. Find the area of the triangle.

7. The Torres Company has a circular pool at its entrance. If the radius of the pool is 42 ft., find the area.
8. A cross section of a wooden beam is a square 4 inches on a side. Find the area of this cross section.
9. Find the area of a parallelogram whose base is 7 yds. and whose height is 5 yds.
10. A forest ranger can see a distance 5 miles in all directions from the top of his fire tower. Find the area, in square miles, that he can observe from the tower. (Hint: the area he can see is a circle.)
11. A rubber disk is circular in shape with an area of 28.26 sq. in. Find the radius of the disk to the nearest inch.
12. A civil engineer needs a triangular brace for a wall of a building. If the height of the brace is 7 yds. and the base is 6 ft., find the area of the brace in square feet. (Hint: 1 yard = 3 feet)
13. The top of a bookcase is rectangular in shape. If the bookcase is 4 feet long and 18 inches wide, find the area of the top in square inches. (Hint: 1 foot = 12 inches)

## 14.3 *Volumes of Geometric Solids*

The earlier sections of this chapter dealt with plane geometric figures; that is, figures which can be drawn entirely on a flat surface. This section will deal with some three dimensional figures and the measure of the space enclosed within them. This space is called the volume.

It is often helpful to visualize these shapes as common figures used in everyday life, since drawing a sketch of a three dimensional figure on a plane (flat) surface requires the use of perspective.

**The following is a list of the more common geometric solids and the formulas to find their volumes:**

1. A Rectangular solid (sometimes called a box)

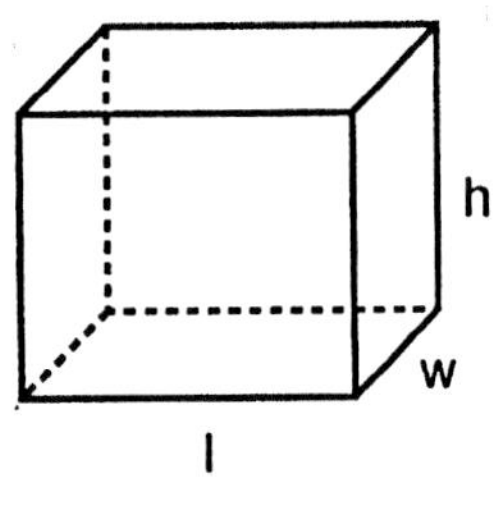


Volume = length · width · height
$V = lwh$

2. A Cube (a rectangular solid all of whose edges are equal in length)

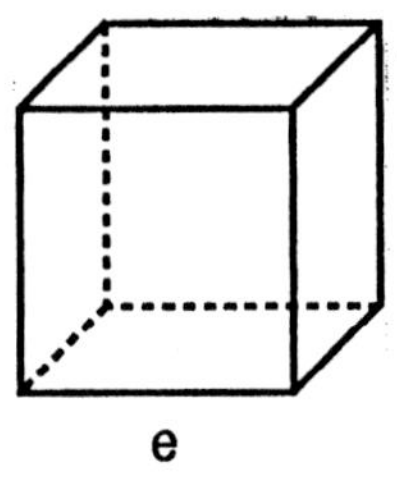


$$\text{Volume} = \text{edge} \cdot \text{edge} \cdot \text{edge}$$
$$V = e^3$$

3. A Circular cylinder

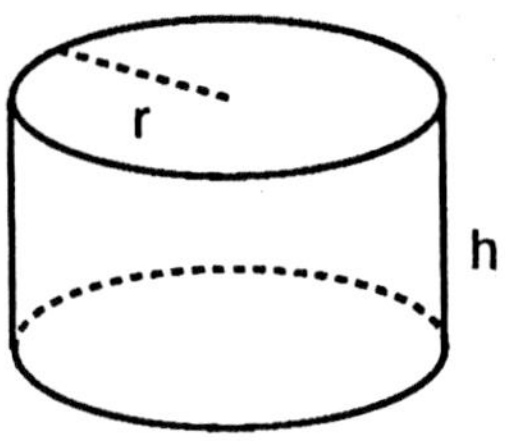


Volume = area of base · height

Since the base is a circle the formula becomes

$$V = \pi r^2 h$$

4. A Cone

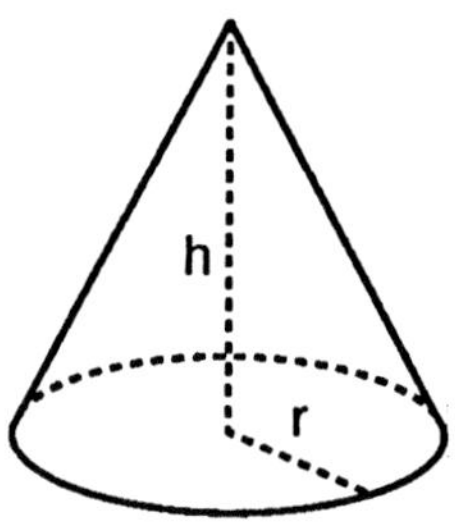


$$\text{Volume} = \frac{1}{3}\text{ area of the base} \cdot \text{height}$$

Since the base is a circle

$$V = \frac{1}{3}\pi r^2 h$$

5. A Sphere (Ball shape)

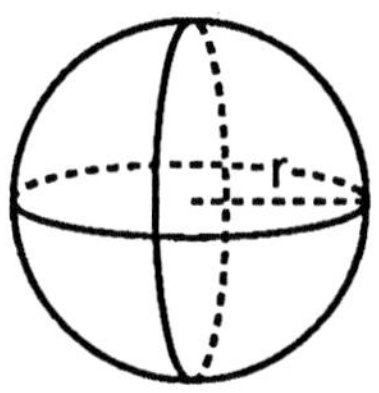


$$V = \frac{4}{3}\pi r^3$$

Notice in all these figures you are multiplying three dimensions. Therefore, when using unit label such as inches, feet, yards, etc. the resulting volume will have the label:

cubic inches ($in^3$),
cubic feet ($ft^3$),
cubic yards ($yds^3$), etc.

In Section 14.2 you saw the rules for area problems. **The following are the rules used for finding volumes.** Notice the similarities.

1. Draw the figure. Represent the given data on it.
2. Select an appropriate formula.
3. Substitute the given data into the formula.
4. Solve the resulting equation.
5. Label your answer in cubic units.

*Example*: Find the volume of a rectangular solid 14 inches high by 8 inches wide by 5 inches long.

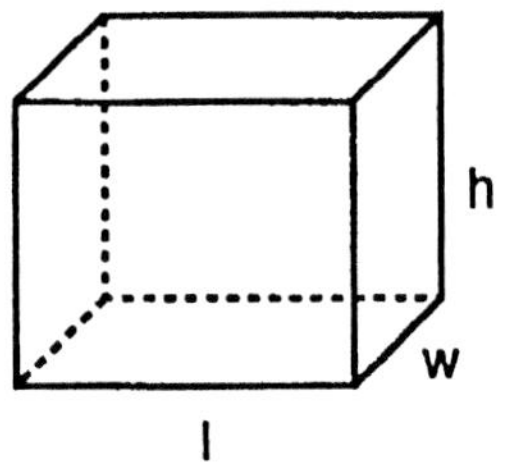


$$\text{Volume} = \text{length} \cdot \text{width} \cdot \text{height}$$
$$V = lwh$$
$$V = 5(8)(14)$$
$$V = 560$$
$$\text{Volume} = 560 \text{ cubic inches}$$

*Example*: Find the volume of a circular cylinder 12 ft high with a radius of 7 ft. Use $\pi = 3.14$.

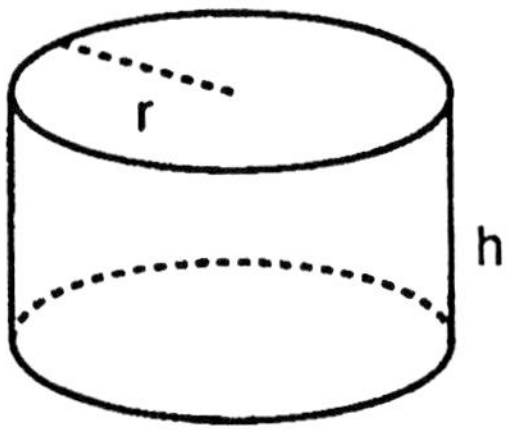


$$\text{Volume} = \text{Area of base} \cdot \text{height}$$
$$V = \pi \cdot r^2 \cdot h$$
$$V = \pi \cdot 7^2 \cdot 12$$
$$V = \pi \cdot 49 \cdot 12$$
$$V = 3.14 \cdot 5.88$$
$$V = 1846.32 \text{ ft.}^3$$

*Example*: Find the volume of a rubber ball whose radius is 9 cm. Use $\pi = 3.14$.

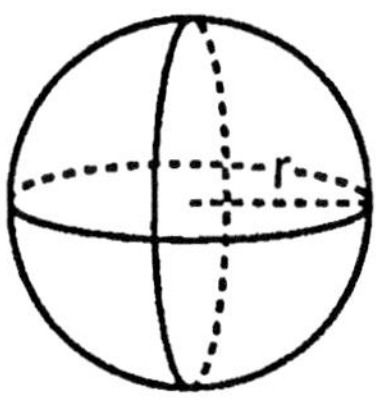


$$V = \frac{4}{3}\pi r^3$$
$$V = \frac{4}{3}(3.14)(9^3)$$
$$V = \frac{4}{3}(3.14)(729)$$
$$V = \frac{9156.24}{3}$$

V = 3052.08 cubic centimeters.

**Exercise Set 14.3**

1. Find the volume of a rectangular solid whose length is 40 inches, width is 10 inches and height is 10 inches.

2. An ice cube is 2 inches on an edge. Find its volume.

3. Find the volume of a cylinder if its height is 12 inches and the radius of its base is 3 inches.

4. A cone has a circular base with radius of 7 ft. and height of 10 ft. Find the volume of the cone.

5. A sphere has a radius of 14 yds. Find its volume.

6. Find the volume of a steel bar 12 ft. long, 2 ft. wide and 1 ft. high.

7. A rail car is in the shape of a rectangular solid 40 ft. long by 10 ft. high by 8 ft. wide. Find its capacity (volume) in cubic feet.

8. A cone shape pile of sand has a radius of 8 ft. and a 4 ft. height. Find its volume.

9. How many cubic feet of water would a cylindrical tank hold if its height is 15 ft. and its radius is 6 ft?

10. A gambling casino has a model of a die above its entrance. If the edge of the die is 8 ft., find its volume.

11. A basketball has a diameter of 22 inches. Find the volume enclosed.

12. The volume of a rectangular solid is 1232 cubic feet. If its length is 22 feet and its width is 7 feet, find its height.

13. A circular wading pool has a diameter of 70 feet and a height of 1 ft. How many cubic feet of water will it hold?

14. A steel rod is 4 inches in diameter and 8 feet long. Find the number of cubic inches in its volume. (Hint: 1 foot = 12 inches)

15. A classroom is 20 ft. by 25 ft. by 8 ft. Find the volume of air, in cubic feet, enclosed by the room.

## 14.4 *Summary of Formulas*

The following is a list of the formulas used in this chapter:

### Perimeter

The sum of all the sides = perimeter

| | |
|---|---|
| Triangle | $p = a + b + c$ |
| Square | $p = 4s$ |
| Rectangle | $p = 2l + 2w$ |
| Parallelogram | $p = 2l + 2w$ |
| Rhombus | $p = 4s$ |
| Trapezoid | $p = s + s + s + s$ |
| Circle (Circumference) | $C = 2\pi r$ or<br>$C = \pi d$ |

### Area

| | |
|---|---|
| Triangle | $A = \frac{1}{2}bh$ |
| Square | $A = s^2$ |
| Rectangle | $A = bh$ |
| Parallelogram | $A = bh$ |
| Rhombus | $A = bh$ |
| Trapezoid | $A = \frac{1}{2} \cdot h(b_1 + b_2)$ |
| Circle | $A = \pi r^2$ |

### Volumes

| | |
|---|---|
| Rectangular Solid | $V = lwh$ |
| Cube | $V = e^3$ |
| Cylinder | $V = \pi r^2 h$ |
| Cone | $V = \frac{1}{3}\pi r^2 h$ |
| Sphere | $V = \frac{4}{3}\pi r^3$ |

## CHAPTER EXERCISES

### Section 14.1

**Answers**

1. A rectangular table top is 4 ft. by x ft. Find its dimensions if its perimeter is 14 ft. 1. ________

2. A rectangular picture frame has a perimeter of 36 inches. If the width of the frame is 10 inches, find its length. 2. ________

3. A triangle has one of its sides 4 ft. longer than a second side. If the third side is 10 ft. and the perimeter is 30 ft, find the sides of the triangle. 3. ________

4. The length of a rectangle is twice its width. Find its dimensions if the perimeter is 240 yards. 4. ________

5. A bike wheel has a diameter of 26 inches. Find its circumference. 5. ________

6. The length of a parallelogram is three times its width. If the perimeter is 80 ft., find the length of the sides. 6. ________

7. A trapezoid has two legs that are equal in length. The other two sides are 10 yds. and 20 yds. If the perimeter is 70 yds., find the length of the legs. 7. ________

8. A rodeo barrel has a circumference of 46 inches. Find its diameter to the nearest inch. 8. ________

9. A living room skylight is square in shape. Each side measures 3.5 ft. Find its perimeter. 9. ________

10. A gardener plants a circular tulip bed with a diameter of 21 ft. He plans to edge the outside with stone. What is the distance around the bed? 10. ________

11. A home decorator has 31 feet of molding and is undecided as to whether to create a rectangular or a square design on a wall. If the rectangle is to have dimensions of 7 ft. by 8 ft. and the square is 8 ft. on a side, does she have enough molding to build either one? 11. ________

12. A farmer needs 310 ft. of wire to fence in a triangular field. Two sides of the field are 124 ft. and 100 ft. Find the length of the third side of the triangle. 12. ________

13. A horse is tied to one end of 35 ft. rope. The other end is tied to a stake in the ground. If the horse keeps the rope stretched as far as possible and walks around the stake, how far can it walk without retracing its steps? 13. ________

14. A cowboy needs 244 ft. of barbed wire to fence a rectangular pen. If the dimensions of the pen are 70 yds. by x yds., find the width of the pen. 14. ________

15. The parks commission plans to fence the border of a playground which has the shape of a rhombus. Find the size of the playground if its perimeter is 530 yds. 15. ________

16. Ms. Kim wishes to create a portion of a basketball court in her driveway. She wants to use a circle 12 ft. in diameter. What is the perimeter of the circle? 16. ________

**Section 14.2**

**Answers**

17. Find the area of a rectangle 14 ft. by 11 ft. 17. ________

18. Find the area of a circle whose radius is 14 yds. 18. ________

19. Find the area of a parallelogram one of whose sides is 20 ft. and whose height is 6 ft. 19. ________

20. A right triangle has legs of 16 ft. and 30 ft. Find its area. 20. ________

21. A city block is in the shape of a rectangle, 400 ft. by 200 ft. Find the area of the block. 21. ________

22. Kitchen floor tiles are in the shape of a square 9 inches on a side. Find the area covered by one tile. 22. ________

23. The base of the Great Pyramid in Egypt is a square 746 feet on a side. Find the area of the base. 23. ________

24. A television station can clearly broadcast its signal a distance of 50 miles in all directions from the station. Find the area the signal covers in square miles. 24. ________

25. A decorator requires a piece of marble in the shape of a trapezoid with bases of 6 ft. and 3 ft. and a height of 3 ft. Find the area of the required piece of marble. 25. ________

26. A carpet installation firm is going to cover the floor of a room in the shape of a parallelogram. The base of the parallelogram is 15 yards and its height is 12 yards. How many square yards of carpet will be needed? 26. ________

27. A Wyoming rancher has a cattle feeding pen in the shape of a rectangle 25.5 rods by 63.5 rods. Find the area of the enclosed pen. 27. ________

28. A Mineola gardener wants to plant grass in a rectangular lawn 200.25 ft. by 75.5 ft. The package of grass seed she purchased will cover 15,000 square feet. Does she have enough seed to cover the lawn? 28. ________

29. A chef has a circular cake pan with a 20 in. radius. He must cover the bottom of the pan with a sheet of parchment paper. What is the area to be covered? 29. ________

30. A greenhouse manager has a triangular plant bench that must be covered with creosote. If the bench is in the shape of a right triangle whose legs are 10 ft. and 15 ft., find the area of the triangle to be covered. 30. ________

31. The area of a square is 64 sq. ft. Find the length of the sides of the square. 31. ________

32. A rectangle is twice as long as it is wide. If its area is 98 square inches, find the dimensions of the figure. 32. ________

33. The base of a triangle is four times its height. If the area of the triangle is 72 sq. meters, find the height of the triangle. 33. ________

34. Which is larger, the area of a circle 6 inches in diameter or the area of a square 6 inches on a side? 34. ________

35. The area of the face of a clock is 800 sq. in. Find its radius to the nearest inch. 35. ________

36. A felt washer is a circle 4 inches in diameter. It has a circular hole 2 inches in diameter removed from its center. Find the area of the washer. 36. ________

37. The area of a rhombus is 126 sq. yds. If the height of the rhombus is 6 ft., find the length of its base. 37. ________

38. A trapezoid has bases of 17 ft. and 24 ft. Find the height if the area is 205 sq. ft. 38. ________

**Section 14.3**

**Answers**

39. Find the volume of a rectangular solid 2 ft. by 3 ft. by 4 ft. 39. ________

40. A room is 18 ft. long, 10 ft. wide and 8 ft. high. Find its volume. 40. ________

41. A vegetable can is 6 inches high and has a radius of 2 inches. Find the volume of the can. 41. ________

42. Find the volume of a handball if its radius is 2 inches. 42. ________

43. Find the volume of a cube 10 inches on an edge. 43. ______

44. A swimming pool is 75 ft. long by 30 ft. wide by 3 ft. deep. Find the volume of the pool. 44. ______

45. How many yards of soil must be removed from a hole in the shape of a cylinder 8 feet deep and 6 feet in diameter? 45. ______

46. What is the capacity, in cubic feet, of a freight car 30 feet long, 12 feet high and 8 feet wide? 46. ______

47. Wheat is stored in a conical pile 30 ft. high and with a radius of 20 ft. Find the volume in cubic feet. 47. ______

48. An oil tank is cylindrical in shape. Its height is 35 feet and its diameter is 10 ft. Find its volume. 48. ______

49. A balloon has a diameter of 15 inches. Find its volume. 49. ______

50. An efficiency expert needs to design a storage tank. She is considering one in the shape of a cylinder 14 feet high with a diameter of 14 feet. The other choice is a sphere with a diameter of 14 feet. Which shape has the greater volume? 50. ______

51. A steel bolt is cylindrical in shape, with a radius of 0.5 inches and a length of 2.5 inches. Find its volume. 51. ______

52. A rectangular solid is 3 times as long as its width. If its height is 5 inches and its volume is 240 cubic inches, find its length and width. 52. ______

53. The volume of a cube is 216 cubic inches. Find the length of an edge. 53. ______

54. A large size juice can is 9 inches high and has a radius of 4 inches. Find its capacity in cubic inches. 54. ______

55. The volume of a cylinder is 1230.88 cubic inches. Find its height if the radius is 7 inches. Use $\pi = 3.14$. 55. ________

56. The volume of a cone is 4.2 cubic ft. Find its height to the nearest foot if the radius is 2 ft. Use $\pi = 3.14$. 56. ________

57. A rectangular solid has a volume of 325 cubic yds. If the length of the solid is 13 ft., and its width is equal to its height, find its height. 57. ________

58. A cube has a volume of 1331 cubic meters. Find the length of one of its edges. 58. ________

59. A box is 7′ long and 6′ wide. If the volume is 336 cubic ft., find the height of the box. 59. ________

60. A conical funnel has a volume of 16.7 cu. in. Find its height to the nearest tenth of an inch if it has a radius of 4 inches. Use $\pi = 3.14$. 60. ________

61. A storage tank for natural gas is cylindrical in shape. If the tank is 24 ft. high and the diameter of its circular base is 4 ft., find its volume to the nearest cubic foot. 61. ________

**CHAPTER TEST** *Form 14 A*

**Directions:** Show all work for the following problems. Be sure to state the formula used and label your answer.

1. Find the perimeter of a rectangular table top 5 ft. by 3 ft.

2. Find the circumference of a bicycle wheel whose radius is 9 inches.

3. The perimeter of a square is 100 ft. Find the length of a side of the square.

4. A parallelogram has a base 4 ft. longer than its width. If its perimeter is 52 ft., find its length and width.

5. Find the perimeter of rhombus one of whose sides is 9.2 feet.

6. Find the area of a circular patio whose diameter is 12 ft.

7. The area of a triangle is 72 sq. inches. The base of the triangle is 12 inches. Find the height.

8. A modern sculpture is in the shape of a rhombus. If its base is 17 feet and its area is 255 sq. ft., find its height.

9. The area of a square is 81 sq. inches. Find the length of its sides.

10. Find the area of a parallelogram whose base is 17 and whose height is 10.

11. The area of a rectangular floor is 180 sq. yds. If the length of the floor is 12 yds., find the width.

12. Find the volume of a ball whose radius is 5 inches.

13. Find the volume of a cone whose height is 5 inches and whose diameter is 4 inches.

14. The volume of a cube is 27 cu. inches. Find the length of an edge.

15. Find the volume of a box 2 ft. by 3 ft. by 4 ft.

## CUMULATIVE EXERCISES Chapters 1–14 *Set A*

1. Evaluate: $-6|(-2)^2| - (-2)^2$
2. Evaluate: $-4^2$
3. Evaluate: $-5[6 - (8 - 2)^2]$
4. Evaluate: $\sqrt{371}$

Perform the indicated operations and simplify:

5. $(4x - 1)^2$
6. $(3x - 1)(4x + 5)$
7. $4xy(2x^2y^2 - 3xy^2 + xy)$
8. $6x(x + 4) - 7(x + 3) - 4x$
9. $-[4 - (x + 3) + 2(x + 6)]$
10. $(-23)^2$

Solve:

11. $x + 2 = 5$
12. $3x - 1 = 5$
13. $6x + 2 = x - 8$
14. $4(x + 3) = x - 6$
15. $(x + 2)(x + 4) = 0$
16. $x(2x - 1) = 0$

Solve by factoring:

17. $36y^2 - 4 = 0$
18. $2x^2 - 8 = 0$
19. $x^2 - 2x - 3 = 0$
20. $x^2 + 5x + 6 = 0$
21. $x^2 - 5x = 0$
22. $x^2 + 5x = 0$

Solve by using the quadratic formula:

23. $x^2 - 6x - 16 = 0$
24. $x^2 + 6x - 27 = 0$

Perform the indicated operations and simplify:

25. $\frac{6x^2y^3z}{25} \cdot \frac{-15}{xyz^3}$
26. $\frac{3xy}{8} \div \frac{x^2y}{16}$
27. $\frac{3x}{x + 2} - \frac{4}{5}$
28. $\frac{y + 4}{3} - \frac{6}{5} + \frac{y}{3}$

29. Solve: $\frac{x + 7}{3} = -11$

30. Solve: $\frac{x}{5} - 3 = \frac{x}{2}$

Use the table of trig values to find:

31. sin 70°

32. tan 30°

33. cos 40°

34. sin 10°

35. Right triangle ABC, with angle C = 90°, has side c = 10 and side a = 2. Find side b.

36. In right triangle ABC, angle A measures twice angle B. Angle C measures 60°. Find the measures of angles A and B.

37. In triangle ABC, angle A measures 70° and angle B measures 35°. Find the measure of angle C.

38. A square measures 4 cm. on a side. Find its perimeter.

39. What is the area of a circle with radius of 5 units?

40. The base of a triangle is 4 in. and the height is 6 in. What is the area of the triangle?

41. If a rectangular box has a length of 10 units, a width of 7 units and a height of 3 units, what is its volume?

42. The area of a rectangle is 70 square inches. If the length is 3 inches more than the width, what are the measurements of the length and width.

43. Find the volume of a sphere if its radius is 7 units.

44. What is the area of a square if it measures 6″ on a side?

45. A cone has a height of 9 units and a radius of 3 units. What is its volume?

## CUMULATIVE TEST Chapters 1–14 *Form A*

1. Evaluate: $-5(-53 + 2) + (3^2 + |-4|)^2$
2. Evaluate: $-4^2 + (-4)^2 + (-4)^3$
3. What is the additive inverse of 274.3?
4. Solve: $8x + 3 = 4(x + 2) + 7$
5. Solve: $5(3x - 2) - x + 5 = 13x - 8$
6. The sum of a number and 3 is multiplied by 6. The result is 48. What is the number?
7. Solve for a: $\dfrac{2a - 3c}{5} = -2$
8. Solve: $\dfrac{3c}{4} - 5 = 16$

Factor completely:

9. $x^2 + 4x - 21$
10. $18x^2 - 2y^2$
11. $a^2 - 2ab + b^2$
12. Solve by factoring: $x^2 + 8x + 16 = 0$
13. Solve by using the quadratic formula: $2x^2 - 3x - 2 = 0$.
14. In the right triangle ABC, the hypotenuse is 12 and the shortest side is 3. Find the other side. (Leave answer in radical form.)
15. In a triangle, the sum of two angles is 123°. What is the measure of the third angle?
16. If the two legs of a right triangle are 50 and 120, what is the length of the hypotenuse?
17. The perimeter of a rectangle is 38 inches and the length is 5 inches longer than the width. Find the length and width.
18. Find the area of a circle with radius 12.
19. Find cos 24°.
20. Find tan 51°.

CHAPTER

# 15 Systems of Linear Equations

## 15.0 *Vocabulary Preview*

Define these words without using a dictionary. Formulate their meanings from past experience or knowledge.

| | | |
|---|---|---|
| **inconsistent** | **intersection** | **parallel** |
| **substitution** | **system** | |

## 15.1 *Systems of Linear Equations*

Recall from Chapter Nine that a linear equation represents the graph of a line. A **system** of linear equations is a group of equations which represent lines; that is, a group of equations which are written in one of the forms:

$$y = mx + b$$

or

$$ax + by + c = 0$$

In this text, we will consider only systems of two equations in two variables.

*Example:* The equations $y = 3x + 2$
and $y = 4x - 1$
represent a system of linear equations.

*Example:* The equations $3x + 4y - 8 = 0$
and $x - 3y + 3 = 0$
represent a system of linear equations.

*Example:* The equations $y = 3x - 7$
and $y = x$
represent a system of linear equations.

*Example:* The equations $4x - 2y - 6 = 0$
and $y = 4$
represent a system of linear equations.

## 15.2 *Solving Systems of Linear Equations by Graphing*

**The solution of a system of linear equations is their point of intersection; that is, the point which satisfies both equations.**

**To Solve a System of Linear Equations by Graphing**

1. **Graph the two equations on the same set of axes.**
2. **Locate the point of intersection of the two lines.**
3. **Check the solution by substituting the x and y values of the point of intersection in the original equations.**

*Example:* Solve the system of linear equations:
$y = 2x + 1$
$y = x + 2.$

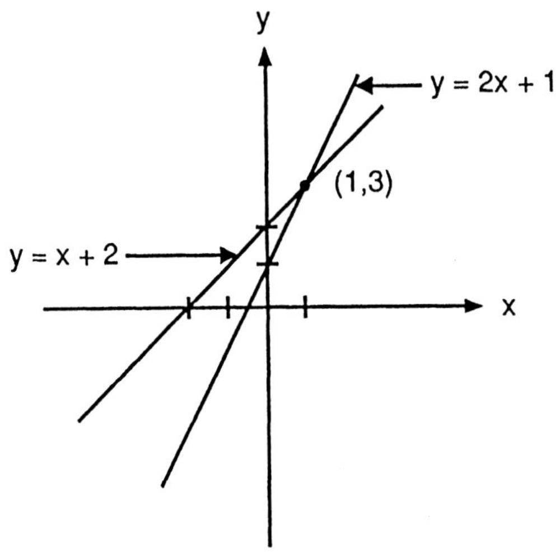


Both graphs pass through the point (1,3). It is the solution of the system of equations.

*Check:*

| | |
|---|---|
| $y = 2x + 1$ | $y = x + 2$ |
| $3 = 2(1) + 1$ | $3 = 1 + 2$ |
| $3 = 2 + 1$ | $3 = 3$ |
| $3 = 3$ | |

*Example:* Solve graphically: $y = 2x - 5$
$y = x$

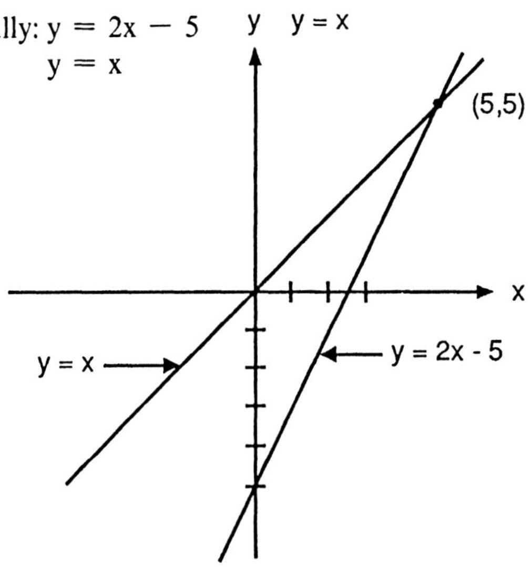


Both lines go through the point (5,5). It is the solution of the system of equations.

*Check:*

| | |
|---|---|
| $y = 2x - 5$ | $y = x$ |
| $5 = 2(5) - 5$ | $5 = 5$ |
| $5 = 10 - 5$ | $3 = 3$ |
| $5 = 5$ | |

*Example:* Find the graphic solution for
$$2x + y = 2$$
$$x + 2y = -2$$

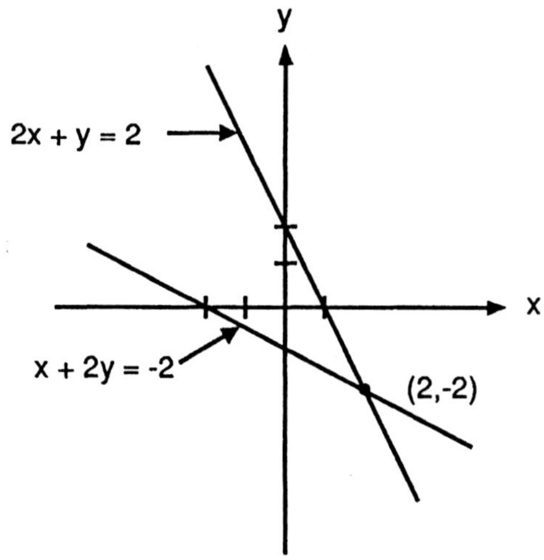


The graphic solution of the system of linear equations is (2,−2).

*Check:*

| | |
|---|---|
| $2x + y = 2$ | $x + 2y = -2$ |
| $2(2) + (-2) = 2$ | $2 + 2(-2) = -2$ |
| $4 + (-2) = 2$ | $2 + (-4) = -2$ |
| $4 - 2 = 2$ | $2 - 4 = -2$ |
| $2 = 2$ | $-2 = -2$ |

*Example:* Solve the following system of linear equations:
$$2x - y = 1$$
$$4x - 2y = -6$$

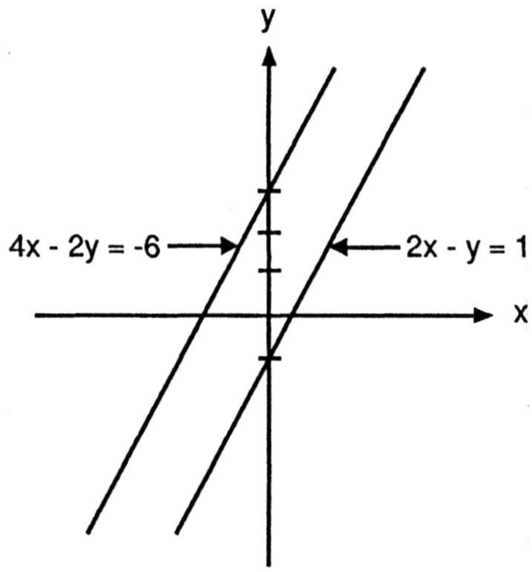


The two lines in the previous example are **parallel.** Parallel lines are lines having the same slope, so they do not intersect. There is NO solution for this system of linear equations. We say that the system is **inconsistent.**

*Example:* Solve the system of linear equations graphically: $4y = 12 - 2x$
$2y = 6 - x$.

Rewriting the first equation in the form
$y = mx + b$ results in $\quad 4y = -2x + 12$

which is the same as $\quad y = \frac{-1}{2}x + 3$

The second equation becomes $\quad 2y = -x + 6$

which is the same as $\quad y = \frac{-1}{2}x + 3$

The graphs of these lines results in the same line with slope of $\frac{-1}{2}$ and y intercept of (0,3).

Equations such as these are called **dependent.** Every point satisfying one line, satisfies the other

**Exercise Set 15.2**

Solve the following systems by graphing the equations.

| | | |
|---|---|---|
| 1. $y = x - 2$<br>$y = 2x - 2$ | 2. $4x - y = 10$<br>$3x - y = 7$ | 3. $x + y = 3$<br>$2x - y = -6$ |
| 4. $3x - 2y = 4$<br>$2x + 3y = -19$ | 5. $4x + 5y = 3$<br>$8x + 10y = 11$ | 6. $5x - y = 8$<br>$y = x - 8$ |
| 7. $x + y = -3$<br>$y = 1$ | 8. $3x - 5y = 8$<br>$3x - 5y = 11$ | 9. $4x + 3y = 2$<br>$3x - 4y = 2$ |
| 10. $4x + 3y = 2$<br>$x - y = 1$ | 11. $y = 3x - 7$<br>$y = x$ | 12. $4x - 2y = 0$<br>$8x - 4y = 0$ |

## 15.3 *Solving Linear Systems by Substitution*

In the previous section, solutions for linear equations were found by graphing the lines and locating their point of intersection.

In cases where it is difficult to locate the point of intersection exactly, graphic solutions have limited accuracy. In those cases, an algebraic solution is preferable.

One method of algebraic solution is by **substitution,** where the y (or x) value from one equation is substituted in the other, thereby eliminating one variable.

*Example:* Solve the following system of linear equations algebraically:

$$y = 2x + 1$$

and $$y = x + 2.$$

Replace the y in the second equation with its equivalent value from the first equation: $2x + 1 = x + 2$

Solve for x:

$$\begin{array}{rcl} 2x + 1 & = & x + 2 \\ -1 & & -1 \\ \hline 2x & = & x + 1 \\ -x & & -x \\ \hline x & = & 1 \end{array}$$

Replace x with 1 in either of the given equations:

$$y = 2x + 1$$
$$y = 2(1) + 1$$
$$y = 2 + 1$$
$$y = 3$$

The solution for the system of equations is (1,3). This is the same solution obtained graphically in the example on page 266.

*Example:* To solve the following system of linear equations:

$$3x - y = 1$$

and $$y = 4 - 2x$$

substitute the value $4 - 2x$ for y in the first equation:

$$3x - (4 - 2x) = 1$$

solve for x:

$$3x - 4 + 2x = 1$$
$$5x - 4 = 1$$
$$5x = 5$$
$$x = 1$$

substitute 1 for x in

$$3x - y = 1$$
$$3(1) - y = 1$$
$$3 - y = 1$$
$$-y = -2$$
$$y = 2$$

The ordered pair (1,2) is the solution of the system of linear equations.

*Check:*

| | |
|---|---|
| $3x - y = 1$ | $y = 4 - 2x$ |
| $3(1) - 2 = 1$ | $2 = 4 - 2(1)$ |
| $3 - 2 = 1$ | $2 = 4 - 2$ |
| $1 = 1$ | $2 = 2$ |

*Example:* Find the solution for the system of equations: $y = 5x + 7$ and
$y = 5x - 1$.

Replace y in the second equation with its value from the first equation:

$$\begin{aligned} 5x + 7 &= 5x - 1 \\ -5x \quad &\quad -5x \\ \hline 7 &= -1 \end{aligned}$$

Solve for x:

This is NOT TRUE, so there is no solution for this system. Look at the graph which follows:

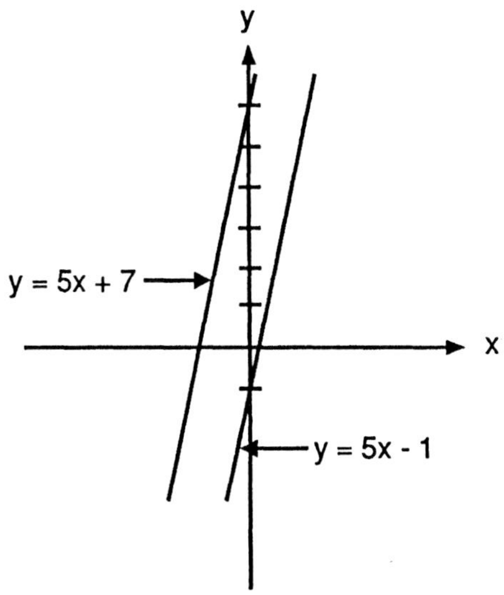


The lines are parallel and do not intersect. **The system is inconsistent.**

*Example:* Find the solution for the system of linear equations using the substitution method.
$2x + 4y = 12$
$x = 6 - 2y$

Replacing x in the first equation with its equivalent from the second equation gives: $2(6 - 2y) + 4y = 12$

which equals: $12 - 4y + 4y = 12$

which results in: $12 = 12$.

**Since this is always true, EVERY point on one of the lines is also on the other; that is, they are the same line.** (See the graphic solution on page 268).

These are **dependent** equations.

**Exercise Set 15.3**

Solve the following systems of linear equations by substitution:

| | | |
|---|---|---|
| 1. $y = x + 2$<br>$3x + y = 6$ | 2. $x + y = 6$<br>$y = x - 3$ | 3. $y = 4x - 6$<br>$2y = 8x + 3$ |
| 4. $9x + 6y = -9$<br>$y = 6x - 4$ | 5. $7x - y = 1$<br>$y = x - 3$ | 6. $6x + y = 5$<br>$y = 8x - 1$ |
| 7. $3x - y = -3$<br>$y = x - 3$ | 8. $2x - 2y = 8$<br>$y = x - 4$ | 9. $y = 3x - 13$<br>$4y - 12x = 5$ |
| 10. $y = x + 6$<br>$y = 2x$ | 11. $2x - 3y = 3$<br>$y = 4x + 8$ | 12. $5x - 8y = -2$<br>$y = 3x + 5$ |

## 15.4 *Solving Systems of Equations by the Addition Method*

Systems of linear equations may also be solved by what is called the **addition method.** In this method, one of the variables is "eliminated" by adding the two equations. The value of the other variable is then found by algebraic solution as in the previous section.

*Example:* To solve the system of linear equations by addition:

$$3x + y = 10 \quad \text{and}$$
$$2x - y = -5.$$

Adding:

$$\begin{array}{l} 3x + y = 10 \quad \text{and} \\ \underline{2x - y = -5} \end{array}$$

results in:

$$5x = 5$$

Solving:

$$x = 1$$

Substituting 1 for x in one of the equations:

$$3x + y = 10$$
$$3(1) + y = 10$$

Solving for y:

$$3 + y = 10$$
$$y = 7$$

The solution (1,7) may be checked in the other equation:

$$2x - y = -5$$
$$2(1) - 7 = -5$$
$$2 - 7 = -5$$
$$-5 = -5$$

*Example:* Solve by the addition method:

$$4x + y = 1$$
$$x + y = 4$$

Here it will be necessary to change one of the equations in order to make it possible to eliminate a variable by addition.

Multiplying the second equation by $-1$ results in the system:

$$4x + y = 1 \text{ and}$$
$$\underline{-x - y = -4}$$

Now, addition results in: $3x = -3$

Solving: $x = -1$

Substituting $x = -1$ in the first equation:

$$4x + y = 1$$
$$4(-1) + y = 1$$
$$-4 + y = 1$$
$$y = 5$$

$(-1,5)$ is the solution of the system.

*Check:*

| | |
|---|---|
| $4x + y = 1$ | $x + y = 4$ |
| $4(-1) + 5 = 1$ | $-1 + 5 = 4$ |
| $-4 + 5 = 1$ | $4 = 4$ |
| $1 = 1$ | |

*Example:* Solve the following system of linear equations by the addition method:

$3x + 2y = 5$ and

$2x + y = 1.$

For this system of equations, addition alone will not result in the elimination of a variable. Neither will multiplying by $-1$ as in the previous example. However, if the second equation is multiplied by $-2$, and then the two equations are added, y will be eliminated and we can solve for x.

$$3x + 2y = 5$$
$$\underline{-2[2x + y = 1]}$$

Makes the system:

$$3x + 2y = 5$$
$$\underline{-4x - 2y = -2}$$

Now adding:

$$-x = 3$$
$$x = -3$$

To solve for y, substitute the value 3 for x in the first equation:

$$3(-3) + 2y = 5$$
$$-9 + 2y = 5$$
$$2y = 14$$
$$y = 7$$

The solution $(-3,7)$ may be checked in each equation:

$$\begin{aligned} 3x + 2y &= 5 \\ 3(-3) + 2(7) &= 5 \\ -9 + 14 &= 5 \\ 5 &= 5 \end{aligned} \qquad \begin{aligned} 2x + y &= 1 \\ 2(-3) + (7) &= 1 \\ -6 + 7 &= 1 \\ 1 &= 1 \end{aligned}$$

*Example:* To solve the system of linear equations
$2x + 3y = 2$ and
$3x + 4y = -4$

both equations must be changed before addition will eliminate a variable.

One way of changing the equations is to multiply the first by 3 and the second by $-2$:

$$\begin{aligned} 3[2x + 3y &= \phantom{-}2] \\ -2[3x + 4y &= -4] \end{aligned}$$

results in:

$$\begin{aligned} 6x + 9y &= 6 \quad \text{and} \\ -6x - 8y &= 8 \\ \hline y &= 14 \end{aligned}$$

Addition gives $y = 14$

Substituting in the first equation:

$$\begin{aligned} 2x + 3y &= 2 \\ 2x + 3(14) &= 2 \\ 2x + 42 &= 2 \\ 2x &= -40 \\ x &= -20 \end{aligned}$$

The solution $(-20,14)$ may be checked in the equations:

$$\begin{aligned} 2x + 3y &= 2 \\ 2(-20) + 3(14) &= 2 \\ -40 + 42 &= 2 \\ 2 &= 2 \end{aligned} \qquad \begin{aligned} 3x + 4y &= -4 \\ 3(-20) + 4(14) &= -4 \\ -60 + 56 &= -4 \\ -4 &= -4 \end{aligned}$$

**Exercise Set 15.4**

Solve the systems of linear equations by the addition method:

1. $2x - y = 8$
   $x + y = 3$
2. $x + y = 7$
   $x - y = -1$
3. $-5x - 2y = 2$
   $3x + y = 3$
4. $3x - y = 11$
   $3x + y = -1$
5. $6x + 3y = 7$
   $-12x - 6y = -3$
6. $3x - 4y = 0$
   $2x - y = -10$
7. $9x - 3y = 39$
   $4x + y = 15$
8. $2x - 3y = 5$
   $6x - 9y = 15$
9. $3x + y = 4$
   $2x + y = 2$
10. $x - y = 7$
    $x + 3y = 5$
11. $2x + y = 2$
    $x + 2y = -2$
12. $x - y = 1$
    $x - 2y = 10$

## CHAPTER EXERCISES

### Section 15.2

Solve the following systems of linear equations graphically. Label each line and clearly indicate the point of intersection.

1. $y = 4x + 2$
   $y = 2x + 4$

2. $y = 5 - x$
   $y = x + 4$

3. $y = 3x$
   $y = x - 12$

4. $y = 2x + 1$
   $y = x + 2$

5. $3x - y = 1$
   $y = 4 - 2x$

6. $y = 5x + 7$
   $y = 5x - 1$

7. $y = x + 2$
   $3x + y = 6$

8. $7x - y = 1$
   $y = x + 7$

9. $6x + y = 5$
   $y = 8x - 1$

### Section 15.3

Solve the systems of linear equations by the substitution method.

10. $y = 2x - 8$
    $y = 3 - x$

11. $y = 7 - x$
    $y = x + 1$

12. $y = 3x - 11$
    $y = x + 1$

13. $y = x + 1$
    $y = 2x + 6$

14. $y = 2 - 2x$
    $y = x + 2$

15. $y = x + 5$
    $y = 3x - 5$

16. $y = 1 - 4x$
    $y = 4 - x$

17. $4x - 2y = 0$
    $y = 4$

18. $y = 3x - 7$
    $y = 1 - 6x$

### Section 15.4

Solve the systems of linear equations by the addition method.

19. $x + y = 3$
    $2x - y = -6$

20. $4x - y = 10$
    $3x - y = 7$

21. $y - x = -2$
    $y - 2x = -2$

22. $3x - 2y = 4$
    $2x + 3y = -19$

23. $4x + 5y = 4$
    $8x + 10y = 11$

24. $5x - y = 8$
    $x - y = 8$

25. $4x + 3y = 2$
    $x - y = 1$

26. $y - 3x = -7$
    $y - x = 0$

27. $4x - 2y = 0$
    $y = 4$

**CHAPTER TEST** *Form 15 A*

Solve graphically. Label each line and clearly indicate the point of intersection.

1. $y = x + 4$
   $y = 2x + 5$

2. $2y + 3x = 7$
   $y - 2x = 0$

Solve by the substitution method:

3. $y = 5x + 14$
   $y = 3x - 6$

4. $y = 12 - x$
   $y = 15 + 2x$

5. $y = x - 5$
   $5y - 3x = -23$

Solve by the addition method:

6. $2x + 6y = 10$
   $3x - 3y = 15$

7. $5y - x = 2$
   $y - 3x = 8$

8. $3x - 4y = 5$
   $6x - 8y = 1$

## CUMULATIVE EXERCISES Chapters 1–15 *Set A*

1. Evaluate $3[4 - 6(7 - 2)^2]$
2. Evaluate $-4|3 - 15| - 12^2$
3. Evaluate $5\sqrt{124.6}$
4. Evaluate $(-3\sqrt{27})^2$

Perform the indicated operation and simplify.

5. $(3x - 1)(6x + 5)$
6. $7x(3x - 1) - (x + 7)$
7. $2ab(3a^2 - 4b)$
8. $3(4x + 1)^2 - 2x$
9. $-[5x + 3(2x - 4)^2 - 7x^2]$
10. $5xy^2(3x^2y - 2xy^2)$

Solve each of the following equations.

11. $3x - 5 = -13$
12. $\frac{1}{2}x + 8 = 0$
13. $9z + 3 = z + 5$
14. $-2(x + 3) = x + 6$
15. $(x + 8)(2x + 3) = 0$
16. $x(4x - 9) = 0$

Solve each of the following equations by factoring.

17. $x^2 + 9x + 14 = 0$
18. $3x^2 = 27$
19. $x^2 - 11x = 0$
20. $k^2 - 3k - 28 = 0$
21. $25x^2 - 36 = 0$
22. $49x^2 - 14x = 0$

Solve each of the following by using the quadratic formula:

23. $x^2 - 3x - 6 = 0$
24. $2x^2 + x - 5 = 0$

Perform the indicated operation and simplify.

25. $\frac{3x^2y^3}{14} \cdot \frac{49}{-x^3y^2z}$
26. $\frac{4ab}{5} \div \frac{a^2b}{20}$
27. $\frac{5y}{2 - y} - \frac{7}{8}$
28. $\frac{3a}{5} + \frac{a}{12} - \frac{5}{8}$

Solve the following equations.

29. $\frac{x + 5}{3} = -9$

30. $\frac{x}{7} - 12 = \frac{x}{3}$

Use the table of trigonometric values to answer #31–34:

31. cos 80°

32. sin 70°

33. tan 50°

34. cos 20°

35. In right triangle ABC (angle C = 90°), side a = 15 and side c = 25. Find the measure of side b.

36. In triangle ABC angle A measures 14° and angle C measures 28°. Find the measure of angle B.

37. In triangle ABC, angle C is 8 times the measure of angle A, and angle B measures 45°. What is the measure of angles A and C?

38. A room measures 15 feet by 12 feet, find the area of this room.

39. The base of a triangle is 14 cm and its height is 20 cm. What is the area of this triangle?

40. What is the volume of a sphere with a radius of 15 cm?

41. A cube measures 2 inches on a side, what is the volume of the cube?

42. Find the volume of a circular cylinder with radius of 12 cm and a height of 20 cm. ($\pi = 3.14$)

Solve graphically

43. $y = 3x - 4$
$y = 6 - 2x$

44. $y = x - 5$
$y + 3x = 7$

Solve by substitution method

45. $y = 2x - 3$
$x - 2y = -15$

46. $x - 2y = 12$
$x = 3y + 13$

Solve by addition method

47. $2x - 3y = -8$
$5x + 3y = -6$

48. $5x + 4y = 12$
$2x + 6y = 7$

## CUMULATIVE TEST **Chapters 1–15** *Form A*

1. Evaluate $-3\{5 - 2[7 + (15 - 4)^2]\}$
2. Perform the indicated operation and simplify.
   $4(x^3y^2 - xy^2) + x(x^2y^2 + 9y^2)$
3. Solve for x: $5x - 2 = 3(x - 6) - 9x$
4. Solve for c: $\dfrac{4a - 3}{c} = b$
5. Solve for x: $\dfrac{3x}{2} - 11 = \dfrac{23}{3}$
6. Factor completely: $3x - 12x^2$
7. Factor completely: $a^2 - 16a + 48$
8. Solve by factoring: $x^3 - 49x = 0$
9. Perform the indicated operation: $\dfrac{3xy^4z}{7} \div \dfrac{9x^2yz^3}{-14}$
10. Solve by using the quadratic formula: $x^2 - 3x - 9 = 0$
11. In right triangle ABC, the hypotenuse is 20 and the shortest side is 8. Find the length of the longest side.
12. In a certain triangle the measure of angle A is twice the measure of angle B, and angle C is thirty degrees more than angle A. What is the measure of all three angles?
13. Solve graphically: $y = 4x + 1$
    $y = x - 5$
14. Solve the system by any method except graphing:
    $2x - y = 8$
    $4x + 3y = 6$
15. Solve the system by any method except graphing:
    $3x + 2y = 12$
    $5x + 3y = 1$

CHAPTER

# 16 Linear Inequalities

## 16.0 *Vocabulary Preview*

Define these words without using a dictionary. Formulate their meaning from past experience or knowledge.

| | | |
|---|---|---|
| **coordinate** | **inequality** | **integer** |
| **intercept** | **linear** | **non-integer** |
| **plane** | **rational** | **real** |
| **slope** | **variable** | |

## 16.1 *Graphing Real Numbers*

In Chapter Two, a number line was used to illustrate the set of numbers called integers. That same line can be used to locate all real numbers, non-integers as well as integers. However, since it is impossible to locate the exact point that represents a number such as 3.45 or $\sqrt{2}$, it is also impossible to locate them on the number line exactly, so their location can only be estimated.

*Example:* Locate the numbers $-3$, 6, 4.5, $\sqrt{3}$, $-1.235$ on the real number line.

$-3$ and 6 are easily located. $-3$ is 3 units left of zero.

6 is located 6 units to the right of zero.

4.5 is located to the right of zero, halfway between 4 and 5.

$\sqrt{3}$ is approximately 1.73 so is located to the right of zero, close to 2.

$-1.235$ is about one-quarter unit left of $-1$.

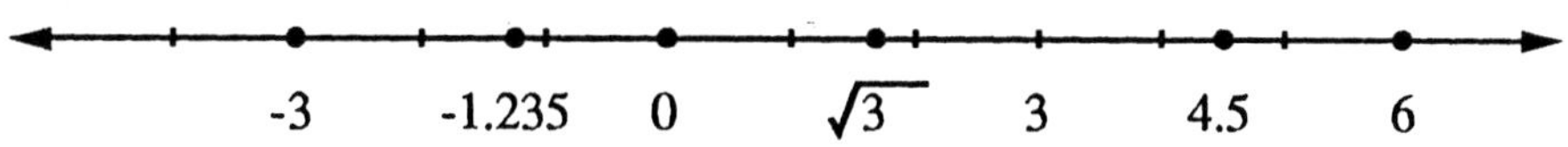


**Exercise Set 16.1**

Graph the real numbers on a number line.

1. $-3.5$
2. 0.45
3. 2.25
4. $-0.5$
5. $\sqrt{16}$
6. $\sqrt{5}$
7. $\frac{-3}{5}$
8. $\frac{1}{6}$

## 16.2 *Graphing Sets of Real Numbers*

In Chapter Eight, we dealt with algebraic equations, that is, algebraic expressions containing equal signs. The equation $2x = 10$ has only one solution: $x = 5$.

The expression $2x \neq 10$ says "twice x IS NOT EQUAL TO 10". This means x can be any number EXCEPT 5, since if x IS 5, 2x IS 10.

If $2x \neq 10$, $x \neq 5$, which means that x is either greater than $5 (x > 5)$, or x is less than $5 (x < 5)$.

Such expressions are called **inequalities.**

*Example:*

| **Equation** | | | **Inequality** | | |
|---|---|---|---|---|---|
| $x = 2$ | $x \neq 2$ | so | $x > 2$ | or | $x < 2$ |
| $x + 3 = 7$ | $x + 3 \neq 7$ | so | $x + 3 > 7$ | or | $x + 3 < 7$ |
| $x - 5 = 2$ | $x - 5 \neq 7$ | so | $x - 5 > 2$ | or | $x - 5 < 7$ |

In addition to the symbols $=$, $\neq$, $>$, and $<$, two other symbols are used to indicate relationships with numbers:

the symbol $\geq$ denotes "greater than or equal to", while the symbol $\leq$ denotes "less than or equal to".

*Example:* Suppose we were to consider the set of integers greater than or equal to 3, but less than or equal to 6. This would include the integers 3, 4, 5, and 6 and would be represented with dots on the number line.

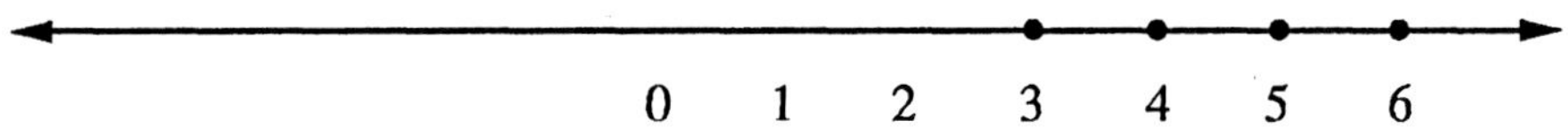


The above set is described: $\{x | x \in I, 3 \leq x \leq 6\}$ This is read "The set of x values, such that x is an integer which is greater than or equal to 3 but less than or equal to 6".

*Example:* To represent all the **real** numbers which are greater than 3 but less than 6, shade the line between the numbers and place an open circle at 3 and 6 to indicate that 3 and 6 are NOT INCLUDED in the set described.

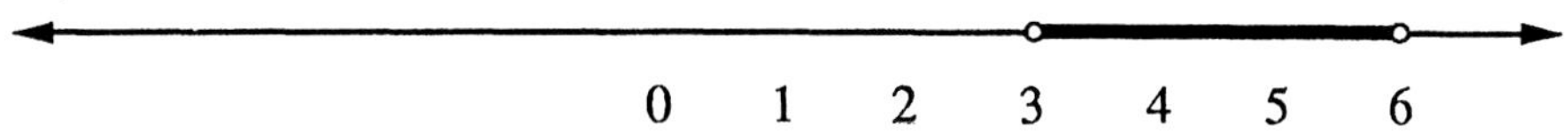

The above set is described: $\{x|x \varepsilon R, 3 < x < 6\}$ which is read, "The set of x values such that x is a real number which is less than 6 but greater than 3".

*Example:* The expression $\{x|x \varepsilon I, 3 < x < 6\}$ describes the set of **integers** "such that the integers are greater than 3 but less than 6". The values 4 and 5 satisfy the inequality, which is graphed by placing a dot on the line above the 4 and 5:

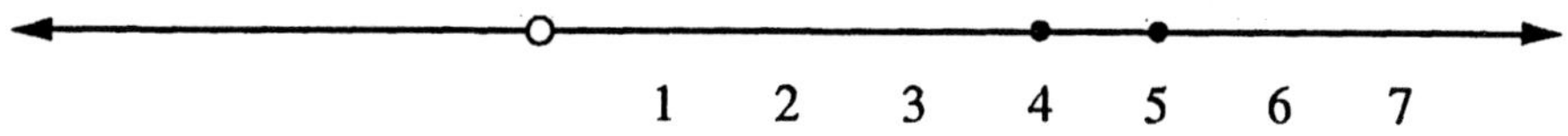


*Example:* The expression $\{x|x \varepsilon R, 3 < x < 6\}$ describes the set of **real numbers** "such that the real numbers are greater than 3 but less than 6". The solution is all numbers between 3 and 6.

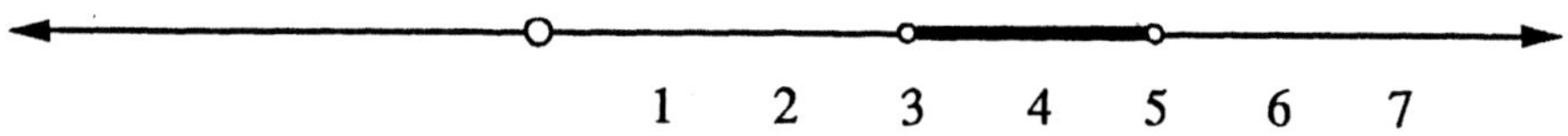


**Open dots** are placed above the numerals 3 and 6 to show that they ARE NOT included in the set. The line between them is shaded to show that all values **between** 3 and 6 are in the set.

*Example:* $\{x|x \varepsilon R, x < 3\}$ describes the set of real numbers which are less than 3.

The open dot above 3 indicates that it is not included in the solution set. The entire line to the left of 3 is shaded because all values less than 3 are in the solution set.

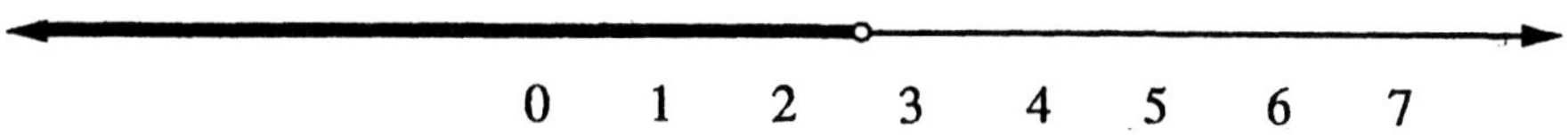


*Example:* $\{x|x \varepsilon R, x > 3\}$ means "the real numbers greater than or equal to 3." This includes 3 and all values to the right of 3:

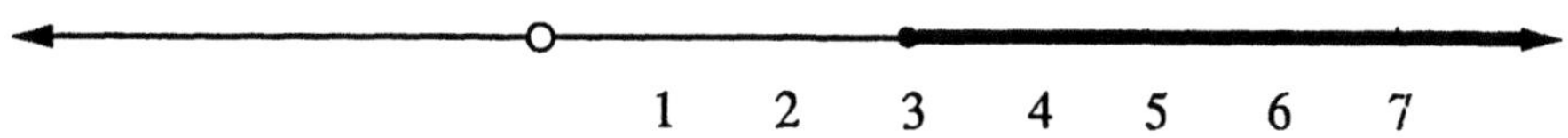

*Example:* $\{x | x \in R, x \leq 3\}$ means "the real numbers less than or equal to 3." This includes 3 and all values to the left:

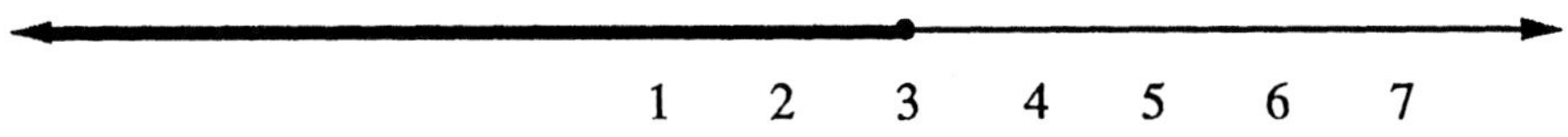


**Exercise Set 16.2**

Complete the following:

| | Set Description | Graphic Representation |
|---|---|---|
| 1. | $\{x | x \in R, 0 \leq x \leq 2\}$ | |
| 2. | $\{x | x \in I, 0 \leq x \leq 2\}$ | |
| 3. | | 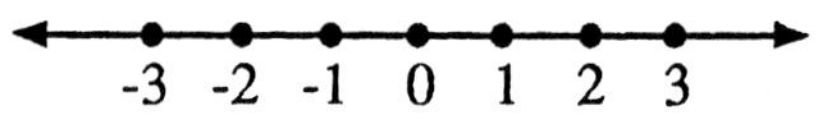  |
| 4. | |   |
| 5. | $\{x | x \in R, 6 \leq x \leq 8\}$ | |
| 6. | $\{x | x \in I, 6 \leq x \leq 8\}$ | |
| 7. | | 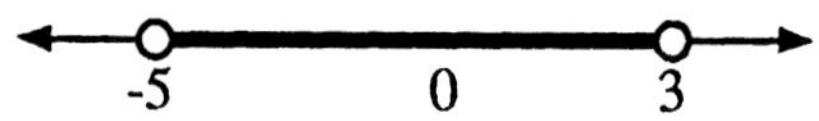  |
| 8. | | 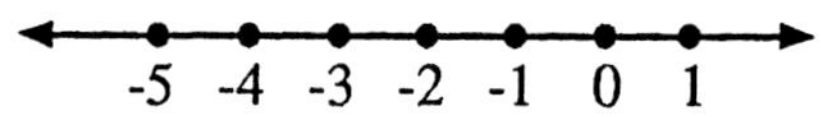  |
| 9. | $\{x | x \in R, x > -2 \text{ or } x < -5\}$ | |
| 10. | $\{x | x \in R, x \geq -2 \text{ or } x \leq -5\}$ | |

## 16.3 *Graphing Linear Inequalities*

Before attempting this section, it would be well to review Chapter Nine. Techniques similar to those used to graph linear equations such as $y = 2x + 3$ or $3y + 6x = 9$, are used to solve linear inequalities such as $y < 2x + 3$ or $3y + 6x \geq 9$.

Recall that the graph of a linear equation in one or two variables is a STRAIGHT LINE.

The graph of a linear inequality in two variables results in a REGION called a HALF-PLANE.

Remember that the solution or graph of an equation or inequality is the set of all points (ordered pairs) that satisfy the relationship.

*Example:* When graphing y = 2x + 3 using the slope-intercept method, you know that the slope is 2 and the y intercept is (0,3). The graph of the line represents all the points which satisfy that equation.

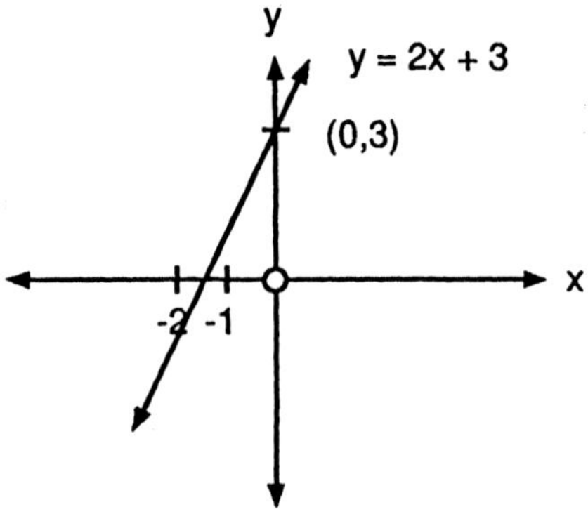


We can show that the points (1,5), (0,3) and (−1,1) "satisfy" the equation:

Using (1,5) we have

$$\begin{aligned} y &= 2x + 3 \\ 5 &= 2(1) + 3 \\ 5 &= 2 + 3 \\ 5 &= 5 \end{aligned}$$

Using (0,3) we have

$$\begin{aligned} y &= 2x + 3 \\ 3 &= 2(0) + 3 \\ 3 &= 0 + 3 \\ 3 &= 3 \end{aligned}$$

Using (−1,1) we have

$$\begin{aligned} y &= 2x + 3 \\ 1 &= 2(-1) + 3 \\ 1 &= -2 + 3 \\ 1 &= 1 \end{aligned}$$

We can also show that the point (1,0) DOES NOT lie on the line since that point doesn't satisfy the equation:

$$\begin{aligned} y &= 2x + 3 \\ 0 &= 2(1) + 3 \\ 0 &= 2 + 3 \\ 0 &= 5 \end{aligned}$$

NO!

As a matter of fact, the point (1,0) lies BELOW the line $y = 2x + 3$, so really satisfies the inequality $y < 2x + 3$.

*Example:* To graph the inequality $y < 2x + 3$:

1. Graph the "boundary line" determined by the equation $y = 2x + 3$. This is drawn as a dotted line to indicate that the points on the line are *not* part of the solution.

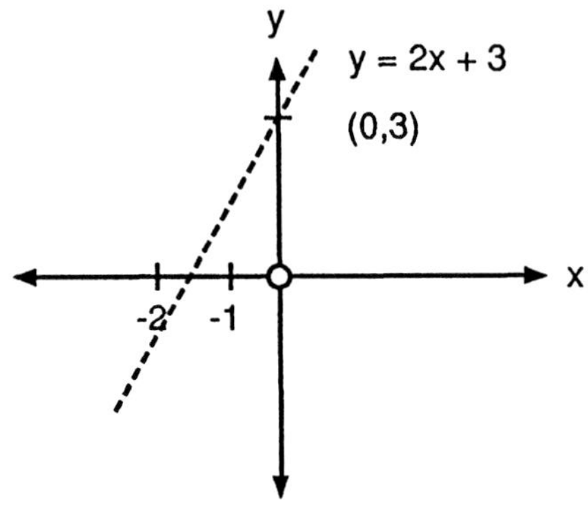


The line divides the region into two parts called half-planes. To determine in which half-plane the points that satisfy the inequality $y < 2x + 3$ lie:

2. Choose any point which is NOT on the line and "test it" in the given inequality. If the point $(-3,0)$ is chosen, then for

$$y < 2x + 3$$
$$0 < 2(-3) + 3$$
$$0 < -6 + 3$$
$$0 < -3 \qquad \text{NOT true.}$$

The half-plane in which the point $(-3,0)$ is located DOES NOT satisfy the inequality.

3. Test a point in the other half-plane which has $y < 2x + 3$ as a boundary; say, (2,1):

$$y < 2x + 3$$
$$0 < 2(1) + 3$$
$$0 < 2 + 3$$
$$0 < 5 \qquad \text{true.}$$

All points in the half-plane containing the point (2,1) satisfy the inequality.

4. Shade the region containing the point (2,1) to represent the solution of $y < 2x + 3$.

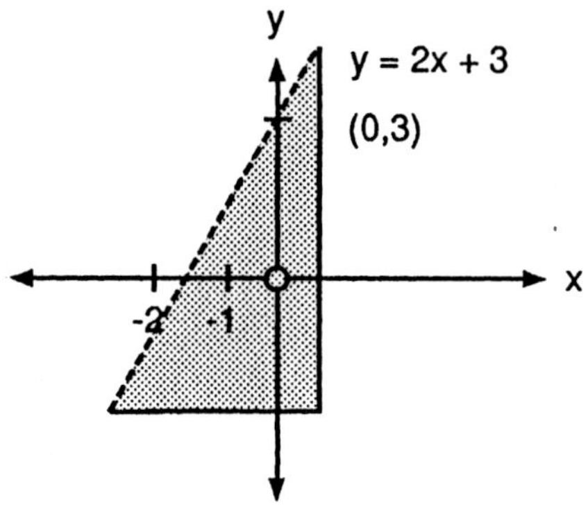


NOTE that the points on the line itself are not included in the solution, since they do not satisfy the inequality, for example: using the point (0,3) gives

$$y < 2x + 3$$
$$3 < 2(0) + 3$$
$3 < 3$ which is not true.

*Example:* Graph $3y + 6x \geq 9$.

1. Graph the boundary line determined by the *equation* $3y + 6x = 9$:

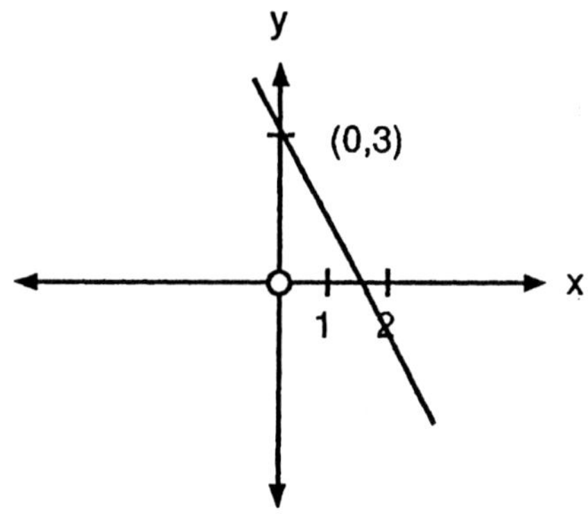


2. Determine which half-plane satisfies the inequality. Choose any point in either half-plane and "test" it. Testing (0,0) which is on the left side of the line:

$$3y + 6x \geq 9$$
$$3(0) + 6(0) \geq 9$$
$$0 + 0 \geq 9$$
$0 \geq 9$ not valid.

Testing the point (3,1) on the right of the border line:

$$\begin{aligned} 3y + 6x &\geq 9 \\ 3(1) + 6(3) &\geq 9 \\ 3 + 18 &\geq 9 \\ 21 &\geq 9 \end{aligned} \qquad \text{valid.}$$

The graphic solution of the inequality consists of all the points in the half-plane to the right of the border line. This is shown by shading that portion of the plane.

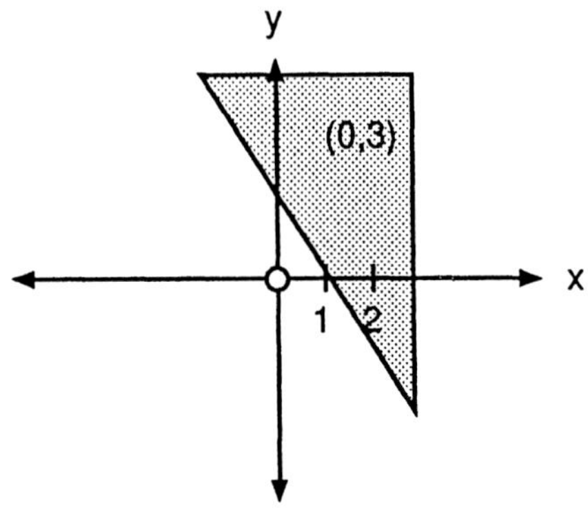


NOTE that the line is drawn as a solid line for $3y + 6x \geq 9$ because any point ON the line satisfies the inequality.

For example, using the point (0,3):

$$\begin{aligned} 3y + 6x &\geq 9 \\ 3(3) + 6(0) &\geq 9 \\ 9 + 0 &\geq 9 \\ 9 &\geq 9 \end{aligned}$$

The result is valid, since $9 = 9$. (Recall that the symbol $\geq$ means "greater than OR EQUAL TO".)

**When Graphing Inequalities:**

**a. the border line will be dotted if the points on the line do not satisfy the inequality. This will occur when the signs $<$ and $>$ are used.**

**b. the border line will be solid if the points on the line do satisfy, that is, when the symbols $\geq$ and $\leq$ are used.**

**c. the half-plane which represents the points that satisfy the inequality is shaded.**

**Exercise Set 16.3**

Graph the following inequalities in the Cartesian plane:

1. $y < 2x - 1$
2. $y > 3x + 1$
3. $y \geq \frac{2x}{3} + 1$
4. $y \leq -2x + 2$
5. $y < 3x$
6. $x + y \geq 4$
7. $2y < 6x + 4$
8. $5x + 2y \geq 10$
9. $4y < 3x$
10. $7x + 2y + 14 < 0$
11. $12 + 2x \geq 6y$
12. $2x > 8$
13. $x - y < 0$
14. $y \leq -2$

## 16.4 *Graphing Systems of Linear Inequalities*

Techniques for graphing systems of inequalities are similar to those used in Chapter Fifteen for graphing systems of linear equalities. It is important to fully understand the material from that chapter and section 16.3 before attempting this section.

A **system of inequalities** is a set of inequalities which are graphed or solved together (simultaneously). The solution of a system of linear inequalities is the set of all points (ordered pairs) which satisfy ALL of the inequalities at the same time. Graphically, this solution will be the region in the Cartesion plane that contains all the points which make the inequalities true at the same time.

**To Graph Systems of Inequalities:**

1. **Graph the boundary line for the first inequality and using vertical lines, shade the region that represents the solution.**
2. **On the same set of axes, graph the border line for the second inequality and using horizontal lines shade the region that represents the solution.**
3. **The solution to the system of inequalities is the region which is cross-hatched.**
4. **Check your solution by choosing a point from the cross-hatch and testing it in each inequality.**

*Example:* Solve the system of inequalities graphically:

$$y \geq 2x - 3 \text{ and } y < -3x + 1$$

1. Graph the border line (solid) for the first inequality and shade the solution with vertical lines.

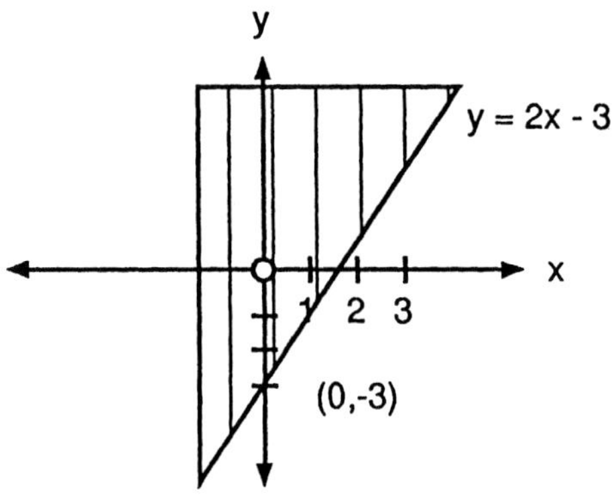


2. On the same set of axes, graph the border line (dotted) for the second inequality and shade the solution with horizontal lines.

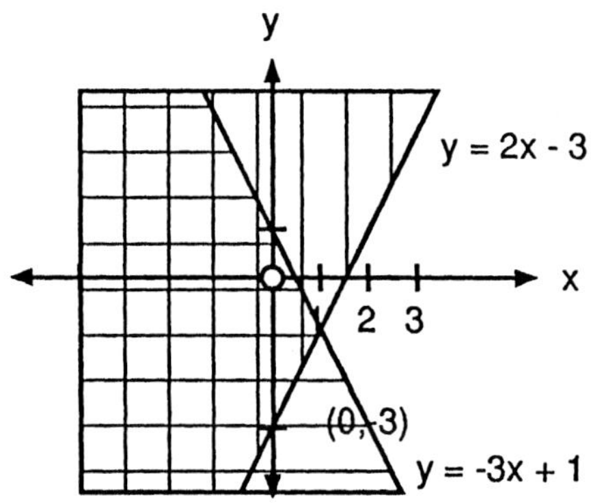


3. The solution is the cross-hatched region on the left, including the line $y = 2x - 3$. The line $y = -3x + 1$ is NOT part of the solution of the system of inequalities.

4. Check the solution by choosing a point in the cross-hatched region and test it in both inequalities. If we choose $(-2,0)$:

| | |
|---|---|
| $y \geq 2x - 3$ | $y < -3x + 1$ |
| $0 \geq 2(-2) - 3$ | $0 < -3(-2) + 1$ |
| $0 \geq -4 - 3$ | $0 < 6 + 1$ |
| $0 \geq -7$ | $0 < 7$ |
| valid | valid |

*Example:* Solve the system of inequalities graphically:

$x - y > 2$
$3x + 2y < 2$

1. The graph of the first inequality is:

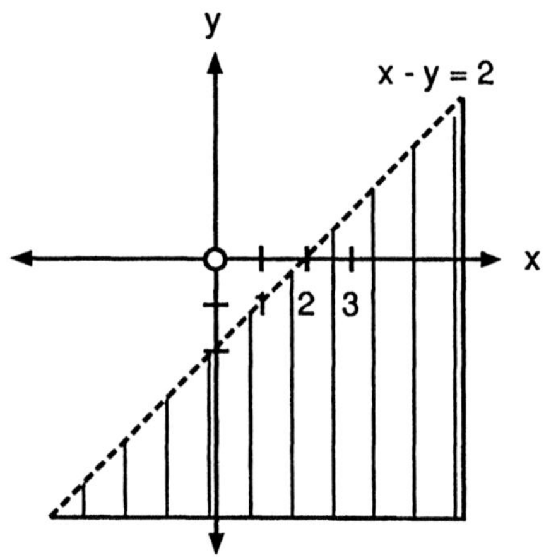


2. Using the same set of axes to graph the second inequality results in the graph:

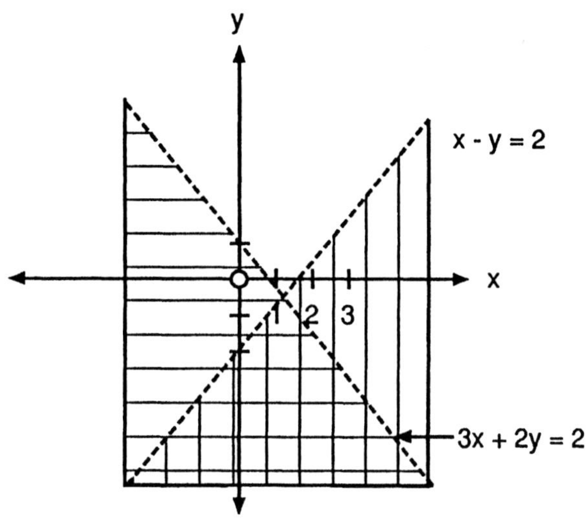


3. The solution is the cross-hatched region at the bottom of the graph. The solution of the system does not contain the points on either of the boundary lines since they were not included in the solution of the original inequalities.

4. To test the solution, choose a point in the crossed section. Choose (0,−4):

| | |
|---|---|
| $x - y > 2$ | $3x + 2y < 2$ |
| $0 - (-4) > 2$ | $3(0) + 2(-4) < 2$ |
| $0 + 4 > 2$ | $0 + (-8) < 2$ |
| $4 > 2$ | $-8 < 2$ |
| valid | valid |

**Exercise Set 16.4**

Graph the solution to the following systems of linear inequalities:

1. $y \le 2x$
   $y > -2x + 2$

2. $3x + 2y < 6$
   $y > x$

3. $x + 2y < -2$
   $x - y \ge 4$

4. $3x - 2y < 8$
   $y \le x + 2$

5. $3y > 9 + x$
   $3y + 2x \ge 7$

6. $2x + y < 1$
   $x \ge -2$

7. $y + x \ge 0$
   $2x > y + 2$

8. $3x + 5y < 15$
   $x - 2y > 6$

9. $x < 2$
   $x \le 2y + 3$

10. $2y + x < 0$
    $4x - 7y \le 28$

11. $x > 2$
    $y \le -3$

12. $2x < 3y$
    $4x + 3y \ge 12$

13. $x + 2y > 1$
    $2x + 4y < -4$

14. $x - y < 4$
    $2x > y + 2$

## CHAPTER EXERCISES

### Section 16.1

Graph the following numbers on the real number line:

1. 3.2
2. $-3.0$
3. $-7.1$
4. 2.71
5. $\frac{-2}{3}$
6. $-0.75$
7. 3.14
8. $\frac{5}{15}$
9. $\frac{-7}{6}$
10. $\sqrt{3}$
11. $-\sqrt{5}$
12. $-2\sqrt{2}$
13. $\sqrt{.04}$
14. $\sqrt{25}$
15. $-\sqrt{49}$

### Section 16.2

Graph each of the following sets:

16. $\{x|x \varepsilon I, x \geq 3\}$
17. $\{x|x \varepsilon I, -1 \leq x \leq 7\}$
18. $\{x|x \varepsilon R, 2 \leq x \leq 4\}$
19. $\{x|x \varepsilon R, x \geq -3\}$
20. $\{x|x \varepsilon I, -4 \leq x < 1\}$
21. $\{x|x \varepsilon R, x < 3\}$
22. $\{x|x \varepsilon R, x < -4\}$
23. $\{x|x \varepsilon I, x < 5\}$
24. $\{x|x \varepsilon I, 2 < x, \text{ or } x < -1\}$
25. $\{x|x \varepsilon R, x < 2 \text{ or } x \geq 3\}$

Write a set description for each of the following graphs:

26. 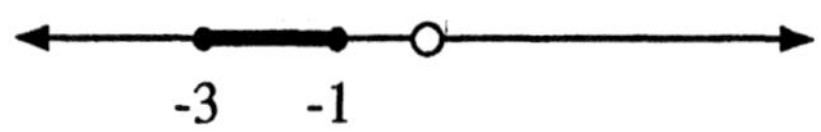


27. 

28. 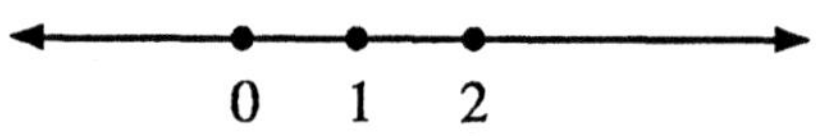


29. 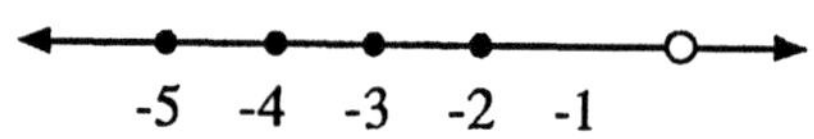


30. 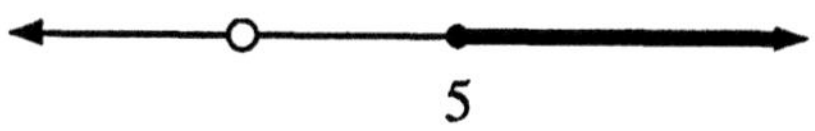


31. 

32. 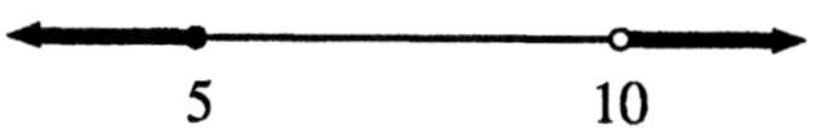


33. 

### Section 16.3

Graph the following inequalities in the Cartesian plane:

34. $y < 3x + 2$
35. $y \geq \frac{2}{3}x + 4$
36. $y \leq 4x - 3$
37. $y < 5x - 1$
38. $y \geq \frac{-4}{3}x$
39. $y \leq 8x - 2$
40. $x + y > 3$
41. $2y + x < -3$
42. $x + 3 < 1$

43. $y - 7 \geq 0$

44. $2x - y \leq 6$

45. $y - 7x > 0$

46. $9x < y$

47. $5x - 7 \geq 2y$

48. $0 > y - 3$

49. $y - 5 < 6x$

50. $-3 > 3x$

51. $5y - 3x \leq 12$

52. $4y - 8 + x \leq 0$

53. $7 \leq y$

**Section 16.4**

Graph the solution for the following systems of inequalities:

54. $y > 3x$
$y \leq x - 1$

55. $y \leq 2x + 4$
$y > -x$

56. $x + y < 2$
$x - y \geq -4$

57. $2 \leq y$
$y < 7 - 2x$

58. $3x + 5y > 15$
$-3x + y < 6$

59. $x < -8$
$x < y - 3$

60. $3y - x < 9$
$y > -x$

61. $3x - 4 \geq 0$
$2x + y + 1 < 0$

62. $5x < 7 - y$
$4y \geq 12 - 2x$

63. $2x - 7 < y$
$-3 > y$

64. $12 + 4y < 3x$
$\frac{-2}{3}x + 3 \leq y$

65. $x - y < 0$
$x \geq -3$

66. $y - x \geq -2$
$4y - 8x \leq 12$

67. $x < y$
$5 \geq x$

68. $\frac{x}{4} + \frac{y}{3} < 1$
$y \geq 2x$

69. $\frac{y}{5} \geq x + 1$
$2y + 4 < -2x$

70. $5y + 4x < 20$
$y \geq \frac{-1}{3}x + 1$

## CHAPTER TEST *Form 16A*

Graph the following sets:

1. $\{x|x \in R, 2 \leq x \leq 4\}$
2. $\{x|x \in I, -2 < x \leq 4\}$
3. $\{x|x \in R, x > 1 \text{ or } x < -2\}$
4. $\{x|x \in R, x > -3 \text{ and } x \leq 0\}$

Graph each of the following inequalities in the Cartesian plane:

5. $y > 2x + 1$
6. $y \leq \frac{3}{4}x - 2$
7. $y > 5x$
8. $x \leq -1$
9. $3x + 4y \leq 12$
10. $x \geq y - 3$

Graph the solution to the following systems of linear inequalities:

11. $y > x + 1$ and
    $y \leq 2x + 3$
12. $x < 3$ and
    $y \geq -2$
13. $x + y \geq 5$ and
    $x - y < -3$
14. $3x + 2y + 4 < 0$ and
    $y > x - 2$

## CUMULATIVE EXERCISES **Chapters 1–16** *Set A*

Perform the indicated operations and simplify your answers:

1. $2(x - 5) - 3(2x + 4) - 7$
2. $8ab[5bc - 3] - 2b[3bca - 5a]$
3. $(3x + 2)(7x - 1)$
4. $(9x - 2)(8x - 7)$
5. $(a - 9)(4 + a)$
6. $(2 - y)^2$
7. $(3a + 2)(a^2 - 2a + 4)$
8. $(4x - 2)^2$
9. $\dfrac{x^2y}{70z^2} \div \dfrac{4xz^2}{35yz}$
10. $\dfrac{8}{4x} - \dfrac{2 - x}{8x}$
11. $\dfrac{a}{4} - \dfrac{b}{3} + \dfrac{2a}{12}$
12. $\dfrac{x - 4}{9} + \dfrac{2x}{3} - 1$
13. $\dfrac{22x^2}{3y} \cdot \dfrac{-6y^2z}{-55xz}$
14. $\dfrac{5x^3 - 20x^2 - 15x}{-5x}$
15. $(y - 1)^2 - (y + 1)^2$
16. $(z - 3)(z + 1) - z(z - 2)$
17. $(6x^2 - 7y) - (3x^2 - 2y)$
18. $\dfrac{x}{x - 4} - \dfrac{3x}{x - 4}$
19. $(-7a^3b)^2$
20. $\dfrac{10y^2}{x^2 - 16} \div \dfrac{2y}{x - 4}$

Solve the following equations:

21. $2x - 3(x - 5) = .7$
22. $22y - 7(y - 1) = 4 + 8y$
23. $(x - 3)(x + 2) = 0$
24. $\dfrac{1}{3}x^2 + \dfrac{2x}{3} = \dfrac{-2}{3}$
25. $9z - 7 = 12.8$
26. $x^2 = 49$
27. $x^2 - 2x = 8$
28. $\dfrac{x}{4} = \dfrac{x}{8} + 3$
29. $18 - 3y + y^2 = 0$
30. $3a - 7.2 = 9a$
31. $15y - 11 = 23$
32. $7(y - 3) = 35(y + 1)$
33. $(a - 2)^2 = 9$
34. $\dfrac{-y}{8} = \dfrac{3y}{24} - 42$
35. $z^2 - 4z = 77$
36. $2x^2 + 5x = 12$
37. $x^2 = .49$
38. $a(a - 4) = 0$
39. $3y^2 - 2y = 0$
40. $11 = \dfrac{8y - 3}{2}$
41. $\dfrac{x - 5}{7} = \dfrac{3x - 7}{35}$

42. Solve for l: $p = 2l + 2w$

43. Solve for r: $V = \frac{1}{3}\pi r^2$

44. Find tan 16°

45. Find cos 87°

Solve the systems of linear equations:

46. $2x - 6y = 4$ and
$x + 6y = 8$

47. $y = 3x - 7$ and
$2x - y = 5$

48. $y = 4x$ and
$y = 2x - 1$

49. $y - x = 3$ and
$y + 2x = 9$

50. $3x + 4y = 6$ and
$x - 2y = 7$

51. $2y - 5x = 4$ and
$3y - 2x = -5$

Graph the solutions for the systems of linear inequalities:

52. $\frac{y}{3} \leq 2x - 4$ and
$y \geq -3x + 1$

53. $y > 5x$ and
$x + y \leq 6$

Solve the linear inequalities graphically:

54. $4x + 3 \geq -13$

55. $\frac{2}{3}x - 6 \leq 0$

Solve the following problems:

56. The circumference of a circle is 43 inches. Find the diameter of the circle to the nearest foot. (Use $\pi = 3.14$)

57. The length of the two legs of a right triangle are 10 feet and 24 feet. Find the length of the hypotenuse to the nearest foot.

58. A pole casts a shadow 3.5 feet long when the angle of elevation of the sun is 63°. Find to the nearest tenth of a foot, the length of the pole.

59. A game has a total of 40 jacks. Patty has five jacks more than Liz. Chris has three times as many jacks as Liz. How many jacks has each girl?

60. One angle of a triangle is three times as large as a second angle. The third angle is 28°. Find the measures of the angles.

61. The base of a triangle is five feet more than its height. If the area of the triangle is 42 square feet, find its height.

## CUMULATIVE TEST **Chapters 1–16** *Form A*

1. Find the multiplicative inverse of 3a.
2. Simplify: $4a[3 - 2a(a^2 - 3b)]$

Factor completely:

3. $x^3 - 9x$
4. $ax^2 + 2ax - 3a$
5. $x^2 - 64$
6. $x^2 - 4x - 21$
7. Find the slope and y intercept of the line $y = \frac{2}{3}x - 5$
8. Graph the equation $y = 2x$

Solve for x:

9. $x - 7 = 8x + 8$
10. Solve for x: $\frac{2x - 3}{5} = \frac{6x - 4}{3}$
11. Multiply: $\frac{4x^2y}{25} \cdot \frac{75}{16xy^3}$
12. Divide: $\frac{22a^2bc}{27xc} \div \frac{33ab^2}{36x^3}$
13. The product of two consecutive odd integers is 99. Find the integers.
14. In right triangle ABC, where the hypotenuse is 17 and one of the legs is 15, find the other leg.
15. Find $\sin 37°$
16. Find $\cos 47°$.
17. In a triangle, angle A is 16°. If angle B is three times angle C, find angles B and C.
18. Multiply: $(3a - 6b)(2a + 5b)$
19. Find the area of a rectangle 7 ft. by 25 ft.
20. Find the area of a circle whose diameter is 22 feet. Use $\pi = 3.14$.
21. Find the perimeter of a rhombus whose side is 3.25.
22. Find the volume of a cylinder 15 feet high if the radius of the base is 6 feet.

Graph the following:

23. $y \leq 4 - 2x$
24. $y > 6x$
25. $y \leq \frac{3}{5}x - 2$
    $x + y < 4$
26. $y = 2x - 5$
    $y = 7 - 4x$

CHAPTER

# 17 Factoring and Algebraic Fractions

## 17.0 *Vocabulary Preview*

Define these words without using a dictionary. Formulate their meaning from past experience or knowledge.

| | | |
|---|---|---|
| **algebraic** | **factor** | **reciprocal** |
| **binomial** | **fraction** | **solution** |
| **coefficient** | **inverse** | **square** |
| **denominator** | **numerator** | **trinomial** |
| **divisor** | **product** | |

## 17.1 *Factoring Algebraic Expressions*

Before attempting this chapter, review the sections on factoring from Chapter Ten. That material is summarized as follows:

*General Rules for Factoring Algebraic Expressions*

**1. *Look for the greatest common factor.* Divide the entire expression by that factor, in order to obtain the second factor.**

*Example:* In the expression $2xz + 6x^2yz^3 - 14xz^3$, 2xz is common to ALL the terms, so it is "factored out". Dividing each term by 2xz results in the factored form:

$$2xz + 6x^2yz^3 - 14xz^3 = 2xz(1 + 3xyz^2 - 7z^2)$$

*Example:* In the expression $3t^2r^3s^4 + 6trs$, the common factor is 3trs. If that is factored out and divided through the expression, the result is:

$$3trs(tr^2s^3 + 2)$$

**2. *Look for a "difference of two squares".* The factors will be the sum and difference of the square roots of the two terms.**

*Example:* $4x^2 - z^2 = (2x + z)(2x - z)$

*Example:* $4x^2 - 36z^2 = 4(x^2 - 9z^2) = 4(x + 3z)(x - 3z)$

*Example:* $a^4 - b^4 = (a^2 + b^2)(a^2 - b^2) = (a^2 + b^2)(a + b)(a - b)$

**3.** ***To factor trinomials with leading coefficient of 1,***

**a. Find the factors of the last term. Their sum must equal the coefficient of the middle term.**

**b. If the sign of the last term is positive, then the signs of both factors will be the same as the sign of the middle term.**

**c. If the sign of the last term is negative, the signs of the factors will be different, and will be determined by the sign of the middle term, that is, the "inner" and "outer" products of the factors of the first and last terms must sum to the middle term.**

*Example:* $x^2 + 5x + 6 = (x + 3)(x + 2)$
because $(+3)(+2) = +6$ and $(+3) + (+2) = +5$

*Example:* $x^2 - 5x + 6 = (x - 3)(x - 2)$
because $(-3)(-2) = +6$ and $(-3) + (-2) = -5$

*Example:* $x^2 + 5x - 6 = (x + 6)(x - 1)$
because $(+6)(-1) = -6$ and $(+6) + (-1) = +5$

*Example:* $x^2 - 5x - 6 = (x - 6)(x + 1)$
because $(-6)(+1) = -6$ and $(-6) + (+1) = -5$

**Exercise Set 17.1**

Factor the following:

1. $12st^2 + 4st - 6t^3$
2. $rst - r^2s + 4r$
3. $s^2 - 16t^2$
4. $c^2d^4 - e^6$
5. $2t^2 - 8s^2t^2$
6. $8y^4 - 800$
7. $s^2 + 4s + 3$
8. $c^2 + 5c + 4$
9. $r^3 + 7r^2 + 6r$
10. $x^2 + 2xy + y^2$
11. $x^4 - y^2$
12. $b^2 + 3b + 2$
13. $x^4y^2 - x^2y^4$
14. $x^2 - 3x - 4$
15. $3x^2 + 6x + 3$
16. $c^2 + 6c - 7$
17. $2c^2 + cd - d^2$
18. $-g^2 + 9g - 20$
19. $49x^2 - 16$
20. $32x^2y^2 - 18$
21. $64a^2 - 36b^2$
22. $x^2 - 2x - 3$
23. $a^2 - 2ab - 3b^2$
24. $6x^3 - 24x^2 + 18x$

## 17.2 *Factoring Trinomials with Leading Coefficient Not 1*

The procedure for factoring a trinomial with leading coefficient NOT one is similar to that for factoring when the leading coefficient IS one, but the factors of the first term must also be considered.

**1. Find the factors of the first and last terms.**

**2. Consider the signs of the last term and the middle term.**

**3. The "inner" and "outer" products of the factors of the first and last terms must sum to the middle term.**

*Example:* To factor $2x^2 + 5x + 3$:

1. The factors of the first term are $(2x)(x)$.
2. The factors of the last term are $(3)(1)$.
3. The sign of the last term is +, so the factors will have LIKE signs.
4. The sign of the middle term is +, so the signs of both factors will be positive.
5. The factors of the first and last terms must sum to $+5x$.

Try the factors $(2x + 1)(x + 3)$
$(2x + 1)(x + 3) = 2x^2 + 6x + x + 3 = 2x^2 + 7x + 3$
This is not the solution.

Try the factors $(2x + 3)(x + 1)$. Here the products are $2x^2 + 2x + 3x + 3$ which does equal $2x^2 + 5x + 3$, so this solution is correct.

*Example:* To factor $3x^2 + 11x + 6$, consider:

1. The factors of $3x^2$ are $3x \cdot x$.
2. The factors of 6 are $6 \cdot 1$ or $3 \cdot 2$.
3. The sign of the last term is positive, so the signs of the factors will be the same.
4. The sign of the middle term is +, so the signs of both factors will be positive.
5. The products of the factors of the first and last terms must sum to $+11x$.

Try the factors $(3x + 6)(x + 1)$.
The "inner" product is $6x$.
The "outer" product is $3x$.
They combine to $9x$. This is NOT the middle term of the original expression.

Try $(3x + 1)(x + 6)$.
The "inner" product is $1x$.
The "outer" product is $18x$.
They combine to $19x$, not the middle term of the original expression.

Try $(3x + 3)(x + 2)$.
The "inner" product is $3x$.
The "outer" product is $6x$.
They combine to $9x$, which is not the middle term of the original expression.

Try $(3x + 2)(x + 3)$.
The "inner" product is $2x$.
The "outer" product is $9x$.
They combine to $11x$, which is the middle term, so $3x^2 + 11x + 6$ factors to $(3x + 2)(x + 3)$.

*Example:* To factor $3x^2 + 2x - 5$

1. The factors of the first term are $(3x)(x)$.
2. The factors of the last term are $(-5)(1)$ or $(5)(-1)$.
3. The sign of the last term is negative, so the factors will have different signs.
4. The "inner" product and "outer" product must combine to $+2x$.

Try: $(3x - 5)(x + 1)$. The "inner" product is $-5x$.
The "outer" product is $3x$.
They combine to $-2x$.
This is not a solution.

Try: $(3x - 1)(x + 5)$. The "inner" product is $-x$.
The "outer" product is $15x$.
They combine to $14x$.
This is not a solution.

Try: $(3x + 5)(x - 1)$. The "inner" product is $5x$.
The "outer" product is $-3x$.
They combine to $2x$, so this is the solution.

The factors of $3x^2 + 2x - 5$ are $(3x + 5)(x - 1)$

**Check by multiplying:**

$(3x + 5)(x - 1) = 3x^2 - 3x + 5x - 5 = 3x^2 + 2x - 5$

*Example:* To factor $4a^2 - 6a + 2$,

FIRST factor the common factor 2: $2(2a^2 - 3a + 1)$

THEN factor the trinomial $2a^2 - 3a + 1$:

1. The factors of $2a^2$ are $(2a)(a)$.
2. The factors of 1 are $(1)(1)$ or $(-1)(-1)$.
3. The sign of the last term is $+$, so the factors will have the same signs.
4. The sign of the middle term is $-$ so the signs of the factors will be negative.
5. The products of the factors of the first and last terms must sum to $-3a$.

In factored form: $$4a^2 - 6a + 2 = 2(2a^2 - 3a + 1)$$
$$= 2(2a - 1)(a - 1)$$

*Example:* Factor $6a^2 - 3ab - 3b^2$

1. Take out the common factor and divide it through the expression: $3(2a^2 - ab - b^2)$.
2. Factor the trinomial $2a^2 - ab - b^2$: $2a^2$ factors into $2a \cdot a$ and $b^2$ factors into $b \cdot b$.
3. The sign of the last term is $-$ so the signs of the factors will be different.
4. The "inner" and "outer" products must sum to $-ab$.

Try $(2a - b)(a + b)$. The "inner" and "outer" products of $-ab$ and $2ab$ combine to $+ab$.

This is not a solution.

Try $(2a + b)(a - b)$. The "inner" and "outer" products, $ab$ and $-2ab$, sum to $-ab$. This is the solution, so in factored form:

$$6a^2 - 3ab - 3b^2 = 3(a^2 - ab - b^2) = 3(2a + b)(a - b).$$

**Check by multiplication:**

$$3(2a + b)(a - b) = (6a + 3b)(a - b) = 6a^2 - 6ab + 3ab - b^2$$
$$= 6a^2 - 3ab - b^2.$$

*Example:* To factor $4y^2 + 6y + 2$:

1. Divide out the common factor 2 and then factor the trinomial.
   $4y^2 + 6y + 2 = 2(2y^2 + 3y + 1)$.
2. The factors of $2y^2$ are $2y \cdot y$.
3. The factors of the last term are $1 \cdot 1$.
4. The sign of the last term is + so the signs of the factors will be alike.
5. The sign of the middle term is + so the signs of both factors will be +.
6. The "inner" and "outer" products must combine to 3y.

   Try: $(2y + 1)(y + 1)$. The "inner" product is y.
   The "outer" product is 2y.
   They combine to 3y.
   This is the solution.

$4y^2 + 6y + 2$ factors completely to $2(2y + 1)(y + 1)$.

This can be checked by multiplication.

*Example:* Factor $6x^3 - 9x^2y + 3xy^2$

Factoring out the common factor 3x gives $3x(2x^2 - 3xy + y^2)$.

Factoring the trinomial:

1. The factors of $y^2$ are $y \cdot y$.
2. The factors of $2x^2$ are $2x \cdot x$.
3. The sign of the last term is + so the signs of both terms will be alike.
4. The sign of the middle term is − so the signs of both terms will be negative.
5. The "inner" and "outer" products must combine to $-3xy$.

$6x^3 - 9x^2 + 3xy^2$ factors completely to $3x(2x - y)(x - y)$

This can be checked by multiplication.

**Exercise Set 17.2**

Factor completely.

1. $2x^2 - 7x + 3$
2. $4y^2 - 8y + 4$
3. $6a^2 - 7ab + b^2$
4. $2x^2 + 3x - 14$

5. $6x^2 - 18xy + -24y^2$

6. $3c^2 - 5c - 2$

7. $2y^2 - 5y + 3$

8. $7x^2 + 10x + 3$

9. $2x^2 + 3x - 14$

10. $3x^2 - 6x + 3$

11. $7a^2 + 14ab - 21b^2$

12. $6d^2 - d - 2$

13. $3x^2 + 8x + 5$

14. $2t^2 + 6t - 20$

15. $5m^2 - 13m + 6$

16. $2x^2 - x - 28$

17. $6s^2 + 13s + 2$

18. $2a^2 + 7a + 6$

19. $4q^2 + 8q + 3$

20. $6x^3 + 3x^2 - 3x$

## 17.3 *Reducing Algebraic Fractions*

Before attempting sections 17.3, 17.4, 17.5 and 17.6, review the material on algebraic fractions from Chapter Eleven.

To reduce an algebraic fraction, factor the numerator and denominator completely. Then divide off the common factors.

*Example:* To reduce the fraction $\frac{8x^2}{4x^3}$

factor the numerator and denominator: $\frac{4 \cdot 2 \cdot x^2}{4 \cdot x \cdot x^2}$

and divide off the common factors: $\frac{8x^2}{4x^3} = \frac{2}{x}$

*Example:* To reduce the fraction $\frac{10x + 50}{10}$

factor the numerator: $\frac{10x + 50}{10} = \frac{10(x + 5)}{10}$

divide off the common factor 10: $\frac{10x + 50}{10} = x + 5$

*Example:* To reduce the fraction $\dfrac{3x^2 + 6x}{x^2 - 4}$

factor the numerator: and the denominator: $\dfrac{3x(x + 2)}{(x - 2)(x + 2)}$

divide off (x + 2). The result is $\dfrac{3x}{(x - 2)}$

*Example:* To reduce the fraction $\dfrac{x^2 - 1}{x^2 + 3x - 4}$

factor the numerator: factor the denominator: $\dfrac{(x + 1)(x - 1)}{(x + 4)(x - 1)}$

divide off (x − 1). The result is $\dfrac{x + 1}{x + 4}$

*Example:* The fraction $\dfrac{2x + 2y}{x^2 - y^2}$ factors to $\dfrac{2(x + y)}{(x - y)(x + y)}$

which reduces to $\dfrac{2}{(x - y)}$

*Example:* $\dfrac{x^2 - 5x + 4}{4x - 16}$ factors to $\dfrac{(x - 4)(x - 1)}{4(x - 4)}$

which reduces to $\dfrac{x - 1}{4}$

**Exercise Set 17.3**

Reduce the fractions:

1. $\dfrac{12}{3x - 6}$
2. $\dfrac{6x - 3}{3}$
3. $\dfrac{64xy}{32x^2y^2}$
4. $\dfrac{x}{x^3y}$
5. $\dfrac{4a + 8b}{6a + 12b}$
6. $\dfrac{49xy^5z}{7x^2y^2z}$
7. $\dfrac{x - 5}{x^2 - 4x - 5}$
8. $\dfrac{y^2 - 49}{y - 7}$
9. $\dfrac{x^2 - 1}{x^2 - 2x + 1}$
10. $\dfrac{x - 5}{x^2 - 7x + 10}$
11. $\dfrac{3x + 6}{3x - 6y}$
12. $\dfrac{x + 1}{x^2 - 1}$
13. $\dfrac{x^2 - y^2}{2x + 2y}$
14. $\dfrac{x + 2}{4x^2 - 16}$
15. $\dfrac{x^2 + 5x + 4}{x^2 - 16}$

16. $\frac{x^2 - 1}{3x^2 + x - 2}$

17. $\frac{6a^2b + 12b^2}{3a^2 + 6b}$

18. $\frac{3x^2 - 27}{2x^2 + 10x + 12}$

19. $\frac{6a^2b + 12b^2}{3a^2 + 6}$

20. $\frac{3x - 3}{3x^2 + 6x - 9}$

21. $\frac{x^2 + 7x - 8}{x - 1}$

22. $\frac{8r^3 + 20r^2 - 12r}{8r^2 - 4r}$

23. $\frac{16a^2 - 1}{4a^2 - 7a - 2}$

## 17.4 *Multiplying Algebraic Fractions*

**To multiply algebraic fractions, factor the numerator and denominator of each fraction. Express the product of the numerators and the product of the denominators as a single fraction. Divide off the common factors.**

*Example:*

$$\frac{3x + 6}{3x - 6y} \cdot \frac{x^2 - 4y^2}{x + 2y} = \frac{3(x + 2)}{3(x - 2y)} \cdot \frac{(x + 2y)(x - 2y)}{(x + 2y)}$$

$$= \frac{3(x + 2)(x + 2y)(x - 2y)}{3(x - 2y)(x + 2y)} = x + 2$$

*Example:*

$$\frac{2x + 4}{x^2 - y^2} \cdot \frac{x - y}{4x^2 - 16} = \frac{2(x + 2)}{(x + y)(x - y)} \cdot \frac{(x - y)}{4(x + 2)(x - 2)}$$

$$= \frac{2(x + 2)(x - y)}{4(x + y)(x - y)(x + 2)(x - 2)}$$

$$= \frac{1}{2(x + y)(x - 2)}$$

*Example:*

$$\frac{3x^2 + 8x + 5}{x^2 - 1} \cdot \frac{x^2 - 2x + 1}{3x + 5} = \frac{(3x + 5)(x + 1)}{(x - 1)(x + 1)} \cdot \frac{(x - 1)(x - 1)}{3x + 5}$$

$$= \frac{(3x + 5)(x + 1)(x - 1)(x - 1)}{(x - 1)(x + 1)(3x + 5)}$$

$$= x - 1$$

*Example:*

$$\frac{2x^2y^2 - 8x^4y^4}{7xy - 14x^2y^2} \cdot \frac{xy}{2x^2y^2} = \frac{2x^2y^2(1 - 4x^2y^2)}{7xy(1 - 2xy)} \cdot \frac{xy}{2x^2y^2}$$

$$= \frac{(2x^2y^2)(1 - 2xy)(1 + 2xy)(xy)}{7xy(1 - 2xy)(2x^2y^2)}$$

$$= \frac{(2x^2y^2)(xy)(1 - 2xy)(1 + 2xy)}{7xy(2x^2y^2)(1 - 2xy)}$$

$$= \frac{1 + 2xy}{7}$$

**Exercise Set 17.4**

Multiply the fractions:

1. $\frac{r - 7}{16} \cdot \frac{6s}{rs - 7s}$
2. $\frac{x - 3}{6} \cdot \frac{36x}{2x^2 - 6x}$
3. $\frac{-b^2c}{b + 2c} \cdot \frac{3b + 6c}{c}$
4. $\frac{16y}{15x^2} \cdot \frac{3x}{8y^2}$
5. $\frac{4a^3}{18} \cdot \frac{9}{a^2 - a}$
6. $\frac{15a + 5b}{3a^2 - 2ab - b^2} \cdot \frac{a^2 - b^2}{5}$
7. $\frac{3x^2 - 12y^2}{x + 6} \cdot \frac{x^2 + 5x - 6}{3x - 6y}$
8. $\frac{2ab}{a^2 - 16b^2} \cdot \frac{a^2 + 5ab + 4b^2}{4a^2b^3}$
9. $\frac{2x + 2y}{x^2 - y^2} \cdot \frac{x + 2}{4x^2 - 16}$
10. $\frac{3x^2 + 8x + 5}{x^2 - 1} \cdot \frac{x^2 - 2x + 1}{3x + 5}$
11. $\frac{x^2 + 2x + 1}{x - 3} \cdot \frac{x^2 - 9}{x + 1}$
12. $\frac{x^2 + 5x + 4}{x - 3} \cdot \frac{3x^2 - 10x - 8}{x^2 - 16}$

## 17.5 *Dividing Algebraic Fractions*

Remember that division is the inverse of multiplication and that $8 \div 4$ is the same as $8 \times \frac{1}{4}$

$18 \div \frac{3}{4}$ is the same as $18 \times \frac{4}{3}$

$\frac{1}{5} \div \frac{4}{15}$ is the same as $\frac{1}{5} \times \frac{15}{4}$

So, when dividing algebraic fractions:

**1. Change the operation from division to multiplication.**

**2. Change the divisor to its reciprocal.**

**3. Factor numerators and denominators completely. Divide off the common factors in numerators and denominators.**

*Example:* Divide $\frac{a^2b^5}{a - b}$ by $\frac{ab}{a^2 - b^2}$

Change the operation and use the reciprocal. Factor the difference of two squares. Divide off the common factors ab and (a − b).

$$\frac{a^2b^5}{a - b} \cdot \frac{a^2 - b^2}{ab} = \frac{a^2b^5(a + b)(a - b)}{(a - b)(ab)}$$

$$= ab^4(a + b)$$

*Example:* Divide $\frac{5x}{x + 5}$ by $\frac{10x^5}{x^2 - 25}$

$$\frac{5x}{x + 5} \cdot \frac{x^2 - 25}{10x^5} = \frac{5x(x + 5)(x - 5)}{(x + 5)(10x^5)}$$

$$= \frac{x - 5}{2x^4}$$

*Example:* Divide $\frac{3x^2 - 2x - 2}{3x^2 + 10x - 12}$ by $\frac{6x - 12}{4x^2 - 64}$

$$\frac{3x^2 - 2x - 2}{3x^2 + 14x + 8} \cdot \frac{4x^2 - 64}{6x - 12} = \frac{(3x + 2)(x - 2)}{(3x + 2)(x + 4)} \frac{[4(x^2 - 16]}{[6(x - 2)]}$$

$$= \frac{4(3x + 2)(x - 2)(x + 4)(x - 4)}{6(3x + 2)(x + 4)(x - 2)}$$

$$= \frac{2(x - 4)}{3}$$

*Example:* Divide $\frac{a^2 - 4b^2}{a^2 + 3ab + 2b^2}$ by $\frac{a^2 - ab}{a^2 + ab}$

$$\frac{a^2 - 4b^2}{a^2 + 3ab + 2b^2} \cdot \frac{a^2 + ab}{a^2 - ab} = \frac{(a - 2b)(a + 2b)[a(a + b)]}{(a + 2b)(a + b)[a(a - b)]}$$

$$= \frac{a - 2b}{a - b}$$

**Exercise Set 17.5**

Divide:

1. $\dfrac{s^2r}{-r^2} \div \dfrac{sr^2}{s^2r^2}$

2. $\dfrac{-x^3}{y^2} \div \dfrac{y^6}{x^2y^2}$

3. $\dfrac{a+b}{ac} \div \dfrac{ac+bc}{ac^2}$

4. $\dfrac{70b^3}{b^2+10b+25} \div \dfrac{14b^2}{5+b}$

5. $\dfrac{x^2+7x-8}{x+3} \div \dfrac{x-1}{x^2-9}$

6. $\dfrac{6a^2}{a^2-1} \div \dfrac{3a}{a+1}$

7. $\dfrac{a^2-4}{15a^3} \div \dfrac{a-2}{5a^2}$

8. $\dfrac{y+6}{3} \div \dfrac{6y^2+36}{2y}$

9. $\dfrac{a^2+2ab+b^2}{4a^2b+2ab^2} \div \dfrac{a+b}{2a+b}$

10. $\dfrac{3x^2-7x-20}{2x^2+3x-2} \div \dfrac{2x-8}{3x+6}$

11. $\dfrac{x^2-2x+1}{6x^2-5x-6} \div \dfrac{x-1}{2x^2-5x+3}$

12. $\dfrac{x^2y^2-16}{2xy+8} \div \dfrac{xy-4}{xy+4}$

## 17.6 *Addition and Subtraction of Algebraic Fractions*

In sections 5.5 and 5.6, you learned the procedures for adding and subtracting fractions:

**1. If the denominators are alike, add or subtract the numerators and express the sum or difference in terms of the common denominator.**

*Example:* $\dfrac{3}{4} + \dfrac{5}{4} = \dfrac{8}{4}$ and $\dfrac{3}{x} - \dfrac{5}{x} = \dfrac{-2}{x}$

**2. If the denominators are not the same, find the LCM and change each fraction to an equivalent one with the LCM as denominator. Then add or subtract as indicated.**

*Example:* $\dfrac{3}{4} + \dfrac{5}{8} = \dfrac{6}{8} + \dfrac{5}{8} = \dfrac{11}{8}$

*Example:* $\dfrac{3}{y} - \dfrac{5}{2y} = \dfrac{6}{2y} - \dfrac{5}{2y} = \dfrac{1}{2y}$

*Example:* $\dfrac{4c}{ab} + \dfrac{7b}{ac} - \dfrac{3a}{bc}$

$\dfrac{4c^2}{abc} + \dfrac{7b^2}{abc} - \dfrac{3a^2}{abc}$

**In the case of algebraic fractions:**

1. **Factor the denominators and find their least common multiple.**
2. **Change each fraction to an equivalent one, having the LCM as common denominator.**
3. **Add or subtract as indicated.**

*Example:* Add: $\frac{6}{x^2} + \frac{5}{6xy}$

1. The LCM of the denominators is $6x^2y$
2. Multiply the left fraction by $\frac{6y}{6y}$ and the right fraction by $\frac{x}{x}$
3. Each fraction is written as an equivalent fraction with the denominator of $6x^2y$.

$$\frac{6}{x^2}\frac{(6y)}{(6y)} + \frac{5}{6xy}\frac{(x)}{(x)} = \frac{36y}{6x^2y} + \frac{5x}{6x^2y}$$

4. Combining: $= \frac{36y + 5x}{6x^2y}$

*Example:* Subtract $\frac{3y}{y^2 - 16} - \frac{4y}{y - 4}$

1. The LCM of the denominators is $y^2 - 16$.

$$\frac{3y}{y^2 - 16} - \frac{4y}{y - 4} = \frac{3y}{(y - 4)(y + 4)} - \frac{4y}{(y - 4)}$$

2. Multiply the right fraction by $\frac{(y + 4)}{(y + 4)}$

$$\frac{3y}{(y - 4)(y + 4)} - \frac{4y}{(y - 4)}\frac{(y + 4)}{(y + 4)} = \frac{3y - [4y(y + 4)]}{(y - 4)(y + 4)}$$

$$= \frac{3y - 4y^2 - 16y}{y^2 - 16}$$

$$= \frac{-4y^2 - 13y}{y^2 - 16}$$

*Example:* Add $\frac{4}{x+1} + \frac{x}{x^2+2x+1}$

1. The factors of $x^2 + 2x + 1$ are $(x + 1)(x + 1)$, so the LM is $(x + 1)(x + 1)$ or $(x + 1)^2$.

2. Multiply the fraction on the left by $\frac{x+1}{x+1}$

$$\frac{4}{(x+1)}\frac{(x+1)}{(x+1)} + \frac{x}{(x+1)(x+1)} = \frac{4x+4+x}{(x+1)(x+1)}$$

3. Combining: $$= \frac{5x+4}{x^2+2x+1}$$

*Example:* Combine the fractions $\frac{4t}{t+2} + \frac{8}{t^2-4} - \frac{6t^2}{t-2}$

1. $t + 2$ and $t - 2$ are factors of $t^2 - 4$, so the LCM for the three fractions is $t^2 - 4$.

2. Multiply the first fraction by $\frac{(t-2)}{(t-2)}$

3. Multiply the last fraction by $\frac{(t+2)}{(t+2)}$

4. Combine as indicated.

$$\frac{4t}{(t+2)}\frac{(t-2)}{(t-2)} + \frac{8}{t^2-4} - \frac{6t^2}{(t-2)}\frac{(t+2)}{(t+2)} =$$

$$\frac{4t^2-8t}{t^2-4} + \frac{8}{t^2-4} - \frac{6t^3+12t^2}{t^2-4} =$$

$$\frac{4t^2-8t+8-6t^3-12t^2}{t^2-4} = \frac{-6t^3-8t^2-8t+8}{t^2-4}$$

*Example:* To subtract $\frac{x-5}{x+2} - \frac{x}{x+4}$

Multiply both numerator and denominator of the left fraction by $x + 4$, and the right fraction by $x + 2$.

$$\frac{(x-5)(x+4)}{(x+2)(x+4)} - \frac{x(x+2)}{(x+4)(x+2)} = \frac{x^2-x-20}{(x+2)(x+4)} - \frac{x^2+2x}{(x+4)(x+2)}$$

$$= \frac{-3x-20}{(x+2)(x+4)}$$

**Exercise Set 17.6**

Add or subtract as indicated:

1. $\frac{7}{8} + \frac{2}{x} + \frac{1}{2x}$

2. $\frac{7}{x} - \frac{6}{3x}$

3. $\frac{10}{5x - 15} - \frac{5}{x + 3}$

4. $\frac{1}{a + 3b} + \frac{5}{4a + 6b}$

5. $\frac{1}{ax^2} - \frac{2}{bx}$

6. $\frac{10}{3x + 9} - \frac{8}{x + 3}$

7. $\frac{8}{3x + 6} + \frac{14}{x + 2}$

8. $\frac{7}{x + 2} - \frac{8}{x^2 - 4}$

9. $\frac{x + 2}{x^2 + 2x + 1} + \frac{7}{x + 1}$

10. $\frac{5}{3a + b} - \frac{a + b}{3a^2 - 2ab - b^2}$

11. $\frac{5a}{2a + 5} + \frac{2b}{3a - 2} - \frac{a + b}{6a^2 + 11a - 10}$

12. $\frac{4}{x + 3} - \frac{5}{x - 3} + \frac{7}{x^2 - 9}$

13. $\frac{4}{t^2 + 6t} - \frac{2}{3t}$

14. $\frac{x - 1}{x^2 + 4x} + \frac{x - 6}{x + 4} + \frac{3x + 7}{x + 4}$

## 17.7 *Solving Equations Involving Fractions*

Review Chapter 11 section 4 before attempting this section.

**To solve equations containing fractions:**

1. **Find the LCM of the denominators.**
2. **Multiply every term of the equation by the LCM.**
3. **Collect terms.**
4. **Solve for the variable.**

**CHECK THE SOLUTION BY SUBSTITUTING YOUR ANSWER IN THE ORIGINAL EQUATION**

*Example:* Add $\frac{8a}{3} = \frac{4a}{3} + 12$

Multiplying each term by the LCM 3:

$$3\left(\frac{8a}{3}\right) = 3\left(\frac{4a}{3}\right) + 3(12)$$
$$8a = 4a + 36$$
$$4a = 36$$
$$a = 9$$

*Check:*

$$\frac{8(9)}{3} = \frac{4(9)}{3} + 12$$
$$\frac{72}{3} = \frac{36}{3} + 12$$
$$24 = 12 + 12$$
$$24 = 24$$

*Example:* Solve for x: $\frac{4}{x} - \frac{6}{2x} = 10$

Multiply each term by the LCM, which is 2x.

$$2x \cdot \frac{4}{x} - 2x \cdot \frac{6}{2x} = 2x \cdot 10$$
$$(2)(4) - (1)(6) = (2x)(10)$$
$$8 - 6 = 20x$$
$$2 = 20x$$
$$\frac{1}{10} = x$$

*Check:*

$$\left(4 \div \frac{1}{10}\right) - \left(6 \div \frac{2}{10}\right) = 10$$
$$(4 \cdot 10) - \left(6 \cdot \frac{10}{2}\right) = 10$$
$$40 - 30 = 10$$
$$10 = 10$$

*Example:* Solve for y: $\frac{2}{y^2 - 1} + \frac{3}{y + 1} = \frac{2}{y - 1}$

$y^2 - 1 = (y + 1)(y - 1)$ so the LCM is $(y + 1)(y - 1)$.

Multiply the entire expression by the LCM:

$$(y^2 - 1)\frac{2}{y^2 - 1} + (y^2 - 1)\frac{3}{y + 1} = \frac{2}{y - 1}(y^2 - 1)$$

$$(1)(2) + (y - 1)(3) = (2)(y + 1)$$
$$2 + 3y - 3 = 2y + 2$$
$$3y - 1 = 2y + 2$$
$$3y - 2y = 2 + 1$$
$$y = 3$$

*Check:*

$$\frac{2}{9 - 1} + \frac{3}{3 + 1} = \frac{2}{3 - 1}$$

$$\frac{2}{8} + \frac{3}{4} = \frac{2}{2}$$

$$\frac{1}{4} + \frac{3}{4} = 1$$

$$1 = 1$$

**Exercise Set 17.7**

Solve for the variable:

1. $\frac{2}{x^2 - 1} + \frac{3}{x + 1} = \frac{2}{x - 1}$
2. $\frac{2}{n + 2} - \frac{3}{n - 3} = \frac{2}{n - 3}$
3. $\frac{2a^2 + a - 5}{a - 1} - \frac{4a + 5}{2} = 0$
4. $\frac{2x + 10}{3x + 9} = \frac{2x + 4}{x + 3}$
5. $\frac{6x + 5}{2x + 4} + \frac{4}{x + 2} = 2$
6. $\frac{2}{x^2 + 2x - 3} - \frac{3}{x + 3} = \frac{7}{x - 1}$
7. $\frac{1}{d + 5} = \frac{6}{d}$
8. $\frac{5 + h}{h} = 2$
9. $\frac{2y - 3}{4} - \frac{y + 4}{12} = 1$
10. $\frac{x + 2}{4} + \frac{x - 3}{8} = 4$
11. $\frac{2t}{t + 3} - \frac{8}{t^2 - 9} = 2$
12. $\frac{c - 4}{3} + \frac{c + 5}{9} = 2c$
13. $\frac{4}{d} + \frac{7}{2d} = \frac{5}{d - 3}$
14. $\frac{4t + 9}{t - 3} - \frac{7t}{2t - 6} = 2$

## CHAPTER EXERCISES

### Section 17.1

Factor the following expressions completely:

1. $3x + 6$
2. $4x^2 - 8x$
3. $x^2 + 5x + 4$
4. $28a^2b - 7b$
5. $x^2 + 3x - 4$
6. $x^4 + 7x^3 + 6x^2$
7. $x^4 - 16y^2$
8. $x^2y^2 - 100$
9. $3x^2 - 12$
10. $6t - 4t^2$
11. $81a^2 - 9b^2$
12. $t^2u^3 + 2t^3u^2$
13. $a^2 - .25$
14. $.64b^2 - .16c^2$

### Section 17.2

Factor the following expressions completely:

15. $3x^2 - 2x - 1$
16. $4x^2 - 11x - 3$
17. $2x^2 + 8x + 6$
18. $6 - 13a + 2a^2$
19. $2x^2 - 9x - 5$
20. $7x^2 - 10x - 8$
21. $4y^2 + 20y + 24$
22. $5 - 9y - 2y^2$
23. $3x^2 - 5x - 2$
24. $6 - 4x - 2x^2$
25. $4 - 11c - 3c^2$
26. $2x^2 + 9x - 5$
27. $2z^2 - 8z - 10$
28. $2t^2 + 8t + 6$
29. $12s^2 + 37s + 3$
30. $6r^2 + 29r - 5$
31. $y^2 - 4y - 77$
32. $9ac^2 - 27a^2c + 33ac$
33. $2r^2 + 5r - 3$
34. $2t^3 - 18t^2$
35. $2x^3 - 3x^2y - 2xy^2$

### Section 17.3

Reduce the algebraic fractions:

36. $\dfrac{x + y}{x^2 - y^2}$
37. $\dfrac{4a - 16b}{a^2 - 16b}$
38. $\dfrac{x^2 + 3x + 2}{x^2 - x - 2}$
39. $\dfrac{3x^2 + 4x + 1}{3x + 3}$
40. $\dfrac{5x^2 - 7x - 6}{5x + 3}$
41. $\dfrac{2x^2y^2 - 8}{4x^2y^2 - 16}$
42. $\dfrac{81a^2 - 9b^2}{27a + 9b}$
43. $\dfrac{8t^2 + 10t + 3}{8t^2 - 14t + 3}$
44. $\dfrac{a^2 + 2at + t^2}{a^2 - t^2}$

### Section 17.4

Multiply. Express the products in lowest terms.

45. $\dfrac{4}{x^2} \cdot \dfrac{6x}{12}$
46. $\dfrac{s + 2}{3t} \cdot \dfrac{9t^2}{st + 2t}$
47. $\dfrac{4z^2}{z + 2} \cdot \dfrac{z^2 + 2z}{2z}$
48. $\dfrac{2d^3 + 8d^2 + 6d}{d^3 + 4d^2 + 3d} \cdot \dfrac{d}{2}$
49. $\dfrac{2c^2 + 3c - 14}{4c^2 - 49} \cdot \dfrac{2c + 7}{c - 2}$

50. $\dfrac{y+2}{y+1} \cdot \dfrac{y^2+4y+3}{y^2-4}$

51. $\dfrac{c^3-2c^2}{4c^2-16} \cdot \dfrac{c+2}{c^3}$

52. $\dfrac{3xy}{x-1} \cdot \dfrac{x^2-1}{6x}$

53. $\dfrac{x^2y}{3x+2} \cdot \dfrac{3x^2-13x-10}{x-5}$

54. $\dfrac{t^3r^4}{t+3} \cdot \dfrac{t^2+4t+3}{t^3+t^2}$

55. $\dfrac{3r^2-17r+10}{r-5} \cdot \dfrac{r^2-25}{3r-2}$

56. $\dfrac{8y^2-32}{y+2} \cdot \dfrac{3}{16y}$

57. $\dfrac{2t+4}{2t-2} \cdot \dfrac{5t^2-5}{t^2+3t+2}$

**Section 17.5**

Divide and express answers in lowest terms.

58. $\dfrac{4x}{x+4} \div \dfrac{5x^5}{x^2+5x+4}$

59. $\dfrac{6t^3r^2}{3t+9} \div \dfrac{tr}{9t+27}$

60. $\dfrac{a^2+4ab}{a^2} \div \dfrac{a+4b}{a}$

61. $\dfrac{s^4+3s^3-4s^2}{s+4} \div \dfrac{s^2}{s-1}$

**Section 17.6**

Add or subtract as indicated.

62. $\dfrac{7}{x} + \dfrac{8}{3x}$

63. $\dfrac{x}{x^2+5x+4} + \dfrac{x+2}{x^2+x}$

64. $\dfrac{2}{3} - \dfrac{5}{12y}$

65. $\dfrac{3}{2a+b} + \dfrac{7}{4a+2b}$

66. $\dfrac{5x}{x^2+3x+2} + \dfrac{3x-6}{x^2+4x+4}$

67. $\dfrac{x-3}{x+1} - \dfrac{2x}{x-1}$

68. $\dfrac{3x}{x^2+3x-10} + \dfrac{2x}{x^2+x-6}$

69. $\dfrac{2}{3x^2-2x-1} - \dfrac{4x}{3x^2+7x+2}$

70. $\dfrac{2y}{y^2-9} + \dfrac{7y}{y+3} - \dfrac{5}{y-3}$

71. $\dfrac{5c}{ad} - \dfrac{8d}{ac} + \dfrac{6a}{cd}$

72. $\dfrac{2y}{3y^2-27} - \dfrac{3y}{3y+9}$

73. $\dfrac{8}{x^2-4y^2} - \dfrac{2}{x^2+5xy+6y^2}$

74. $\dfrac{3}{2x^2 + x - 3} + \dfrac{5}{2x^2 + 5x + 3} - \dfrac{13}{x^2 - 1}$

75. $\dfrac{2b - c}{ac} + \dfrac{c + a}{c}$

76. $\dfrac{3}{y - 3} - \dfrac{6}{y - 6}$

77. $\dfrac{x - 5}{x^2 - 2x} + \dfrac{3 - x}{x - 2} - \dfrac{5x - 8}{x^2 - 2x}$

**Section 17.7**

Solve for the variable:

78. $\dfrac{3}{t} + \dfrac{2}{3t} = \dfrac{11}{t + 4}$

79. $\dfrac{k - 12}{k + 3} - \dfrac{5k}{2k + 6} = 6$

80. $\dfrac{5r - 1}{r - 1} - \dfrac{4r}{r - 1} = -9$

81. $\dfrac{9}{z - 5} - \dfrac{12}{z} = \dfrac{2}{z}$

82. $\dfrac{2q - 7}{q + 1} - 6 = \dfrac{9q}{q + 1}$

83. $\dfrac{8}{3} + \dfrac{9}{s} = \dfrac{5}{s + 3}$

84. $\dfrac{5}{4} - \dfrac{c + 2}{c - 3} = \dfrac{-5}{c - 3}$

85. $\dfrac{16}{v} = \dfrac{v - 8}{4} - \dfrac{v}{8}$

86. $\dfrac{1}{d + 5} - \dfrac{6}{d} = 0$

87. $2c - \dfrac{c - 4}{3} = \dfrac{c - 2}{9}$

88. $\dfrac{2y - 3}{4} = 1 + \dfrac{y + 4}{12}$

89. $\dfrac{4x + 9}{x - 3} - 2 = \dfrac{7x}{2x - 6}$

90. $\dfrac{3}{x + 1} + \dfrac{2}{x^2 - 1} = \dfrac{2}{x - 1}$

91. $\dfrac{2x - 3}{4} - 1 = \dfrac{x + 4}{12}$

92. $\dfrac{2x + 10}{3x + 9} - \dfrac{2x + 4}{x + 3} = 0$

93. $\dfrac{2}{x^2 + 2x - 3} - \dfrac{7}{x - 1} = \dfrac{3}{x + 3}$

94. $\dfrac{2y}{y^2 - 2y - 3} = \dfrac{3}{y - 3}$

95. $\dfrac{6x}{x + 2} - \dfrac{4x}{2x + 4} = -8$

96. $\dfrac{x + 4}{9} - \dfrac{x + 2}{3} = 6$

97. $\dfrac{c}{5} - \dfrac{7c}{20} = 3$

## CHAPTER TEST *Form 17 A*

Factor the numerators and denominators of the following fractions and reduce each to lowest terms:

1. $\dfrac{x^2 - y^2}{2x - 2y}$

2. $\dfrac{x^2 - 6x + 5}{x - 5}$

3. $\dfrac{x^2 + 4x + 3}{x^2 - 9}$

4. $\dfrac{3x^2y^2 - 12}{6xy + 12}$

Multiply or divide as indicated:

5. $\dfrac{3x + 6}{4 - x^2} \cdot \dfrac{2 - x}{x^2 - 4}$

6. $\dfrac{2x}{3x^2 - 2x - 1} \div \dfrac{6x^2y}{x^2 + x - 2}$

7. $\dfrac{5a + 10b}{a^2b^3} \cdot \dfrac{ab}{a^2 - 4b^2}$

8. $\dfrac{xy^2 - x^2y}{6y^2} \div \dfrac{x - y}{3x}$

Add or subtract as indicated:

9. $\dfrac{4x}{x^2 + 5x + 6} + \dfrac{7x}{x + 3}$

10. $\dfrac{x^3}{x^4 - a^2} - \dfrac{x}{x^2 + a}$

11. $\dfrac{x^2 + 2}{x^2 + 5x + 4} + \dfrac{x + 1}{x + 4}$

12. $\dfrac{4x}{x + 3} - \dfrac{x}{x - 3} + \dfrac{3}{x^2 - 9}$

Solve for the variable:

13. $\dfrac{x + 4}{3} = 3 + \dfrac{2x}{5}$

14. $\dfrac{x + 1}{2} = \dfrac{x + 6}{5}$

15. $\dfrac{2x}{x + 1} = \dfrac{3x}{2x + 2} - \dfrac{5}{x + 1}$

## CUMULATIVE EXERCISES Chapters 1–17 *Set A*

Perform the indicated operations and simplify your answers:

1. $3(4x - 1) - 2(x + 3) - 9$
2. $8a\{b - a(ab + 4)\}$
3. $(x + 1)(x - 3)$
4. $(2x + 4)(2x - 4)$
5. $(x + 1)^2$
6. $(3 + y)(2 - 6y)$
7. $(a + 1)(a^2 - 3a - 1)$
8. $(2x - 6)^2$
9. $\dfrac{xy^2}{36z^3} \div \dfrac{8x^2z}{y^3}$
10. $\dfrac{4}{x} - \dfrac{3 + x}{3x}$
11. $\dfrac{x}{3} - \dfrac{y}{6} + \dfrac{3z}{9}$
12. $\dfrac{x + 2}{5} + \dfrac{3x}{15} - 5$
13. $\dfrac{11a^2}{-6b} \cdot \dfrac{-3b^2c}{33ac}$
14. $\dfrac{2a^3 - 4ab^2c - 6a}{2a}$
15. $(y - 1) + (y + 1)^2$
16. $(a + b)(b + c) - 2(a - 1)$
17. $(7a^2 + b^3 - 3) - (3a^2 + 4)$
18. $\dfrac{c}{c + 1} - \dfrac{2c}{c + 1}$
19. $(3ab^3c^2)^4$
20. $\dfrac{5z^2}{z^2 - 4} \div \dfrac{2z}{z + 2}$

Solve the following equations:

21. $x - 3(x + 5) = 1$
22. $2y - (y + 1) = 6$
23. $(t + 1)(t - 8) = 0$
24. $x^2 + 2x = 3$
25. $3x - 7 = 2.3$
26. $y^2 = 64$
27. $a^2 + 2a = 8$
28. $\dfrac{x}{8} = \dfrac{x}{4} + 1$
29. $9 - 3y + y = 1$
30. $2a - 3.2 = 4a$
31. $7y - 12 = y$
32. $3(y + 1) = 2(y - 3)$
33. $(a - 1)^2 = 4$
34. $2 - \dfrac{y}{24} = \dfrac{y}{8}$
35. $x^2 + 4x = 5$
36. $2x^2 - 5x = 12$
37. $z^2 = .36$
38. $b(b + 4) = 0$
39. $2y^2 + 4y = 0$
40. $12 = \dfrac{x + 5}{2}$
41. $\dfrac{a + 2}{4} = \dfrac{3 - a}{16}$

42. Solve for P: $PRT = I$

43. Solve for r: $A = \pi r^2$

44. Find tan 45°

45. Find sin 60°

Solve the systems of linear equations:

46. $x + 6y = 8$ and
$2x - 6y = 4$

47. $y - 3x = -7$ and
$y - 2x = -5$

48. $x = 4y$ and
$x = 2y - 1$

49. $y = x + 3$ and
$y + 2x = 9$

Graph the inequalities:

50. $2x + 1 > 6$

51. $3x + 9 \leq 0$

Graph the solution for the systems of linear equations:

52. $y \leq 5x$ and
$y \geq x + 6$

53. $y \geq x - 2$
$y + 1 \leq x$

Reduce the fractions:

54. $\dfrac{x + 2}{6x + 12}$

55. $\dfrac{x^2 + 3x + 2}{x + 1}$

56. $\dfrac{x^2 - 16}{x - 2}$

Multiply or divide as indicated. Reduce answers to lowest terms.

57. $\dfrac{x + 3}{xy} \cdot \dfrac{x^3}{x^2 - 9}$

58. $\dfrac{9 - y^2}{y + 1} \div \dfrac{3 + y}{y^2 - 1}$

59. $\dfrac{a^2 + 7a - 8}{a^2 - 64} \div \dfrac{a - 1}{a - 8}$

60. $\dfrac{2a + 3b}{a - 1} \cdot \dfrac{a^2 - 2a + 1}{4a + 6b}$

Add or subtract as indicated:

61. $\dfrac{3x + 1}{5} + \dfrac{x + 2}{10}$

62. $\dfrac{3}{4} - \dfrac{8x}{3}$

63. $\dfrac{a + b}{4} - \dfrac{3a}{b}$

64. $\dfrac{4}{x^2 - 1} + \dfrac{5}{x + 1}$

Solve the equations:

65. $\dfrac{x}{4} + \dfrac{5}{8} = 2$

66. $\dfrac{7x}{3} - 2 = \dfrac{4}{5}$

67. Find the hypotenuse of a right triangle with legs of 15 and 20 feet.

## CUMULATIVE TEST Chapters 1–17 *Form A*

1. Find the additive inverse of $(2x + 4)$.

Simplify:

2. $3x - 2[4x - 5(x - 1)]$

3. $(y - 5)^2$

4. $(6x - 2)(x + 3)$

5. $\frac{x - 4}{2} - \frac{x - 1}{4}$

6. $3a^2b^3 - 6ab^2 + 4a^2b^3 - 6ab^2 - 4$

7. $7x(4x - 1) + 3x^2 - 5$

8. $\frac{9a^3y^2}{64x} \div \frac{27a^2x}{8x^3y^2}$

9. $(x^2 - 2a)(x + 3a)$

Solve the following equations:

10. $4x - 6 + x = 9x - 30$

11. $-2(y - 3) = y - 9$

12. $4x^2 - 16x = 0$

13. $x^2 - 4x = 21$

14. $\frac{5x}{3} - 8 = \frac{4x}{3}$

15. $\frac{3x - 6}{5} = 6$

Solve the systems of linear equations:

16. $3x + y = 3$ and
    $-5x - 2y = 2$

17. $y = x + 2$ and
    $y = 3x - 4$

Solve by graphing:

18. $y \geq x - 4$ and
    $y < 3 - 2x$

19. $y = 2x + 1$ and
    $x = y - 2$

Combine and simplify:

20. $\frac{x + 3}{x^2 + x - 2} + \frac{2}{x - 1}$

21. $\frac{x^2 - x - 12}{x^2 - 9} \div \frac{2x - 8}{x - 3}$

22. Solve for x: $\frac{x}{3x + 9} = \frac{x + 4}{x + 3}$

23. Find sin 37°.

24. The area of a rectangle is 75 square feet. If the length is 10 feet more than the width, find the dimensions of the rectangle.

25. Find the area of a circular pond if the diameter is 300 meters. Leave answer to nearest square meter.

CHAPTER

# 18 Roots and Radicals

## 18.0 *Vocabulary Preview*

Define these words without using a dictionary. Formulate their meanings from past experience or knowledge.

| | | |
|---|---|---|
| **cube** | **radical** | **square** |
| **index** | **radicand** | **square root** |
| **power** | **root** | |

## 18.1 *Determining Roots*

The expressions $\sqrt{16}$, $\sqrt[3]{8}$ and $\sqrt[4]{64}$ indicate **roots** of numbers.

$\sqrt{16}$ is read "the square root of 16".

$\sqrt[3]{8}$ is read "the cube root of 8".

$\sqrt[4]{64}$ is read "the fourth root of 64".

The root symbol ($\sqrt{\ }$) is called the **radical** symbol, the number inside the radical sign is called the **radicand,** and the small number outside the radical sign is called the **index.**

The index states the root to be found. When no index is stated, square root is indicated.

**Square root is related to squaring a number.**

Since $(3)^2 = 9$ and $(-3)^2 = 9$, we can say that $\sqrt{9} = 3$ and $\sqrt{9} = -3$. The **principal root** is 3.

$\sqrt{-9}$ is not a real number, since $(3)^2$ does not equal $-9$ and $(-3)^2$ does not equal $-9$.

**Cube root is related to cubing a number.**

Since $(2)^3 = 8$ and $(-2)^3 = -8$, we can say that $\sqrt[3]{8} = 2$ and $\sqrt[3]{-8} = -2$.

Notice that it is possible to find the cube root of a negative number, since a negative number cubed remains negative.

**Finding a fourth root is related to raising a number to the fourth power.**

Since $(5)^4 = 625$ and $(-5)^4 = 625$, we can say that $\sqrt[4]{625} = 5$ and $\sqrt[4]{625} = -5$. The principal root is 5.

$\sqrt[4]{-625}$ is not a real number, since $(5)^4$ does not equal $-625$ and $(-5)^4$ does not equal $-625$.

Similarly, since $(2)^5 = 32$ and $(-2)^5 = -32$, we can say that $\sqrt[5]{32} = 2$ and $\sqrt[5]{-32} = -2$.

Since $(3)^6 = 729$ and $(-3)^6 = 729$, we can say that $\sqrt[6]{726} = 3$ and $\sqrt[6]{729} = -3$.

$\sqrt[6]{-729}$ is not a real number, since neither $(3)^6$ nor $(-3)^6$ will result in $-729$.

In general: **It is only possible to find even roots of positive numbers.** No real number can be an even root of a negative number. We say that such roots are **imaginary.**

**Exercise Set 18.1**

Find the indicated roots without using a calculator.

1. $\sqrt[3]{125}$
2. $\sqrt[3]{-125}$
3. $\sqrt[6]{64}$
4. $\sqrt[6]{-64}$
5. $\sqrt[5]{243}$
6. $\sqrt[5]{-243}$
7. $\sqrt[3]{512}$
8. $\sqrt[4]{10000}$
9. $\sqrt{.04}$
10. $\sqrt[4]{81}$
11. $\sqrt[4]{-81}$
12. $-\sqrt[4]{81}$
13. $-\sqrt[3]{216}$
14. $\sqrt[3]{-216}$
15. $\sqrt{2500}$
16. $-\sqrt{100}$
17. $\sqrt[4]{.0016}$
18. $-\sqrt{121}$
19. $\sqrt[4]{16}$
20. $\sqrt[4]{-16}$
21. $-\sqrt[3]{-27}$

## 18.2 *Roots and Powers*

Consideration of the relationship between roots and powers leads to some interesting results. With the use of calculators, we can show that:

$\sqrt{4} = (4)^{1/2}$, $\sqrt[3]{64} = (64)^{1/3}$, $\sqrt[4]{81} = (81)^{1/4}$

$\sqrt[5]{3125} = (3125)^{1/5}$, $\sqrt[6]{64} = (64)^{1/6}$, and so on.

*Example:* Use a calculator to show that $\sqrt{4} = (4)^{1/2}$.

To calculate $\sqrt{4}$:

Enter 4
Enter $\sqrt{\ }$
**The result is 2.**

To calculate $(4)^{1/2}$:

Enter 1 ÷ 2 =
Store the result (0.5) in memory.
Enter 4 and $y^x$.
Recall 0.5 from memory.
Enter =
**The result is 2.**

We have shown that $\sqrt{4} = (4)^{1/2} = 2$.

*Example:* Use your calculator to show that $\sqrt[3]{64} = (64)^{1/3}$.

Your calculator has no cube root key, however, since $(4)^3 = 64$, it follows that $\sqrt[3]{64} = 4$.

To calculate $(64)^{1/3}$:

Enter 1 ÷ 3 =
Store the result ($0.\overline{3}$) in memory.
Enter 64 and $y^x$.
Recall $0.\overline{3}$ from memory.
Enter =
**The result is 4.**

We have shown that $\sqrt[3]{64} = (64)^{1/3} = 4$.

*Example:* Show that $\sqrt[4]{81} = (81)^{1/4}$.

Your calculator has no fourth root key, however, since $(3)^4 = 81$, it follows that $\sqrt[4]{81} = 3$.

To calculate $(81)^{1/4}$:

Enter 1 ÷ 4 =
Store the result (0.25) in memory.
Enter 81 and $y^x$.
Recall 0.25 from memory.
Enter =
**The result is 3.**

We have shown that $\sqrt[4]{81} = (81)^{1/4} = 3$.

*Example:* Use your calculator to show that $\sqrt[5]{3125} = (3125)^{1/5}$.

Your calculator has no fifth root key, however, since $(5)^5 = 3125$, it follows that $\sqrt[5]{3125} = 5$.

To calculate $(3125)^{1/5}$:

Enter 1 ÷ 5 =
Store the result (0.2) in memory.
Enter 3125 and $y^x$.
Recall 0.2 from memory.
Enter =
**The result is 5.**

We have shown that $\sqrt[5]{3125} = (3125)^{1/5} = 5$.

*Exmaple:* Show that $\sqrt[6]{64} = (64)^{1/6}$.

Your calculator has no sixth root key, however, since $(2)^6 = 64$, it follows that $\sqrt[6]{64} = 2$.

To calculate $(64)^{1/6}$:

Enter 1 ÷ 6 =
Store the result in memory.
Enter 64 and $y^x$.
Recall the exponent from memory.
Enter =
**The result is 2.**

We have shown that $\sqrt[6]{64} = (64)^{1/6} = 2$.

**Exercise Set 18.2**

Change each expression from Exercise Set 18.1 to an expression involving powers. Use a calculator to evaluate each. The results must be the same.

## 18.3 *Operations Involving Roots and Radicals*

The exercises from the previous sections should have led the student to the generalization that finding the nth root of a number is the same as raising the number to the 1/n power; that is,

$$\sqrt[n]{b} = (b)^{1/n}$$

It is reasonable to assume, therefore, that the rules which apply to roots are similar to those which apply to powers, where the power is a non-integer.

*Example:* $(xy)^2 = (x)^2(y)^2$ and by the commutative property $(x)^2(y)^2 = (xy)^2$.

Similarly, $(xy)^{1/2} = (x)^{1/2}(y)^{1/2}$ and by the commutative property $(x)^{1/2}(y)^{1/2} = (xy)^{1/2}$.

Since the fractional exponents of the previous expression indicate roots, we may write:

$\sqrt{xy} = [\sqrt{x}][\sqrt{y}]$ and by the commutative property $[\sqrt{x}][\sqrt{y}] = \sqrt{xy}$.

*Example:* $\left(\frac{x}{y}\right)^3 = \frac{(x)^3}{(y)^3}$ and by the commutative property $\frac{(x)^3}{(y)^3} = \left(\frac{x}{y}\right)^3$.

Similarly, $\left(\frac{x}{y}\right)^{1/3} = \frac{(x)^{1/3}}{(y)^{1/3}}$ and by the commutative property $\frac{(x)^{1/3}}{(y)^{1/3}} = \left(\frac{x}{y}\right)^{1/3}$.

Since the fractional exponents of the previous expression indicate roots, we may write:

$\sqrt[3]{\frac{x}{y}} = \frac{[\sqrt[3]{x}]}{[\sqrt[3]{y}]}$ and by the commutative property $\frac{[\sqrt[3]{x}]}{[\sqrt[3]{y}]} = \sqrt[3]{\frac{x}{y}}$.

*Example:* $(x^4)^3 = (x)^{12}$ and by the commutative property $(x)^{12} = (x^4)^3$.

Similarly, $(x^{1/4})^{1/3} = (x)^{1/12}$ and by the commutative property $(x)^{1/12} = (x^{1/4})^{1/3}$.

Since the fractional exponents of the previous expression indicate roots, we may write:

$\sqrt[3]{\sqrt[4]{x}} = [\sqrt[12]{x}]$ and by the commutative property $[\sqrt[12]{x}] = \sqrt[3]{\sqrt[4]{x}}$.

**Summary of Rules**

$$[\sqrt[n]{a}][\sqrt[n]{b}] = \sqrt[n]{ab}$$

$$\frac{\sqrt[n]{a}}{\sqrt[n]{b}} = \sqrt[n]{\frac{a}{b}}$$

$$\sqrt[n]{\sqrt[m]{a}} = \sqrt[mn]{a}$$

*Example:* Using the concepts from the above examples, show that $[\sqrt{81}][\sqrt{4}] = \sqrt{(81)(4)}$.

a. Since $\sqrt{81} = 9$ and $\sqrt{4} = 2$, $[\sqrt{81}][\sqrt{4}] = (9)(2) = 18$.

b. Since $(81)(4) = 324$, $\sqrt{(81)(4)} = \sqrt{324} = 18$.

We have shown that $[\sqrt{81}][\sqrt{4}] = \sqrt{(81)(4)}$.

*Example:* Show that $[\sqrt[4]{16}][\sqrt[4]{625}] = \sqrt[4]{(16)(625)}$.

a. Since $\sqrt[4]{16} = 2$ and $\sqrt[4]{625} = 5$, $[\sqrt[4]{16}][\sqrt[4]{625}] = (2)(5) = 10$.

b. Since $(16)(625) = 10000$, $\sqrt[4]{(16)(625)} = \sqrt[4]{10000} = 10$.

Therefore: $[\sqrt[4]{16}][\sqrt[4]{625}] = \sqrt[4]{(16)(625)}$

*Example:* Show that $\frac{[\sqrt{100}]}{[\sqrt{25}]} = \sqrt{\left(\frac{100}{25}\right)}$.

a. Since $\sqrt{100} = 10$ and $\sqrt{25} = 5$, $\frac{[\sqrt{100}]}{[\sqrt{25}]} = \frac{(10)}{(5)} = 2$.

b. Since $\frac{100}{25} = 4$, $\sqrt{\frac{100}{25}} = \sqrt{4} = 2$.

We have shown that $\frac{[\sqrt{100}]}{[\sqrt{25}]} = \sqrt{\left(\frac{100}{25}\right)}$.

*Example:* Show that $\sqrt{\sqrt{16}} = \sqrt[4]{16}$.

a. Since $\sqrt{16} = 4$ and $\sqrt{4} = 2$, $\sqrt{\sqrt{16}} = 2$.

b. Considering $\sqrt[4]{16}$ to be $(16)^{1/4}$
   Enter $1 \div 4 =$ into your calculator.
   Store the result in memory.
   Enter 16 and $y^x$.
   Recall the exponent from memory.
   Enter =. The result is 2.

   Therefore, $\sqrt{\sqrt{16}} = \sqrt[4]{16}$.

*Example:* Show that $\sqrt[3]{\sqrt{729}} = \sqrt[6]{729}$.

a. $\sqrt{729} = 27$. We know from previous exercises that $\sqrt[3]{27} = 3$.

Therefore, $\sqrt[3]{\sqrt{729}} = \sqrt[3]{27} = 3$.

b. Considering the sixth root to be the same as the $\frac{1}{6}$ power:

Enter $1 \div 6 =$ into your calculator.
Store the result in memory.
Enter 729 and $y^x$.
Recall the exponent from memory.
Enter =. The result is 3.

Therefore, $\sqrt[3]{\sqrt{729}} = \sqrt[6]{729}$.

*Examples:* Show that the following statements are true:

a. $(\sqrt{4})^2 = 4$
b. $(\sqrt[3]{27})^3 = 27$
c. $(\sqrt[4]{64})^4 = 64$
d. $(\sqrt[5]{32})^5 = 32$.

Generalize your observations.

a. $\sqrt{4} = 2$ and $(2)^2 = 4$ so $(\sqrt{4})^2 = 4$

b. $(\sqrt[3]{27}) = 3$ and $(3)^3 = 27$ so $(\sqrt[3]{27})^3 = 27$

c. $(\sqrt[4]{64}) = 4$ and $(4)^4 = 64$ so $(\sqrt[4]{64})^4 = 64$

d. $(\sqrt[5]{32}) = 2$ and $(2)^5 = 32$ so $(\sqrt[5]{32})^5 = 32$.

The generalization which follows is:

$$(\sqrt[n]{a})^n = a$$

**NOTE:** The above expressions may also be written and evaluated as follows:

a. $\sqrt{(4)^2} = \sqrt{16} = 4$
b. $\sqrt[3]{(27)^3} = \sqrt[3]{729} = 27$
c. $\sqrt[4]{(64)^4} = \sqrt[4]{16777216} = 64$
d. $\sqrt[5]{(32)^5} = \sqrt[5]{33554432} = 32$

The generalization which follows is:

$$\sqrt[n]{(a)^n} = a$$

**Exercise Set 18.3**

Perform the indicated operations:

| | | |
|---|---|---|
| 1. $\sqrt{3} \cdot \sqrt{2}$ | 2. $\sqrt{3} \div \sqrt{2}$ | 3. $[\sqrt{3}]^2$ |
| 4. $\sqrt{(3)^2}$ | 5. $\sqrt{7} \cdot \sqrt{7}$ | 6. $\sqrt[4]{12} \div \sqrt[4]{2}$ |
| 7. $\sqrt{4} \cdot \sqrt{4}$ | 8. $(\sqrt[3]{8})^3$ | 9. $\sqrt{(3)^3}$ |
| 10. $(\sqrt[4]{6})^4$ | 11. $\sqrt[3]{3} \cdot \sqrt[3]{3}$ | 12. $\sqrt[3]{3} \div \sqrt[3]{3}$ |
| 13. $\sqrt[4]{12} \cdot \sqrt[4]{2}$ | 14. $\sqrt[4]{12} \div \sqrt[4]{2}$ | 15. $[\sqrt[4]{12}]^4$ |
| 16. $\sqrt[3]{27} \div \sqrt[3]{8}$ | 17. $\sqrt[3]{(8)^3}$ | 18. $\sqrt[3]{64} \div \sqrt[3]{8}$ |

## 18.4 *Simplifying Roots and Radicals*

It might be helpful for the student to review the laws of exponents and compare them to laws governing radicals.

| | |
|---|---|
| $(a^n)(b^n) = (ab)^n$ | $[\sqrt[n]{a}][\sqrt[n]{b}] = \sqrt[n]{ab}$ |
| $\dfrac{(a^n)}{(b^n)} = \left(\dfrac{a}{b}\right)^n$ | $\dfrac{[\sqrt[n]{a}]}{[\sqrt[n]{b}]} = \sqrt[n]{\dfrac{a}{b}}$ |
| $(a^n)(a^m) = (a)^{n+m}$ | $\sqrt[n]{a} \cdot \sqrt[m]{a} = \sqrt[nm]{a^{m+n}}$ |
| $(a^n)^m = (a)^{n \cdot m}$ | $\sqrt[n]{\sqrt[m]{a}} = \sqrt[nm]{a}$ |
| $(a^{1/n})^n = a^1 = a$ | $(\sqrt[n]{a})^n = a$ |
| $(a^n)^{1/n} = (a)^1 = a$ | $\sqrt[n]{(a)^n} = a$ |
| $(a^n)^m = (a^m)^n$ | $(\sqrt[n]{a})^m = \sqrt[n]{a^m}$ |

The following examples show how the laws governing radicals are used to simplify expressions containing radicals.

*Example:* Simplify $\sqrt{18}$.

$$\sqrt{18} = \sqrt{9 \cdot 2} = \sqrt{9} \cdot \sqrt{2} = 3\sqrt{2}$$

*Example:* Simplify $\sqrt[3]{16}$.

$$\sqrt[3]{16} = \sqrt[3]{8 \cdot 2} = \sqrt[3]{8} \cdot \sqrt[3]{2} = 2(\sqrt[3]{2})$$

*Example:* Simplify $\sqrt{\dfrac{9}{4}}$.

$$\sqrt{\frac{9}{4}} = \frac{\sqrt{9}}{\sqrt{4}} = \frac{3}{2}$$

*Example:* Simplify $\sqrt{50x^3}$.

$$\sqrt{50x^3} = \sqrt{25 \cdot 2 \cdot x^2 \cdot x}$$
$$= \sqrt{25} \cdot \sqrt{x^2} \cdot \sqrt{2} \cdot \sqrt{x} = 5x\sqrt{2x}$$

*Example:* Simplify: $\sqrt[5]{x^5y}$

$$\sqrt[5]{x^5y} = \sqrt[5]{x^5} \cdot \sqrt[5]{y} = x(\sqrt[5]{y})$$

*Example:* Simplify $\sqrt{16x^4}$

$$\sqrt{16x^4} = \sqrt{16} \cdot \sqrt{x^2} \cdot \sqrt{x^2} = 4 \cdot x \cdot x = 4x^2$$

*Example:* Simplify $\sqrt[3]{108}$.

$$\sqrt[3]{108} = \sqrt[3]{27} \cdot \sqrt[3]{4} = 3(\sqrt[3]{4})$$

*Example:* Simplify $\sqrt{x^6y^3z^2}$

$$\sqrt{x^6y^3z^2} = \sqrt{x^2} \cdot \sqrt{x^2} \cdot \sqrt{x^2} \cdot \sqrt{y^2} \cdot \sqrt{y} \cdot \sqrt{z^2}$$
$$= x \cdot x \cdot x \cdot y \cdot \sqrt{y} \cdot z$$
$$= x^3yz\sqrt{y}$$

*Example:* Simplify $\dfrac{\sqrt{144x^2}}{2}$

$$\frac{\sqrt{144x^2}}{2} = \frac{\sqrt{144} \cdot \sqrt{x^2}}{2} = \frac{12x}{2} = 6x$$

*Example:* Simplify $\sqrt{x^4y^2z}$

$$\sqrt{x^4y^2z} = \sqrt{x^4} \cdot \sqrt{y^2} \cdot \sqrt{z} = x^2y\sqrt{z}$$

**Exercise Set 18.4**

Simplify the following expressions:

1. $\sqrt{32}$
2. $\sqrt{162}$
3. $\sqrt{75}$
4. $\sqrt{x^5}$
5. $\sqrt{ab^2}$
6. $\sqrt[3]{54}$
7. $\sqrt[4]{64y^5}$
8. $\sqrt{\left(\dfrac{16}{25}\right)}$
9. $\sqrt[3]{27b^3}$
10. $\sqrt[4]{a^5}$
11. $\sqrt{121b^6}$
12. $\sqrt{\left(\dfrac{a^4}{4}\right)}$
13. $\sqrt{36a^5}$
14. $\sqrt[3]{\left(\dfrac{8}{27}\right)}$
15. $\sqrt[3]{108}$
16. $\sqrt[4]{16y^5}$
17. $\sqrt[3]{y^6}$
18. $\sqrt{48x^3}$
19. $\sqrt{12a^3b^4}$
20. $\sqrt{32x^2y^3}$
21. $\sqrt[5]{x^7y^6}$

## 18.5 *Combining Roots and Radicals*

When adding or subtracting ALGEBRAIC EXPRESSIONS, the bases must be alike. Then the coefficients are added or subtracted.

$6a + 3a = (6 + 3)a = 9a$ and $6a - 3a = (6 - 3)a = 3a$.

$6a + 3b$ cannot be added. $6a - 3b$ can't be subtracted.

**Similarly, when adding or subtracting RADICAL EXPRESSIONS, the radicands must be alike. Then the coefficients are added or subtracted.**

$6\sqrt{a} + 3\sqrt{a} = (6 + 3)\sqrt{a} = 9\sqrt{a}$. $6\sqrt{a} - 3\sqrt{a} = (6 - 3)\sqrt{a} = 3\sqrt{a}$

$6\sqrt{a} + 3\sqrt{b}$ cannot be added. $6\sqrt{a} - 3\sqrt{b}$ can't be subtracted.

*Examples:*

a. $\sqrt{7} + 6\sqrt{7} - 2\sqrt{7} = (1 + 6 - 2)(\sqrt{7}) = 5\sqrt{7}$

b. $3\sqrt{x} + 7\sqrt{x} - 4\sqrt{x} = (3 + 7 - 4)(\sqrt{x}) = 6\sqrt{x}$

c. $\sqrt{ab} + \sqrt{a} + \sqrt{b}$ cannot be simplified.

d. $13\sqrt{5} - 7\sqrt{3} + 4\sqrt{5} = (13 + 4)(\sqrt{5}) + (-7\sqrt{3}) = 17\sqrt{5} - 7\sqrt{3}$

e. $7\sqrt{2} - 9\sqrt{7} + 2\sqrt{2} + 12\sqrt{7} = (7 + 2)(\sqrt{2}) + (-9 + 12)\sqrt{7} = 9\sqrt{2} + 3\sqrt{7}$

f. $6\sqrt{ab} + 4\sqrt{a} - \sqrt{ab} = (6 - 1)(\sqrt{ab}) + 4\sqrt{a} = 5\sqrt{ab} + 4\sqrt{a}$

**Exercise Set 18.5 A**

Add or subtract as indicated:

1. $\sqrt{5} + 6\sqrt{5} - 2\sqrt{5}$
2. $\sqrt{7} - 5\sqrt{7} + 9\sqrt{7}$
3. $13\sqrt{y} + 17\sqrt{y} - 4\sqrt{y}$
4. $3\sqrt{x} - 4\sqrt{x} + \sqrt{x}$
5. $\sqrt{xy} + \sqrt{x} + \sqrt{y}$
6. $\sqrt{6} + \sqrt{5} - 8\sqrt{6}$
7. $13\sqrt{5} - 7\sqrt{3} + 4\sqrt{5}$
8. $\sqrt{3} + 4\sqrt{5} - 7\sqrt{3} - 17\sqrt{5}$
9. $7\sqrt{a} - 9\sqrt{b} + 2\sqrt{a} + 12\sqrt{b}$
10. $7\sqrt{2} - \sqrt{11} - 9\sqrt{2} + \sqrt{11}$
11. $6\sqrt{cb} + 3\sqrt{c} - \sqrt{cb}$
12. $2\sqrt{13} - 3\sqrt{3} + 4\sqrt{3}$
13. $5\sqrt{ac} + 4\sqrt{a} - 10\sqrt{ac}$
14. $\sqrt{ax} - 7\sqrt{ay} - 8\sqrt{ax}$

In some cases, the radicands must be simplified before the expressions can be added or subtracted.

*Example:*

$$\sqrt{80} - \sqrt{45} =$$
$$\sqrt{16} \cdot \sqrt{5} - \sqrt{9} \cdot \sqrt{5} =$$
$$4\sqrt{5} - 3\sqrt{5} = (4 - 3)(\sqrt{5}) = \sqrt{5}$$

*Example:*

$$4\sqrt{27} - \sqrt{75} =$$
$$4\sqrt{9}\sqrt{3} - \sqrt{25}\sqrt{3} =$$
$$4 \cdot 3 \cdot \sqrt{3} - 5\sqrt{3} = 12\sqrt{3} - 5\sqrt{3} = (12 - 5)(\sqrt{3}) = 7\sqrt{3}$$

*Example:*

$$\sqrt{180} - \sqrt{80} + \sqrt{20} =$$
$$\sqrt{36}\sqrt{5} - \sqrt{16}\sqrt{5} + \sqrt{4}\sqrt{5} =$$
$$6\sqrt{5} - 4\sqrt{5} + 2\sqrt{5} = (6 - 4 + 2)\sqrt{5} = 4\sqrt{5}$$

*Example:*

$$4\sqrt{18} - \sqrt{72} + 6\sqrt{36} =$$
$$4\sqrt{9}\sqrt{2} - \sqrt{36}\sqrt{2} + 6\sqrt{36} =$$
$$4 \cdot 3\sqrt{2} - 6\sqrt{2} + 6 \cdot 6 = 12\sqrt{2} - 6\sqrt{2} + 12 = 6\sqrt{2} + 12$$

*Example:*

$$4\sqrt{9b^2} - 6\sqrt{4c^4} + 2\sqrt{b^2} + 5\sqrt{c^4} =$$
$$4\sqrt{9}\sqrt{b^2} - 6\sqrt{4}\sqrt{c^4} + 2\sqrt{b^2} + 5\sqrt{c^4} =$$
$$(4 \cdot 3 \cdot b) - (6 \cdot 2 \cdot c^2) + 2 \cdot b + 5 \cdot c^2 =$$
$$12b - 12c^2 + 2b + 5c^2 =$$
$$12b + 2b - 12c^2 + 5c^2 = 14b - 7c^2$$

*Example:*

$$7\sqrt{x^2y} + 4\sqrt{x^2y} - 16\sqrt{x^2y^2} =$$
$$7\sqrt{x^2}\sqrt{y} + 4\sqrt{x^2}\sqrt{y} - 16\sqrt{x^2}\sqrt{y^2} =$$
$$7x\sqrt{y} + 4x\sqrt{y} - 16xy = 11x\sqrt{y} - 16xy$$

*Example:* $3\sqrt{a^2b^4} - 7\sqrt{a^4b^4} + 4\sqrt{a^2b^2} =$

$3ab^2 - 7a^2b^2 + 4ab.$

This cannot be combined further.

**Exercise Set 18.5 B**

Simplify and combine if possible.

1. $6\sqrt{18} - \sqrt{2} + \sqrt{75}$
2. $\sqrt{36x^2} + \sqrt{x^3}$
3. $3\sqrt{a^3} + 17a\sqrt{a} - 2\sqrt{a^3}$
4. $\sqrt{45} + \sqrt{80} - \sqrt{20}$
5. $6\sqrt{98} - \sqrt{72}$
6. $\sqrt{\left(\frac{36a^4}{16b^6}\right)}$
7. $5\sqrt{28} + 2\sqrt{63} + \sqrt{7}$
8. $\sqrt{\left(\frac{81a^6}{4b^4}\right)}$
9. $3\sqrt{6} + 2\sqrt{8} - \sqrt{6}$
10. $y\sqrt{27y^3} - \sqrt{48y^5}$
11. $5\sqrt{24} - 8\sqrt{54} + 6\sqrt{96}$
12. $\sqrt{5t^2} + \sqrt{20t^2}$
13. $5\sqrt{5} - 8\sqrt{45} + 3\sqrt{125}$
14. $\sqrt{12x^3} + \sqrt{3x} - \sqrt{27x}$
15. $\sqrt[5]{2} + \sqrt[5]{64}$
16. $7\sqrt{x^3y^3} - 7xy\sqrt{xy}$

## 18.6 *Multiplication of Radicals*

The following sections on multiplication and division of radicals expand on the concepts already developed in sections 18.3, 18.4, and 18.5

**The rules for multiplying radicals are similar to the rules for multiplying algebraic expressions. The laws of signs are the same.**

*Example:* $(2\sqrt{3})(6\sqrt{7})$.

By the commutative property:
$(2\sqrt{3})(6\sqrt{7}) = 2 \cdot 6 \cdot \sqrt{3} \cdot \sqrt{7} = 12\sqrt{21}$.

Since 21 contains no factor which is a square, the expression cannot be simplified further.

*Exmaple:* Multiply $6\sqrt{2}(-5\sqrt{8})$.

$$\begin{aligned} 6\sqrt{2}(-5\sqrt{8}) &= (6)(-5)(\sqrt{2})(\sqrt{8}) \\ &= -30(\sqrt{16}) \\ &= -30(4) \\ &= -120 \end{aligned}$$

*Example:* Multiply $5\sqrt{7}(2\sqrt{3} - 3\sqrt{5})$.

By the distributive property:

$$\begin{aligned} 5\sqrt{7}(2\sqrt{3} - 3\sqrt{5}) &= 5\sqrt{7}(2\sqrt{3}) - 5\sqrt{7}(3\sqrt{5}) \\ &= 5 \cdot 2 \cdot \sqrt{7} \cdot \sqrt{3} - 5 \cdot 3 \cdot \sqrt{7} \cdot \sqrt{5} \\ &= 10\sqrt{21} - 15\sqrt{35} \end{aligned}$$

*Example:* Multiply and simplify $(2 - \sqrt{x})(3 + \sqrt{6x})$.

By the distributive property:

$$\begin{aligned} (2 - \sqrt{x})(3 + \sqrt{6x}) &= 2(3 + \sqrt{6x}) + (-\sqrt{x})(3 + \sqrt{6x}) \\ &= (2 \cdot 3) + (2 \cdot \sqrt{6x}) + (-\sqrt{x} \cdot 3) + (-\sqrt{x} \cdot \sqrt{6x}) \\ &= (6 + 2\sqrt{6x} - 3\sqrt{x} - \sqrt{6} \cdot \sqrt{x^2} \\ &= 6 + 2\sqrt{6x} - 3\sqrt{x} - x\sqrt{6} \end{aligned}$$

*Example:* Multiply and simplify $(\sqrt{5} + 2\sqrt{3})^2$.

$$\begin{aligned} (\sqrt{5} + 2\sqrt{3})^2 &= [(\sqrt{5} + 2\sqrt{3})][(\sqrt{5} + 2\sqrt{3})] \\ &= \sqrt{5}[\sqrt{5} + 2\sqrt{3}] + 2\sqrt{3}[\sqrt{5} + 2\sqrt{3}] \\ &= \sqrt{25} + 2\sqrt{15} + 2\sqrt{15} + 4\sqrt{9} \\ &= 5 + 4\sqrt{15} + 4 \cdot 3 \\ &= 5 + 4\sqrt{15} + 12 \\ &= 4\sqrt{15} + 17 \end{aligned}$$

*Example:* Multiply and simplify $(\sqrt{5} + \sqrt{3})(\sqrt{5} - \sqrt{3})$.

$$\begin{aligned}(\sqrt{5} + \sqrt{3})(\sqrt{5} - \sqrt{3}) &= \sqrt{5}(\sqrt{5} - \sqrt{3}) + \sqrt{3}(\sqrt{5} - \sqrt{3}) \\ &= \sqrt{25} - \sqrt{15} + \sqrt{15} - \sqrt{9} \\ &= 5 + 0 - 3 \\ &= 2\end{aligned}$$

*Example:* Multiply and simplify $(2\sqrt{3} + \sqrt{7})^2$.

$$\begin{aligned}(2\sqrt{3} + \sqrt{7})^2 &= [(2\sqrt{3} + \sqrt{7})][2\sqrt{3} + \sqrt{7}] \\ &= 2\sqrt{3}(2\sqrt{3} + \sqrt{7}) + \sqrt{7}(2\sqrt{3} + \sqrt{7}) \\ &= 4\sqrt{9} + 2\sqrt{21} + 2\sqrt{21} + \sqrt{49} \\ &= 4 \cdot 3 + 2\sqrt{21} + 2\sqrt{21} + 7 \\ &= 12 + 4\sqrt{21} + 7 \\ &= 19 + 4\sqrt{21}\end{aligned}$$

**Exercise Set 18.6**

Multiply and simplify:

1. $2\sqrt{5}(\sqrt{5} + 3\sqrt{6})$
2. $\sqrt{x}(3\sqrt{x} + \sqrt{y})$
3. $4\sqrt{2}(5\sqrt{3} - 1)$
4. $2\sqrt{x}(\sqrt{x} + 3\sqrt{y})$
5. $x\sqrt{5}(\sqrt{5} + x)$
6. $3\sqrt{5}(5\sqrt{5} - 2\sqrt{3})$
7. $(2\sqrt{b} - \sqrt{3})(b + 5\sqrt{3})$
8. $(\sqrt{t} + 2)(\sqrt{t} - 2)$
9. $(3\sqrt{s} - \sqrt{3})(\sqrt{s} - 2\sqrt{3})$
10. $(5\sqrt{b} - \sqrt{c})(\sqrt{b} + 2\sqrt{c})$
11. $(\sqrt{5} + 5)^2$
12. $(2\sqrt{a} + \sqrt{b})(2\sqrt{a} - \sqrt{b})$
13. $(\sqrt{a} - \sqrt{b})^2$
14. $(\sqrt{x} + \sqrt{y})^2$
15. $(6\sqrt{x} + 2)(6\sqrt{x} - 2)$
16. $(2\sqrt{y} + 6)(6\sqrt{y} + 2)$

## 18.7 *Division of Radicals*

Dividing radicals such as $\dfrac{\sqrt{8x^6y^5}}{\sqrt{2x^2y}}$ involves the

property of radicals which says that $\dfrac{\sqrt{a}}{\sqrt{b}} = \sqrt{\dfrac{a}{b}}$ and the process of simplification.

*Example:* $\dfrac{\sqrt{8x^6y^5}}{\sqrt{2x^2y}} = \sqrt{\dfrac{8x^6y^5}{2x^2y}} = \sqrt{4x^4y^4} = 2x^2y^2$

*Example:* To divide $\sqrt{9x^3y^5}$ by $\sqrt{3xy^2}$ write the expression as a single square root:

$$\frac{\sqrt{9x^3y^5}}{\sqrt{3xy^2}} = \sqrt{\frac{9x^3y^5}{3xy^2}} = \sqrt{3x^2y^3} = xy\sqrt{3y}$$

*Example:* To divide $\sqrt{3xy^2}$ by $\sqrt{9x^3y^5}$, write the expression as a single square root:

$$\frac{\sqrt{3xy^2}}{\sqrt{9x^3y^5}} = \sqrt{\frac{3xy^2}{9x^3y^5}} = \sqrt{\frac{1}{3x^2y^3}} = \frac{1}{xy\sqrt{3y}}$$

If an expression has a radical in the denominator, it is usual to **rationalize** it; that is, **to write the expression as an equivalent one WITHOUT a radical in the denominator.**

When rationalizing denominators, it is important to remember two properties: $\sqrt{a} \cdot \sqrt{a} = a$ and $\frac{\sqrt{a}}{\sqrt{a}} = 1$.

The latter property indicates that **multiplying the numerator and denominator of a fraction by the same number is the same as multiplying by 1, which is the identity element for multiplication.**

*Example:* To rationalize the denominator of the solution in the previous example, multiply numerator and denominator by $\sqrt{3}$:

$$\frac{1}{xy\sqrt{3}} = \frac{(1)(\sqrt{3})}{(xy\sqrt{3})(\sqrt{3})} = \frac{\sqrt{3}}{xy\sqrt{9}} = \frac{\sqrt{3}}{3xy}$$

*Example:* Rationalize the denominator of the fraction $\frac{5}{\sqrt{7}}$.

$$\frac{5}{\sqrt{7}} \cdot \frac{\sqrt{7}}{\sqrt{7}} = \frac{5\sqrt{7}}{\sqrt{49}} = \frac{5\sqrt{7}}{7}$$

*Example:* Rationalize the denominator of $\frac{\sqrt{6}}{\sqrt{3}}$.

Multiplying numerator and denominator by $\sqrt{3}$ gives:

$$\frac{\sqrt{6}}{\sqrt{3}} \cdot \frac{\sqrt{3}}{\sqrt{3}} = \frac{\sqrt{18}}{3} = \frac{3\sqrt{2}}{3} = \sqrt{2}$$

The same results are obtained with simple division: $\frac{\sqrt{6}}{\sqrt{3}} = \sqrt{\frac{6}{3}} = \sqrt{2}$

*Example:* Rationalize the denominator of $\frac{\sqrt{ab}}{\sqrt{a}}$

Multiplying numerator and denominator by $\sqrt{a}$ gives: $\frac{\sqrt{ab}}{\sqrt{a}} \cdot \frac{\sqrt{a}}{\sqrt{a}} = \frac{\sqrt{a^2b}}{a}$

$$= \frac{a\sqrt{b}}{a} = \sqrt{b}$$

Simple division gives the same result: $\frac{\sqrt{ab}}{\sqrt{a}} = \sqrt{\frac{ab}{a}} = \sqrt{b}$

*Example:* Rationalize the denominator of $\frac{\sqrt{a+b}}{\sqrt{a}}$.

Multiplying numerator and denominator by $\sqrt{a}$ gives: $\frac{\sqrt{a+b}}{\sqrt{a}} \cdot \frac{\sqrt{a}}{\sqrt{a}} = \frac{\sqrt{a^2+ab}}{a}$

**This result cannot be simplified.**

*Example:* Rationalize the denominator of $\frac{\sqrt{32x^3y^7z^8}}{\sqrt{4x^6y^8z^4}}$

Multiplying by $\frac{\sqrt{xy}}{\sqrt{xy}}$ gives: $\frac{2z^2\sqrt{2}\,\sqrt{xy}}{x\sqrt{xy}\sqrt{xy}} = \frac{2z^2\sqrt{2xy}}{x\sqrt{x^2y^2}} = \frac{2z^2\sqrt{2xy}}{x^2y}$

In cases where a denominator contains more than one term, it is necessary to multiply by the **conjugate** of the denominator in order to eliminate the radical.

The conjugate of $\sqrt{a} + \sqrt{b}$ is $\sqrt{a} - \sqrt{b}$ since
$(\sqrt{a} + \sqrt{b})(\sqrt{a} - \sqrt{b}) = \sqrt{a^2} + \sqrt{ab} - \sqrt{ab} - \sqrt{b^2} = a - b$

The conjugate of $3 - \sqrt{x}$ is $3 + \sqrt{x}$ since
$(3 - \sqrt{x})(3 + \sqrt{x}) = 9 + 3\sqrt{x} - 3\sqrt{x} - \sqrt{x^2} = 9 - x$.

| **Expression** | **Conjugate** | **Expression** | **Conjugate** |
|---|---|---|---|
| $\sqrt{a} - \sqrt{b}$ | $\sqrt{a} + \sqrt{b}$ | $\sqrt{u} + \sqrt{u+v}$ | $\sqrt{u} - \sqrt{u+v}$ |
| $33 + \sqrt{y}$ | $33 - \sqrt{y}$ | $\sqrt{cd} - \sqrt{e}$ | $\sqrt{cd} + \sqrt{e}$ |
| $\sqrt{7} + \sqrt{xy}$ | $\sqrt{7} - \sqrt{xy}$ | $3\sqrt{x} + 2\sqrt{y}$ | $3\sqrt{x} - 2\sqrt{y}$ |
| $6\sqrt{y} + 4$ | $6\sqrt{y} - 4$ | $5\sqrt{a} - 3\sqrt{c}$ | $5\sqrt{a} + 3\sqrt{c}$ |

*Example:* Rationalize the denominator of the fraction $\dfrac{6x}{\sqrt{x} + 5}$

Multiplying by the conjugate:

$$\frac{6x}{\sqrt{x}+5} = \frac{6x(\sqrt{x}-5)}{(\sqrt{x}+5)(\sqrt{x}-5)} = \frac{6x\sqrt{x}-30x}{\sqrt{x}\sqrt{x}+5\sqrt{x}-5\sqrt{x}-25}$$
$$= \frac{6x\sqrt{x}-30x}{x^2-25}$$

*Example:* Divide and simplify: $\dfrac{3x+5}{\sqrt{x}-\sqrt{5}}$

$$\frac{3x+5}{\sqrt{x}-\sqrt{5}} = \frac{(3x+5)(\sqrt{x}+\sqrt{y})}{(\sqrt{x}-\sqrt{y})(\sqrt{x}+\sqrt{y})} = \frac{(3x+5)(\sqrt{x}+\sqrt{y})}{\sqrt{x^2}+\sqrt{xy}-\sqrt{xy}+\sqrt{y^2}}$$
$$= \frac{(3x+5)(\sqrt{x}+\sqrt{y})}{x+y}$$

*Example:* Divide and simplify: $\dfrac{\sqrt{x}+\sqrt{3}}{\sqrt{x}-\sqrt{3}}$

$$\frac{\sqrt{x}+\sqrt{3}}{\sqrt{x}-\sqrt{3}} = \frac{(\sqrt{x}+\sqrt{3})(\sqrt{x}+\sqrt{3})}{(\sqrt{x}-\sqrt{3})(\sqrt{x}+\sqrt{3})} = \frac{\sqrt{x^2}+\sqrt{3x}+\sqrt{3x}-\sqrt{9}}{\sqrt{x^2}+\sqrt{3x}-\sqrt{3x}-\sqrt{9}}$$
$$= \frac{x+2\sqrt{3x}-3}{x-3}$$

*Example:* Divide and simplify: $\dfrac{3-5\sqrt{x}}{4-3\sqrt{x}}$

$$\frac{3-5\sqrt{x}}{4-3\sqrt{x}} = \frac{(3-5\sqrt{x})\,(4+3\sqrt{x})}{(4-3\sqrt{x})\,(4+3\sqrt{x})} = \frac{12+9\sqrt{x}-20\sqrt{x}-15\sqrt{x^2}}{16+12\sqrt{x}-12\sqrt{x}-9\sqrt{x^2}}$$
$$= \frac{12-11\sqrt{x}-15x}{16-9x}$$

**An expression containing a square root is in simplest form:**

**1. if the radicand contains no factor (other than 1) which is a perfect square**

**2. if the radicand does not contain a fraction**

**3. if a denominator does not contain a square root.**

**Exercise Set 18.7**

Rationalize denominators, divide and simplify the following:

1. $\frac{\sqrt{36}}{\sqrt{4}}$

2. $\frac{\sqrt{25}}{\sqrt{49}}$

3. $\frac{1}{\sqrt{100}}$

4. $\frac{4}{\sqrt{6}}$

5. $\frac{7a}{\sqrt{a}}$

6. $\frac{\sqrt{5}}{\sqrt{20b}}$

7. $\frac{a}{\sqrt{7}}$

8. $\frac{\sqrt{x^5}}{\sqrt{4x}}$

9. $\frac{3x}{\sqrt{x}}$

10. $\frac{5\sqrt{3}}{\sqrt{2}}$

11. $\frac{7\sqrt{xy}}{\sqrt{z}}$

12. $\frac{\sqrt{6} + \sqrt{2}}{\sqrt{4}}$

13. $\frac{\sqrt{24x^5}}{\sqrt{x^2}}$

14. $\frac{\sqrt{75a^5b}}{\sqrt{3b}}$

15. $\frac{\sqrt[4]{16x^4}}{\sqrt[4]{81x^8}}$

16. $\frac{\sqrt{x} + 1}{\sqrt{x}}$

17. $\frac{1 - \sqrt{xy}}{\sqrt{y}}$

18. $\frac{4x}{\sqrt{2}}$

19. $\frac{3}{\sqrt{3} - x}$

20. $\frac{2y}{1 - \sqrt{y}}$

21. $\frac{x}{\sqrt{x} + 1}$

22. $\frac{2}{\sqrt{y} + 3}$

23. $\frac{\sqrt{5}}{\sqrt{5} + \sqrt{x}}$

24. $\frac{\sqrt{5} - \sqrt{x}}{\sqrt{5} + x}$

25. $\frac{\sqrt{y}}{\sqrt{y} + 3}$

26. $\frac{\sqrt{3y}}{\sqrt{3} + \sqrt{y}}$

27. $\frac{\sqrt{3}}{\sqrt{3} + y}$

28. $\frac{\sqrt{x} + \sqrt{2x}}{\sqrt{x} - \sqrt{2x}}$

29. $\frac{3xy}{\sqrt{x} + \sqrt{y}}$

30. $\frac{3xy}{\sqrt{x} + y}$

## CHAPTER EXERCISES

### Section 18.1

Find indicated roots without the aid of a calculator. Write NO SOLUTION if such is the case.

1. $\sqrt{144}$
2. $-\sqrt{144}$
3. $\sqrt{-144}$
4. $\sqrt[3]{64}$
5. $\sqrt[3]{-64}$
6. $-\sqrt[3]{64}$
7. $\sqrt[4]{81}$
8. $\sqrt[4]{-81}$
9. $-\sqrt[4]{81}$
10. $\sqrt{0.64}$
11. $\sqrt[3]{0.008}$
12. $\sqrt{3600}$

### Section 18.2

Use a calculator to find the roots. (Hint: Use the property $\sqrt[n]{a} = (a)^{1/n}$.)

13. $\sqrt[3]{125}$
14. $\sqrt{50625}$
15. $\sqrt[3]{-216}$
16. $\sqrt{1296}$
17. $\sqrt[4]{1296}$
18. $\sqrt[3]{91125}$
19. $\sqrt{529}$
20. $\sqrt{2025}$
21. $\sqrt[4]{4096}$
22. $\sqrt{-10000}$
23. $\sqrt[5]{16807}$
24. $\sqrt[4]{0.0625}$

### Section 18.3

Perform the indicated operations. Simplify all answers.

25. $\sqrt{3} \cdot \sqrt{7}$
26. $\sqrt{9} \cdot \sqrt{2}$
27. $\sqrt[3]{8} \cdot \sqrt[3]{27}$
28. $\sqrt{x} \cdot \sqrt{y}$
29. $\sqrt{7} \cdot \sqrt{10}$
30. $\sqrt[3]{x^2} \cdot \sqrt[3]{y}$
31. $(\sqrt{3})^2$
32. $(\sqrt[3]{7})^3$
33. $\sqrt[4]{a} \cdot \sqrt[3]{a}$
34. $\sqrt{b} \cdot \sqrt{c}$
35. $\dfrac{\sqrt{18}}{\sqrt{2}}$
36. $\dfrac{\sqrt[3]{8}}{2}$
37. $\dfrac{\sqrt{39}}{\sqrt{3}}$
38. $\dfrac{\sqrt{x^3y^2}}{\sqrt{xy}}$
39. $\dfrac{\sqrt[4]{16}}{\sqrt[4]{4}}$
40. $(\sqrt{xyz})^2$
41. $\dfrac{\sqrt{xyz}}{\sqrt{xy}}$
42. $\dfrac{[\sqrt{5} \cdot \sqrt{10}]}{\sqrt{2}}$

### Section 18.4

Simplify:

43. $\sqrt{72}$
44. $\sqrt[3]{81}$
45. $\sqrt{108}$
46. $\sqrt[3]{x^4y^3}$
47. $\sqrt{18x^3y^7}$
48. $\sqrt{27s^4t^5}$
49. $\sqrt[3]{27a^4y^6}$
50. $\sqrt{0.04z^{12}}$
51. $\sqrt{144c^5d^3}$
52. $\sqrt[4]{16s^4t^8}$
53. $\sqrt{32a^8b^9}$
54. $\sqrt{x^5y^6z^2}$

**Section 18.5**

Combine by addition or subtraction.

55. $3\sqrt{8} + 5\sqrt{8}$

56. $4\sqrt{7} - 8\sqrt{7}$

57. $-6\sqrt{x} + 4\sqrt{x}$

58. $7\sqrt{a^5} + 6a^2\sqrt{a}$

59. $x^2\sqrt{7} - \sqrt{7x^4}$

60. $\sqrt[3]{16} + 7$

61. $\sqrt{x^2y^4} + xy + \sqrt[3]{x^3y}$

62. $3\sqrt{a^3} - 6a\sqrt{a}$

63. $\sqrt[3]{27m^3} + \sqrt{9m^2}$

**Section 18.6**

Multiply and simplify:

64. $\sqrt{14} \cdot \sqrt{3}$

65. $\sqrt{8} \cdot \sqrt{2}$

66. $\sqrt[3]{8} \cdot \sqrt[3]{27}$

67. $3(\sqrt{x} - 5)$

68. $x(\sqrt{x} + y)$

69. $2\sqrt{5}(\sqrt{5} + 1)$

70. $(\sqrt{x} + \sqrt{y})^2$

71. $3\sqrt{a}(b + 3\sqrt{a})$

72. $(\sqrt{x} - \sqrt{y})(\sqrt{x} + \sqrt{y})$

**Section 18.7**

Rationalize denominators, divide and simplify:

73. $\dfrac{\sqrt{7xy}}{\sqrt{x}}$

74. $\dfrac{\sqrt[3]{8x^4y^3}}{\sqrt[3]{x^2}}$

75. $\dfrac{3x}{\sqrt{y}}$

76. $\dfrac{4x}{\sqrt{x + 2}}$

77. $\dfrac{4x}{\sqrt{x} + 2}$

78. $\dfrac{4x}{\sqrt{x} + \sqrt{2}}$

79. $\dfrac{x + 5}{\sqrt{x} + 5}$

80. $\dfrac{3x - 1}{3\sqrt{x} + 1}$

81. $\dfrac{4x + 2}{\sqrt{2} + 4x}$

## CHAPTER TEST *Form 18A*

Use a calculator to find the indicated root, if it exists:

1. $\sqrt{676}$
2. $-\sqrt{225}$
3. $\sqrt{-100}$
4. $\sqrt{0.64}$
5. $-\sqrt{-400}$
6. $-\sqrt{196}$

Use a calculator and the fact that $\sqrt[n]{a} = (a)^{1/n}$ to find the indicated roots:

7. $\sqrt[3]{125}$
8. $\sqrt[4]{256}$
9. $\sqrt[3]{1728}$
10. $\sqrt[5]{32}$
11. $\sqrt[4]{81}$
12. $\sqrt[5]{-3125}$

Perform the indicated operations. Simplify all answers.

13. $3\sqrt{5} + 4\sqrt{7} - 8\sqrt{5}$
14. $a^2\sqrt{a} - \sqrt{a^5}$
15. $(\sqrt{5})^2 + \sqrt[3]{8}$
16. $x\sqrt{y^2} + x\sqrt{y^3} - 3\sqrt{x^2y^3}$
17. $\sqrt{\dfrac{8x^3y^5}{2xy^3}}$
18. $\dfrac{\sqrt{90}}{\sqrt{5}}$
19. $\sqrt{5}(\sqrt{5} + \sqrt{20})$
20. $\sqrt{x}(\sqrt{x} + 5)$
21. $(\sqrt{a} + b)^2$
22. $(\sqrt{a} + 5)(\sqrt{a} - 5)$
23. $\dfrac{3xy}{\sqrt{xy}}$
24. $\dfrac{7}{\sqrt{x} + 7}$
25. $\dfrac{(3 - x)}{\sqrt{3} - x}$

## CUMULATIVE EXERCISES Chapters 1–18 *Set A*

Perform the indicated operations. Simplify answers.

1. $2(3x + 5) - 5(x - 1) + 3$
2. $3b\{c - a(bc + 2)\}$
3. $(y + 2)(y - 5)$
4. $(3p + 1)(3p - 1)$
5. $(s + t)^2$
6. $(2 + t)(3 - 3t)$
7. $(b - 1)(b^2 + 2b + 3)$
8. $(3x - 1)^2$
9. $\frac{s^2t^3}{12v^3} \div \frac{4s^2t^4}{3v^6}$
10. $\frac{3}{6} - \frac{2 + y}{2y}$
11. $\frac{s}{2} - \frac{t}{6} + \frac{3s}{4}$
12. $\frac{x + 1}{10} - \frac{3x - 2}{5}$
13. $\frac{5b^2}{3a} \cdot \frac{-6a^2c}{2bc}$
14. $\frac{6a^2b^2c - 3bc^3 + 9}{3bc}$
15. $(y - 1)^2 + y(y + 1)$
16. $(x + y)(y + z) - 2(x - z)$
17. $(6b^2 - 4c^3d) + (4b^2 - 3c^3d)$
18. $(3xy^2z^3)^3$
19. $\frac{x}{x - 1} + \frac{3x}{x + 1}$
20. $\frac{7x^3}{x^2 - 16} \div \frac{14x^5}{x + 4}$

Solve the following equations:

21. $x + 2(3 - x) = 6$
22. $3x - (x + 1) = 7$
23. $(c + 2)(c - 3) = 0$
24. $x^2 + 2x = 3$
25. $2x + 3.5 = 1$
26. $a^2 = 36$
27. $b^2 - 2b = 8$
28. $\frac{x}{3} = \frac{x - 2}{6}$
29. $2 + 3y = y$
30. $3a - 3.5 = 3.1$
31. $6y + 4 = 2y$
32. $2(x - 1) = 3(x + 5)$
33. $(x + 1)^2 = 4$
34. $6 - \frac{2y}{3} = \frac{y}{2}$
35. $x^2 - 4x = 5$
36. $3x^2 - 2x - 1 = 0$
37. $c^2 = .81$
38. $z(z - 1) = 0$
39. $3x^2 + 6x = 0$
40. $\frac{6}{5} = \frac{x + 5}{3}$

41. Solve for r: $V = \pi r^2$

42. Solve for t: $d = st$

43. Find tan 30°

44. Solve the system of linear equations: $x = 3y$
$y = x + 4$

45. Solve for x and y: $2x - y = 5$
$3x - 2y = 17$

46. Graph the inequality: $3x - 1 \geq 2$

47. Graph the solution for the system of linear inequalities:
$y \leq 3$
$2y \geq 4x + 2$

Perform the indicated operations. Reduce fractions to lowest terms.

48. $\frac{6x}{6x + 12} \cdot \frac{x + 2}{x^5}$

49. $\frac{a^2 - 2a + 1}{a} \div \frac{a - 1}{a^4}$

50. $\frac{3x + 5}{3} + \frac{5x - 3}{12}$

51. $\frac{b}{a + b} - \frac{a}{a - b}$

Solve for the variable:

52. $\frac{x + 5}{2} = 4$

53. $\frac{x + 1}{4} = \frac{x + 6}{2}$

54. $\frac{6x}{4} - 7 = \frac{5}{4}$

55. Find the side b of a right triangle if the hypotenuse is 100 units long and the other side is 60 units long.

Perform the indicated operations. Simplify all answers. Rationalize all denominators.

56. $3\sqrt{a} + 7\sqrt{b} - 4\sqrt{a}$

57. $(7\sqrt{c})(\sqrt{a} - \sqrt{c})$

58. $\sqrt{\frac{18x^2y^5}{4xy}}$

59. $\sqrt{\frac{4a^5}{b^4}}$

60. $\frac{3}{\sqrt{x - 1}}$

61. $\frac{4\sqrt{x}}{\sqrt{y}}$

62. $\sqrt{a^3b^2} + 3\sqrt{a} - 7a\sqrt{a}$

63. $(\sqrt{a} - \sqrt{b})^2$

## CUMULATIVE TEST Chapters 1–18 *Form A*

1. Simplify: $-7[x + (y - 8) - 3(x + y)]$
2. Multiply: $(3y + 4)(3y - 4)$
3. Simplify: $(x + 3y)^2 - (x^2 + 9y^2)$
4. Divide: $\frac{2(x + 5)}{x^2y^3} \div \frac{x^2 - 25}{4xy^5}$
5. Combine: $\frac{3x}{7} - \frac{4x}{14} + \frac{2xy}{28}$
6. Simplify: $(4x^2y^5z^7)^3$
7. Simplify: $(3x^2y^3 - 6xy^2 - 5xy) + (7xy^2 - 2xy + 3)$

Solve for the variable:

8. $3x - 2 = 7$
9. $2x - (x + 5) = 2$
10. $(4 - y)(4 + y) = 0$
11. $x^2 + 4x - 5 = 0$
12. $\frac{x + 2}{5} = 3$
13. $\frac{4}{x + 2} - \frac{3x}{x + 2} = 2$
14. $2x - (x + 1) = 15$
15. Solve for h: $P = 2b + 2h$
16. Find the measure of a third angle of a triangle if the other two measure 42° and 36°.
17. In right triangle ABC, the hypotenuse c is 16 units and side b is 7 units. Find side a to the nearest hundredth of a unit.
18. Graph the system and label the point of intersection:
    $3x - y = 2$
    $2x + y = 3$
19. Use the substitution method to find x:
    $2x - y = 7$
    $y = -1$
20. Graph the inequality: $y \leq 3x + 1$
21. Graph the solution for the system:
    $y \geq x + 1$
    $y \leq 3x - 2$

Perform the indicated operations.

22. $\frac{x^2 + 3x - 4}{x^2 - 4} \cdot \frac{x + 2}{x - 1}$
23. $\frac{x + 2}{3x} + \frac{x - 1}{x} - 4$
24. Solve for x: $\frac{x - 1}{3} - \frac{x + 4}{6} = 2$
25. Simplify: $7\sqrt{a^3} - 8\sqrt{a} + 3a\sqrt{a}$
26. Multiply: $(3\sqrt{5} + 2)(3\sqrt{5} - 2)$
27. Divide: $\frac{\sqrt{3x}}{\sqrt{x + 2}}$

# Additional Chapter Tests

## CHAPTER TEST *Form 1B*

Use your calculator to perform the indicated operations:

1. $175 + 365 + 975.3$
2. $144 - 762$
3. $975 \div 85$
4. $51 \times (38 - 19)$
5. $\sqrt{25} \div 5$
6. $17^2 + 34$
7. $\sqrt{49} - \sqrt{9} + 61$
8. $(-123)^2$
9. $-18 \times 5$
10. $175.3 \div 24.7$
11. $-19.1 \times -2.5 \times -.175$
12. $-82 \div -4$
13. $(1.8)^2$
14. $7 \div (-5)$

Change the percents to decimals, and decimals to percents:

15. .75
16. 2.5
17. 33%
18. .3%

Solve the following problems. Check your answers when possible.

19. Maurice bought a new coat originally priced at $225. If the store gave a 20% discount, how much did Maurice pay for the coat?
20. Jill made the following purchases: 3 items at $1.70 each, 5 items at $1.59 each, and 7 items at $1.98 each. What was the total amount she spent?
21. In a group of students, four were thirteen years old, six were nine years old, three were ten years old, and one was eighteen years old. What was the average age?
22. How many items at $1.79 each can be bought, if the largest amount to be spent is $20.00?

## CHAPTER TEST *Form 2B*

1. Is the following statement TRUE or FALSE? $-2 > 5$
2. Is this statement TRUE or FALSE? $17.5 < -1$
3. Which of the symbols $<$ or $>$ should be placed between the numbers 17.5 and 4.39 to make a true statement?

For the following questions, select the letter which best locates the numbers on the number line:

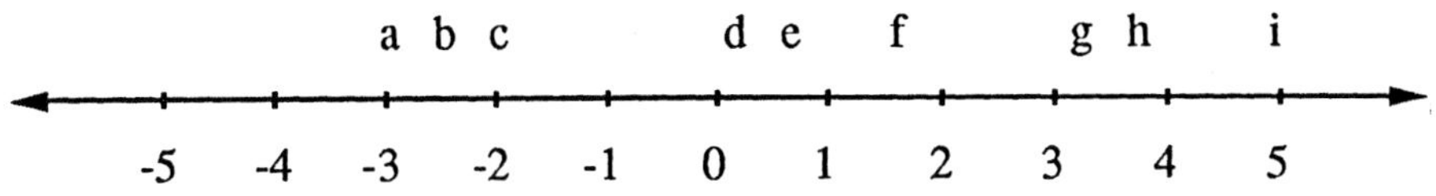


4. $\frac{1}{2}$ 5. $-2.3$ 6. $-3$

7. $\frac{25}{5}$ 8. .05 9. 3.9

10. $1\frac{3}{4}$ 11. $\sqrt{10}$ 12. $-\sqrt{4}$

13. What is the additive inverse of 12?
14. What is the multiplicative inverse of $-6.2$?

*Match the statements below with the letter of the property illustrated:*
a. associative property
b. distributive property
c. commutative property

15. $4.1 + 2 = 2 + 4.1$
16. $13 \times (7.1 + 2.3) = (13 \times 7.1) + (13 \times 2.3)$
17. $(9 \times 7.5) \times 1.1 = 9 \times (7.5 \times 1.1)$
18. $(8 \times 5) + (8 \times 7) = 8 \times (5 + 7)$
19. $(\sqrt{3} \times 17) = (17 \times \sqrt{3})$

Using the property of inverses, supply the value which will make the statement true:

20. [12 × (7 + 3)] + [__?__] = 0

## CHAPTER TEST *Form 3B*

Perform the indicated operations: (some may require the use of a calculator.)

1. $377 - 485 + 16 - 94$
2. $(-7) + (-3) + 15$
3. $-419 - 181$
4. $(18)(-6) + 10$
5. $4 + 13(7 - 8)$
6. $10(6 - 9)^2$
7. $(81)^3$
8. $-|47| + |81|$
9. $7\{|-3| - |-24|\}$
10. $11^2 + 5^2 - 12^2$
11. $|-4|^4$
12. $(0)^2$
13. $(-1)^{14}$
14. $9^2 + (-4)^2 - (16)^2$

For Numbers 15 through 18 insert the sign $<$, $>$, $=$ which will make the statement true:

15. $|-a|$ ____ $|a|$
16. $-|-14|$ ____ $|14|$
17. $(-2)^2$ ____ $2$
18. $(-3)^3$ ____ $27$

Solve the following word problems:

19. In Cold Spot, U.S.A. the temperature at 6 A.M. was $-12$ degrees Fahrenheit. By noon it warmed to 11 degrees Fahrenheit. How many degrees had the temperature risen?
20. If the temperature in Freezon is usually only half that of Cold Spot, what was the temperature in Freezon at 6 A.M. and at noon?

## CHAPTER TEST *Form 4B*

Evaluate the following:

1. $6.7 + 3.26 - 4.37$
2. $-62.7 + 75.1 - 28.4$
3. $-11.8 - (-2.5)$
4. $.05 - (-2.1)$
5. $4.27 - 3.6 - (-5.1)$
6. $-1.75 - (-1.2) + 3.6$
7. $18.6 \div 2.1$
8. $27.3 \div (-3.5)$
9. $-17.4 \div (3.0)$
10. $-26.2 \div (-4.6)$
11. $(2.3)^3$
12. $(.04)^2$
13. $\sqrt{78.6}$
14. $\sqrt{.073}$
15. $2.6(-7.9 + 1.26)$
16. $(4.2)^2 - 1.1(-8.43 - 2.81)$
17. $(-3.1)^2 + 1.2[3(4.5 - 1.9)]$
18. $(-4.8)(6.41)$
19. $(-8.93)(-.15)$
20. $(-2.1)(-.32)(-1.3)^2$
21. $|-3.4| - 2.2(-6.5)$
22. $-4.3[-6.2(-5.7)]$
23. $(38.6) - (16.67)$
24. $(-9.78) - (3.46)$
25. $(-5.9) - (-3.64)$

## CHAPTER TEST *Form 5B*

Write each number as a product of primes:

1. 48
2. 128
3. 418
4. 300
5. 1000
6. 376

Write each fraction in simplest form:

7. $\frac{15}{75}$
8. $\frac{42}{77}$
9. $\frac{48}{120}$
10. $\frac{50}{85}$
11. $\frac{42}{62}$
12. $\frac{103}{206}$

Multiply and simplify:

13. $\frac{5}{8} \cdot \frac{4}{10}$
14. $\frac{27}{15} \cdot \frac{3}{9}$
15. $\frac{-3}{13} \cdot \frac{1}{7}$
16. $\frac{125}{13} \cdot \frac{117}{15}$
17. $\frac{-16}{3} \cdot \left(\frac{-7}{32}\right)$
18. $\frac{-7}{8} \cdot \frac{3}{21} \cdot 6$
19. $\frac{24}{54} \cdot \frac{9}{8}$
20. $\frac{102}{11} \cdot \frac{121}{153}$

Divide and simplify:

21. $\frac{3}{7} \div \frac{2}{11}$
22. $\left(\frac{-3}{7}\right) \div \left(\frac{-5}{3}\right)$
23. $\frac{7}{2} \div 35$
24. $\frac{128}{33} \div \frac{4}{3}$
25. $\left(\frac{-49}{6}\right) \div \left(\frac{-21}{2}\right)$
26. $\frac{121}{-7} \div \frac{11}{14}$
27. $35 \div \frac{7}{2}$
28. $\frac{2}{7} \div \frac{39}{28} \div \frac{5}{65}$

Find the least common multiple for each group of numbers:

29. 12, 15
30. 9, 21, 48
31. 34, 136
32. 36, 72, 120
33. 6, 8, 12
34. 24, 34, 42

## CHAPTER TEST *Form 6B*

Evaluate the following:

1. $10^{-6}$
2. $10^{7}$
3. $8^{-2}$
4. $256^{0}$
5. $14^{2}$
6. $2^{8}$

Supply the value for x which makes each statement true:

7. $3^{X} = 27$
8. $7^{X} = 49$
9. $30^{X} = 900$
10. $5^{X} = \frac{1}{125}$

Rewrite the following using POSITIVE exponents:

11. $(37)^{-2}$
12. $\left(\frac{1}{13}\right)^{-2}$
13. $\left(\frac{4}{11}\right)^{-8}$
14. $\left(\frac{1}{X}\right)^{-3}$

Evaluate:

15. $\left(\frac{1}{4}\right)^{-3}$
16. $\left(\frac{2}{3}\right)^{3}$
17. $(17)^{0}$
18. $(100)^{-2}$
19. $\left(\frac{5}{3}\right)^{-3}$
20. $(17)^{-2}$

Combine like terms to simplify.

21. $21x^3 + 14y^2 - 3y^2$
22. $12y^2 - 8 - 4y^2 + 4$
23. $3a^2 + 4b^2 - 5a^2 + b$

Simplify the following without evaluating:

24. $39^{-6} \div 39^{-2}$
25. $337^{-11} \div 337^{-11}$
26. $5^{0} \div 5^{-6}$
27. $(23^{3})(23^{-2})$
28. $(3.6)^{14} \cdot (3.6)^{-14}$
29. $y^{-1} \cdot y^{0} \cdot y^{9}$

Simplify. Leave answers in exponential form.

30. $(3^{-7})^{-8}$
31. $[(191)^{5}]^{-2}$
32. $(11^{-4})^{4}$
33. $\left[\left(\frac{1}{4}\right)^{-3}\right]^{3}$
34. $(.3^{9})^{-1}$
35. $(12^{0})^{2}$

## CHAPTER TEST *Form 7B*

Classify each as monomial, binomial or polynomial:

1. $a + b^2$
2. $a^2 + ab - 6$
3. $3x^2$
4. $4x^2 + 9y^2$
5. $7x^3 + 4x^2 - 6x + 3$
6. $8x^{10} - 1$
7. $y^3 - 41y - 2$
8. $x^5 - x^4 + x^3 - 6x + 2$
9. $18y^2 - 14y + 3$
10. $2x^3 - 7x + 1$

Combine like terms:

11. $3a^2 + 5c + 2a^2$
12. $-6x + 3y - 4y - 16x$
13. $x^2y^2 + 3x^2y - 4x^2y^2$
14. $9A^2B - 2AB^2 - 4A^2B - 6AB^2 + 3$
15. $3x - 4y + 5z - 6y + 2z - x + 2$
16. $(5a^2b^3 + 3a^2b - 2ab^2) + (4a^2b^3 + 6a^2b - 2a^2b^3)$
17. $(3y^2 - x^2y^3 + x^2y^3) - (4x^3y^2 + 3x^3y^2 - 1)$
18. $7x^2y + 9x^3y - 4x^2y + 13xy + 2x^3y$

Perform the indicated operations and simplify answers:

19. $(2x + 3)(4x - 2)$
20. $(7x^2 - 4x)(4x^2 + 2x)$
21. $(4x^3 - x + 3) + 3x(8x^2 - 2x + 1)$
22. $(5x^5 - 6x + 2) - (3x^5 + 7x - 8)$
23. $(x^2 + 4)^2 - 8x^2(4x^2 + 3)$
24. $\dfrac{28x^3 - 7x^2 - 14x}{7x}$
25. $\dfrac{-(6y^3 - 4y^2 - 18y)}{2y}$

## CHAPTER TEST *Form 8B*

Write the following as algebraic expressions or equations, using x to represent the variable:

1. A number decreased by 68
2. Nine times a number increased by 113
3. A number decreased by $\frac{3}{5}$ of itself
4. Five times a number less nine is thirty seven.
5. Seventy five less one-half of a number is thirty.
6. Fifty percent of a number is six.

Solve each of the following and check your answers:

7. $8x - 2 = 22$
8. $3 - x = 14$
9. $x - 11 = 42$
10. $6x - 7 = -4$
11. $4(x - 1) = -24$
12. $-3(2x - 5) = 12$
13. $\frac{2}{3}x + \frac{1}{3} = 5$
14. $\frac{1}{4}x - 18 = -22$
15. $7x + 5 - 2x = 25$
16. $8x + 6 = 3x - 9$

Write the equation and solve for the variable:

17. Three times a number gives the same result as subtracting twice the number from 110.
18. Five times a number increased by 14 is equivalent to three times the number.
19. Jack's math test grade was 10 points higher than his psychology grade. If he received a 96 in math, what was his psych grade?
20. Maurice purchased his first new car in 1965. A new car in 1988 cost four times the other, plus $3000. If the 1988 car cost $13,000, what did Maurice pay for the car in 1965?

Solve for the indicated variable:

21. $A = \pi r^2$ for r
22. $b^2 - 4ac = 0$ for a

## CHAPTER TEST *Form 9B*

1. Without plotting the points, state the quadrant in which each is located. If the point is on an axis, state if x or y.

   a. $(2,7)$ b. $(-5,0)$

   c. $(3,-8)$ d. $(-3,-11)$

   e. $(-7,36)$

2. Plot the following points on the Cartesian axes and label each with its matching letter:

   a. $(-3,-4)$ b. $(2,0)$

   c. $(-5,3)$ d. $(4,-1)$

3. Does the point $(-3,-13)$ lie on the line $y = 4x - 1$?

4. The equation of a line is $y = \frac{2x}{3} + 4$.

   If the point $(3,y)$ lies on this line, find the value of y.

5. Find any two points which satisfy the following equations, then graph the line represented by each:

   a. $y = 2x + 3$

   b. $x = -2$

   c. $y = 3$

   d. $y = \frac{1}{2}x + 3$

   e. $y = 3x$

6. Find the slope and y intercept of the lines represented by the following equations. Graph the lines.

   a. $y = 3x + 1$

   b. $y = -4x - 3$

   c. $y = x - 5$

## CHAPTER TEST *Form 10B*

Find the greatest common factor of the following:

1. $12x^2$ and $18x$
2. $4a^2b$ and $8ab^2$
3. $2y^2$, $15y$ and $9y^3$
4. $27w^2y$ and $24y^2$
5. $a^2b^2c^3$ and $b^2c^3$

Factor completely:

6. $10x - 16y$
7. $x^2 - 49$
8. $27xy + 6xby^2$
9. $x^2 - 7x + 10$
10. $a^2 + 8a + 15$
11. $64 - x^2$
12. $24 - 2d + d^2$
13. $x^6y^3 + x^4y^6$
14. $ax^2 - ay^2$
15. $y^2 - 9y - 36$
16. $w^2 + 4w - 21$
17. $2y^2 - 50$
18. $4x^2 + 28x - 120$
19. $2.1ax^2 + 4.2ax$
20. $\frac{x^2}{2} + \frac{3x}{2} + \frac{3}{2}$

## CHAPTER TEST *Form 11B*

Reduce the fractions to lowest terms:

1. $\frac{50a^2}{25a}$

2. $\frac{4wz^3}{16wz^2}$

Perform the indicated operations and simplify your answers:

3. $\frac{3y}{7} \cdot \frac{14}{y}$

4. $\frac{8a}{c} \div \frac{2a}{c^2}$

5. $\frac{15}{7y} \cdot \frac{35y^2}{3}$

6. $\frac{9xy^2}{2x} \cdot \frac{4}{6x}$

7. $\frac{x^2y}{16} \div \frac{y}{4}$

8. $\frac{-7y^2z}{2c} \div \frac{28}{c^2}$

9. $\frac{7y}{34} - \frac{2y}{34}$

10. $\frac{7a}{15} + \frac{3a}{5}$

11. $\frac{3}{4t} + \frac{5}{t}$

12. $t - \frac{3t}{5}$

13. $\frac{x}{x+6} - \frac{2}{x+6}$

14. $\frac{y^2}{9} + \frac{2y^2}{3} + 1$

Solve the equations and check your answers:

15. $2x + 3 = 19$

16. $3 + 4y = 2y + 1$

17. $\frac{x}{3} + 2 = -11$

18. $\frac{x+5}{3} = 7$

19. $\frac{3y}{2} - 4 = \frac{5y}{3}$

20. $x + 7 = \frac{3x}{2} - 3$

## CHAPTER TEST *Form 12B*

Solve the following equations:

1. $(x - 6)(x + 7) = 0$
2. $(a + 21)(a - 23) = 0$
3. $(2w - 5)(w - 1) = 0$
4. $a^2 + a - 6 = 0$
5. $z^2 - 2z - 48 = 0$
6. $x^2 + 8x = 0$
7. $r^2 - 8r + 16 = 0$
8. $e^2 - 0.01 = 0$
9. $y^2 - 22y = 0$
10. $x^2 - 3x = 28$
11. $p^2 - 14p + 49 = 0$

Write the following equations in general form and state the values of a, b, and c:

12. $x^2 - 3x + 2 = 0$
13. $2x^2 = 3x - 4$
14. $16 = 3y^2$

Solve the following by using the quadratic formula:

15. $x^2 - 4x - 21 = 0$

**CHAPTER TEST** *Form 13B*

For all questions involving right triangle ABC, angle C is the right angle.

1. In right triangle ABC, the measure of angle B is 34°. What is the measure of angle A?
2. The hypotenuse of the right triangle ABC is 60 units long. The shortest side of the triangle is 27 units long. What is the length of the third side?
3. In triangle ABC, angle B measures 116° and angle C measures twice that of angle A. Find the measures of angles A and C.
4. In right triangle ABC, side a = 10 units and side b = 24 units. Find the length of side c.
5. Using the triangle of question 4, find the cosine of angle B.
6. Use the table on page 233 to find the sine of 45°.
7. Use the table on page 233 to find the tangent of 70°.

For questions 8, 9 and 10, draw and label the triangle in order to see which trig relationship to use for the solution to the problem.

8. In right triangle ABC, the hypotenuse is 64 units and the measure of one angle is 30°. What is the length of the side opposite the 30° angle?
9. In right triangle ABC side c = 110 units and side a = 55 units. What is the measure of angle B?
10. In right triangle ABC, side a = 12 units and side b = 9 units. Find the measure of angle B to the nearest degree.

In questions 11 through 15, use the Pythagorean Theorem to find the third side of right triangle ABC when two sides are given. Answers should be rounded to the nearest tenth.

11. a = 6 and b = 8
12. b = 12 and c = 20
13. a = 12 and b = 5
14. a = 1.5 and b = 1.6
15. a = 11 and b = 11

## CHAPTER TEST *Form 14B*

**Directions:** Show all work for the following problems. Be sure to state the formula used and label the answer.

1. Find the perimeter of a rectangular table top 25 ft by 13 ft.
2. Find the circumference of a wheel whose radius is 27 inches.
3. The perimeter of a square is 144 ft. Find the length of a side of the square.
4. A parallelogram has a base 8 ft longer than its width. If its perimeter is 60 ft, find its length and width.
5. Find the perimeter of a rhombus one of whose sides is 0.2 feet.
6. Find the area of a circular window whose diameter is 2 ft.
7. Find the area of a triangle whose height is 15 inches and whose base is 4 inches.
8. A rhombus has a base of 16 feet and an area of 256 sq. ft. Find its height.
9. The area of a square is 189 sq. inches. Find the length of its sides.
10. Find the area of a parallelogram whose base is 17 and whose height is 10.
11. Find the volume of a ball whose radius is 15 centimeters.
12. Find the volume of a cone whose height is 7 inches and whose diameter is 8 inches.
13. The volume of a cube is 125 cu. inches. Find the length of an edge.
14. Find the volume of a box 4 ft by 5 ft by .5 ft.
15. A triangle has a perimeter of 19.0 yds. If one side is twice as long as the second side, and the third side is 4 yds., find the lengths of the two unknown sides.

## CHAPTER TEST *Form 15B*

For questions 1 and 2, solve graphically. Label each line and clearly indicate the point of intersection.

1. $y = 3x + 2$
   $y = 5x + 6$

2. $3x - 2y = 2$
   $y = 4x - 6$

For questions 3, 4, and 5, use substitution to solve for the variables.

3. $y = 5x + 8$
   $y = x - 4$

4. $y = 9x + 7$
   $2x - y = -7$

5. $y - 4 = x$
   $y = 2x + 5$

For questions 6, 7, and 8, use the addition method to solve for the variables.

6. $2x + 3y = 8$
   $2x - 4y = -6$

7. $4x + 2y = 5$
   $3x - 4y = 1$

8. $3x + y = 11$
   $2x - y = -1$

## CHAPTER TEST *Form 16B*

Graph the following sets of numbers:

1. $\{x|x \varepsilon I, -1 \leq x \leq 5\}$
2. $\{x|x \varepsilon R, -4 < x \leq -1\}$
3. $\{x|x \varepsilon R, x > 0 \text{ or } x \leq -2\}$
4. $\{x|x \varepsilon R, x \geq 1 \text{ and } x \leq 6\}$

Graph each of the following inequalities in the coordinate plane:

5. $y \leq 3x - 4$
6. $y > \frac{2}{3}x - 3$
7. $y \leq 2$
8. $y < 5x$
9. $5x - 2y \leq 10$
10. $x + 4 > 3y$

Graph the solution to the following systems of linear inequalities:

11. $y \leq 2x + 5$
    $y > -x + 1$
12. $x \geq 3$
    $y < -1$
13. $2x + y > 4$
    $x - 3y \geq 6$
14. $4x - 3y \geq 12$
    $y + x + 3 < 0$

## CHAPTER TEST *Form 17B*

Factor the numerators and denominators of the following fractions and reduce each to lowest terms:

1. $\dfrac{x - y}{x^2 - y^2}$

2. $\dfrac{x - 6}{x^2 - 5x - 6}$

3. $\dfrac{x^2 + 5x + 4}{x^2 - 16}$

4. $\dfrac{x^2y^2 - xy}{6xy}$

Multiply or divide as indicated:

5. $\dfrac{3x + 9}{9 - x^2} \cdot \dfrac{3 - x}{x^2 - 9}$

6. $\dfrac{2xy^2}{x^2 - 2x + 1} \div \dfrac{6x^2y}{x - 1}$

7. $\dfrac{4a + 12b}{4a^2b^3} \cdot \dfrac{ab}{a^2 - 9b^2}$

8. $\dfrac{xy^2 - x^2y}{14y^2} \div \dfrac{x - y}{7y}$

Add or subtract as indicated:

9. $\dfrac{7xy}{x + 3} + \dfrac{14x}{2x + 6}$

10. $\dfrac{a}{a^4 - b^2} - \dfrac{b}{a^2 + b}$

11. $\dfrac{x^2 - 25}{x^2 + 6x + 5} + \dfrac{x + 1}{x + 5}$

12. $\dfrac{4x}{x + 3} + \dfrac{x}{x - 3} - \dfrac{3}{x^2 - 9}$

Solve for the variable:

13. $\dfrac{x + 4}{3} = \dfrac{3}{5} - 2x$

14. $\dfrac{x + 1}{3} = \dfrac{x + 6}{7}$

15. $\dfrac{2x}{x + 1} - \dfrac{3x}{2x + 2} = 5$

## CHAPTER TEST *Form 18B*

Use a calculator to find the indicated root, if it exists:

1. $\sqrt{1024}$
2. $\sqrt{22500}$
3. $\sqrt{-900}$
4. $\sqrt{0.0121}$
5. $-\sqrt{289}$
6. $\sqrt{2500}$

Use a calculator and the fact that $\sqrt[n]{a} = (a)^{1/n}$ to find the indicated root:

7. $\sqrt[3]{512}$
8. $\sqrt[4]{50625}$
9. $\sqrt[3]{-343}$
10. $\sqrt[5]{243}$
11. $\sqrt[4]{14641}$
12. $\sqrt[3]{-1331}$

Perform the indicated operations. Simplify all answers.

13. $4\sqrt{a} + \sqrt{16a^2} - \sqrt{64a}$
14. $\sqrt{x^5} - 4x^2\sqrt{x} + 2\sqrt{x}$
15. $(\sqrt{6})^2 + (3\sqrt{7})^3$
16. $a\sqrt{b^3} - ab\sqrt{b} + \sqrt{ab}$
17. $\sqrt{\frac{32c^3d^4}{4c^2d}}$
18. $\sqrt{\frac{500}{5}}$
19. $\sqrt{7}(\sqrt{3} - \sqrt{7})$
20. $\sqrt{x}(\sqrt{x} + \sqrt{y})$
21. $(\sqrt{s} - \sqrt{t})(\sqrt{s} + \sqrt{t})$
22. $(\sqrt{3} - \sqrt{2})^2$
23. $\frac{7ab^2c^3}{\sqrt{a} + 2}$
24. $\frac{6\sqrt{x}}{\sqrt{x} - 6}$
25. $\frac{(3 + \sqrt{y})}{\sqrt{y} - 3}$

## CHAPTER ONE

### Exercise Set 1.1

**1.** 35.21 **3.** 82.7 **5.** 183.28 **7.** 134.1 **9.** 3.18 **11.** 115.93 **13.** −5.96
**15.** −61 **17.** −21 **19.** 147.06 **21.** −54.41

### Exercise Set 1.2

**1.** 2726.4 **3.** 11.268 **5.** 209.28 **7.** 102 **9.** $10.\overline{09}$ **11.** .376 **13.** 281.7
**15.** 20.4375 **17.** 4080 **19.** 48.84 **21.** 99.9

### Exercise Set 1.3

**1.** 16 **3.** 125 **5.** 49 **7.** .25 **9.** 10000 **11.** 1,000,000 **13.** 37.21 **15.** 100
**17.** 0

### Exericse Set 1.4 A

**1.** 6 **3.** 10 **5.** 10.4403 **7.** 19 **9.** 3.6056 **11.** 8

### Exercise Set 1.4 B

**1.** 25 **3.** 16 **5.** 0 **7.** 0 **9.** 0 **11.** error

### Exercise Set 1.5

**1.** .8571 **3.** 4.6667 **5.** 3.75 **7.** 1.6667 **9.** .875 **11.** 5.5 **13.** 1.6667 **15.** 0

### Exercise Set 1.6 A

**1.** 25% **3.** .12 **5.** 130% **7.** 8% **9.** .875 **11.** 12.5% **13.** 290% **15.** .05
**17.** 1100%

### Exercise Set 1.6 B

**1.** .333, 33.3% **3.** .8, 80% **5.** .75, 75% **7.** .25, 25% **9.** .083, 8.3% **11.** .25, 25%
**13.** .125, 12.5% **15.** 3.0, 300%

### Exercise Set 1.7

**1.** 7 **3.** 21 **5.** 21

### Exercise Set 1.8

**1.** $2125 **3.** $104.67 **5.** 24 **7.** $84 **9.** $15, $90

### Chapter One Exercises

**1.** 24.02 **3.** 6.36 **5.** 1.160 **7.** 65.5 **9.** −23.46 **11.** 888 **13.** 4.9092
**15.** 126.9231 **17.** .08 **19.** 170.81 **21.** 65.61 **23.** .0225 **25.** 1331 **27.** 81
**29.** 1,000,000 **31.** 5 **33.** 13 **35.** .7 **37.** 4.6583 **39.** .1581 **41.** .875 **43.** .75
**45.** 1.5 **47.** 1.4286 **49.** .35 **51.** 17% **53.** 3% **55.** 62.5% **57.** 30%
**59.** 710% **61.** 87.5% **63.** 40% **65.** 58.8% **67.** 240% **69.** 400% **71.** 13
**73.** 15.5 **75.** 8.2 **77.** 9 **79.** 11 **81.** $252 **83.** 60 **85.** $.50 **87.** 13 ft, 1 ft.
**89.** 35

### Chapter Test *Form 1A*

**1.** 188.62 **3.** 15.549 **5.** 15 **7.** 1.915 **9.** 98.8 **11.** 19.4481 **13.** 2.2913
**15.** 75.3% **17.** 150% **19.** .05 **21.** 60 **23.** 5 ft. 7 in. **25.** $37.50

## CHAPTER TWO

### Exercise Set 2.1

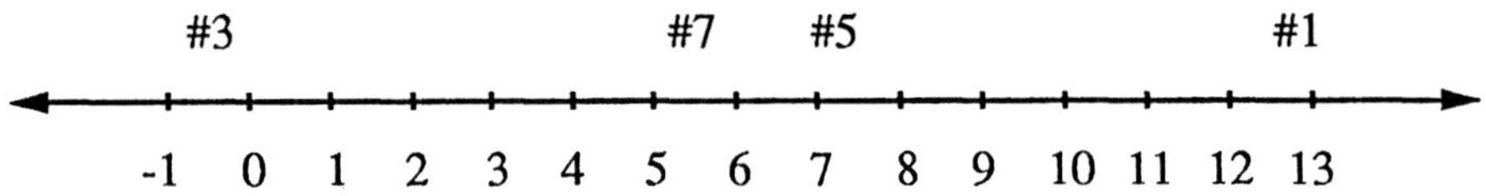


### Exercise Set 2.2

**1.** $4.5 < 5$ **3.** $-3.33 < -3$ **5.** $2.66 > 2.5$ **7.** $5.0 > -5.3$ **9.** $\sqrt{81} = 9$
**11.** $-1 < 0.1$

### Exercise Set 2.7 A

**1.** $-10$ **3.** 75 **5.** 1 **7.** $\frac{1}{3.75}$ **9.** $\frac{1}{-4.2}$

### Exercise Set 2.7 B

**1.** commutative **3.** distributive **5.** commutative **7.** identity **9.** inverse
**11.** commutative **13.** distributive

### Exercise Set 2.7 C

**1.** $(6 - b)$ **3.** $6(5)$ **5.** $(6 + a)$, $6a$

### Chapter Two Exercises

**13.** $10 < 12.5$ **15.** $-13 < 5$ **17.** $2.25 > 0$ **19.** $2\frac{3}{4} > 1.93$ **21.** $\frac{3}{4} > -.75$
**23.** $.01 > .001$ **25.** commutative **27.** distributive **29.** commutative **31.** associative
**33.** distributive **35.** 17.25 **37.** 40 **39.** .208 **41.** $-\frac{1}{3}$ **43.** $-\sqrt{5}$
**45.** $\frac{1}{4.3}$ or $\frac{10}{43}$ **47.** $\frac{-1}{3}$ **49.** $10, \frac{-1}{10}$ **51.** $2.5, \frac{1}{-2.5}$ **53.** $-.8, \frac{1}{.8}$ or $\frac{10}{8}$
**55.** $\sqrt{2}, \frac{1}{-\sqrt{2}}$

### Chapter Test *Form 2A*

**1.** d **3.** a **5.** c **7.** $>$ **9.** $=$ **11.** $=$ **13.** $<$ **15.** $<$ **17.** $<$ **19.** inverse
**21.** identity **23.** associative **25.** commutative **27.** $31, \frac{1}{-31}$ **29.** $-3.5, \frac{1}{3.5}$
**31.** $-1, 1$ **33.** $-3.6, \frac{1}{3.6}$

## CHAPTER THREE

**Exercise Set 3.1**

**1.** +15 **3.** 3 **5.** −8 **7.** −9 **9.** −23 **11.** 9 **13.** −11 **15.** −14 **17.** 4
**19.** −5 **21.** −10 **23.** 0 **25.** −5

**Exercise Set 3.2**

**1.** −4 **3.** −21 **5.** −11 **7.** −11 **9.** 50 **11.** −67 **13.** 25 **15.** 20
**17.** 191

**Exercise Set 3.3 A**

**1.** 36 **3.** −36 **5.** −20 **7.** 20 **9.** −21 **11.** −168 **13.** 168 **15.** 144
**17.** −144 **19.** 30 **21.** −30 **23.** 30 **25.** 96

**Exercise Set 3.3 B**

**1.** −4 **3.** −20 **5.** −3 **7.** 300 **9.** −2 **11.** −40 **13.** −2 **15.** −55
**17.** −8 **19.** −13 **21.** −50 **23.** −50

**Exercise Set 3.4 A**

**1.** 6 **3.** −6 **5.** −5 **7.** 5 **9.** $2.\overline{3}$ **11.** $2.\overline{3}$ **13.** −2.625 **15.** 2.625 **17.** 1
**19.** −1 **21.** −9 **23.** −5 **25.** 10

**Exercise Set 3.4 B**

**1.** −6 **3.** 6 **5.** −1 **7.** 1 **9.** −1 **11.** −1 **13.** −1 **15.** −.5 **17.** 6
**19.** .0675566

**Exercise Set 3.5**

**1.** 4 **3.** −5 **5.** 11 **7.** −2 **9.** 3 **11.** 10 **13.** 5 **15.** −9 **17.** −28
**19.** 12 **21.** 36 **23.** 25 **25.** 0

**Exercise Set 3.6**

**1.** 20,736 **3.** 15625 **5.** 256

**Exercise Set 3.7 A**

**1.** 71 **3.** −6 **5.** −128 **7.** 90 **9.** 300

**Exercise Set 3.7 B**

**1.** −12 **3.** 17 **5.** 88

**Exercise Set 3.7 C**

**1.** 7 **3.** 18 **5.** 15 **7.** −8 **9.** 36

**Exercise Set 3.8**

**1.** −28° **3.** −$158.55 **5.** $51.60 **7.** −$150.00 **9.** +27° **11.** $54.25, $95.75

### Chapter Three Exercises

**1.** 10 **3.** 8 **5.** 16 **7.** −8 **9.** −5 **11.** +8 **13.** −12 **15.** 8 **17.** −9
**19.** 8 **21.** 12 **23.** −9 **25.** 31 **27.** −10 **29.** 4 **31.** −3 **33.** 1 **35.** 18
**37.** −85 **39.** −210 **41.** 24 **43.** −30 **45.** +35 **47.** −30 **49.** −1000
**51.** 48 **53.** −77 **55.** 603 **57.** 90 **59.** 780 **61.** 9 **63.** −20.5 **65.** −21
**67.** −4 **69.** −11 **71.** 3 **73.** −12.5 **75.** 21.6 **77.** 16.4 **79.** $-1.1\overline{6}$
**81.** 115 **83.** −51 **85.** −800 **87.** 4 **89.** −5 **91.** −40 **93.** 9 **95.** −1
**97.** −25 **99.** 8.57 **101.** 7 **103.** 2 **105.** −4 **107.** 6 **109.** 9 **111.** −12
**113.** −12 **115.** 4 **117.** 123 **119.** 13 **121.** 0 **123.** 1 **125.** 28 **127.** −180
**129.** −7 **131.** −23 **133.** −5 **135.** 11 **137.** 95 **139.** −203 **141.** 80°
**143.** $15.00

### Chapter Test *Form 3A*

**1.** 26 **3.** 1 **5.** 58 **7.** −33 **9.** 0 **11.** 84 **13.** 76 **15.** 504 **17.** 102
**19.** $1.\overline{6}$ **21.** 36 **23.** 39 **25.** 256 **27.** 576 **29.** −32

## CHAPTER FOUR

### Exercise Set 4.1

**1.** 92.444 **3.** −20.196 **5.** 177.6 **7.** 106.4 **9.** .088 **11.** −.088 **13.** −15.2
**15.** 0

### Exercise Set 4.2 A

**1.** 20.196 **3.** −92.444 **5.** 106.4 **7.** 177.6 **9.** .108 **11.** −.108 **13.** 0
**15.** −15.2

### Exercise Set 4.2 B

**1.** .84 **3.** −6.6 **5.** −7.07 **7.** −43.52 **9.** .84 **11.** −6.6 **13.** −7.07
**15.** −43.52

### Exercise Set 4.3

**1.** −48.396 **3.** −2.5 **5.** 30.15 **7.** 30.7075 **9.** −.24 **11.** 6.72 **13.** −1.05

### Exercise Set 4.4

**1.** 7.36 **3.** 3.54 **5.** 9.90

### Exercise Set 4.5

**1.** −24.2 **3.** 6.79 **5.** 25.734

**Chapter Four Exercises**

**1.** 20.8 **3.** 14.77 **5.** −16 **7.** −18.4 **9.** 8.37 **11.** .06 **13.** −23.7
**15.** −13.53 **17.** −38.4 **19.** −393.78 **21.** 10.32 **23.** −9.7 **25.** 43.6
**27.** 119.2 **29.** 2.52 **31.** −.45 **33.** −65.86 **35.** −39.36 **37.** 0 **39.** 214.08
**41.** −6.37 **43.** 5.28 **45.** −62.32 **47.** 4.16 **49.** 18.944 **51.** 17.451
**53.** −38.64 **55.** 2.43 **57.** 5.8 **59.** −26.02 **61.** 2 **63.** −5.8 **65.** 138.24
**67.** 4.18 **69.** .89 **71.** 10.23 **73.** 21.83 **75.** 2.85 **77.** error **79.** 7.34
**81.** 3402.07 **83.** 1.07 **85.** 1.40 **87.** 11.2 **89.** −1.5 **91.** −188.15 **93.** −.907
**95.** −50.68 **97.** 4.62 **99.** −248.56

**Chapter Test** *Form 4A*

**1.** 6.21 **3.** −46.4 **5.** −8.67 **7.** 39.7 **9.** 10.88 **11.** 11.29 **13.** 3.6
**15.** −94.3 **17.** .0008 **19.** 19.74 **21.** −21.7 **23.** 5.16 **25.** .148 **27.** 4860
**29.** −28.84 **31.** −6.98 **33.** 174.24 **35.** 1,488,964.1 **37.** 3.48 **39.** −50.430
**41.** −12.000 **43.** 50.661 **45.** −193 **47.** −52 **49.** 300

**CUMULATIVE EXERCISES Chapters 1–4** *Set A*

**1.** commutative **3.** associative **5.** inverse **7.** 72 **9.** −28.8 **11.** −11.6
**13.** −18.22 **15.** −127.92 **17.** −16.3 **19.** 35.2 **21.** .4 **23.** .064 **25.** .76

**CUMULATIVE TEST Chapters 1–4** *Form A*

**1.** d **3.** c **5.** b **7.** 54 **9.** −14 **11.** 255 **13.** −10.61 **15.** −8.8 **17.** 7.9
**19.** 33.58 **21.** −7.3 **23.** 9.96 **25.** −1.4

## CHAPTER FIVE

**Exercise Set 5.1 A**

**1.** 18 · 2 **3.** prime **5.** 9 · 9 **7.** 12 · 6 **9.** 11 · 11

**Exercise Set 5.1 B**

**1.** 2 · 11, 5 · 11, 11 **3.** 2 · 2 · 3 · 5, 3 · 3 · 11, 3
**5.** 2 · 2 · 17, 2 · 3 · 7, 2 **7.** 3 · 3 · 7, 2 · 3 · 3 · 7, 3 · 3 · 3 · 7, 9
**9.** 2 · 2, 3 · 5, 3 · 7, 2 · 2 · 2 · 3 There is no common factor.

**Exercise Set 5.2 A**

**1.** $\frac{1}{20}$ **3.** $\frac{1}{12}$ **5.** $\frac{17}{93}$ **7.** 3 **9.** $\frac{1}{86}$ **11.** $\frac{1}{11}$

**Exercise Set 5.2 B**

**1.** .05, 5% **3.** $.08\overline{3}$, 8.3% **5.** .183, 18.3% **7.** 3.0, 300% **9.** .012, 1.2%
**11.** .091, 9.1%

**Exercise Set 5.3**

**1.** $\frac{1}{4}$ **3.** $\frac{4}{7}$ **5.** $\frac{-7}{3}$ **7.** $-6$ **9.** $\frac{1}{-12}$

**Exercise Set 5.4**

**1.** $\frac{5}{2}$ **3.** $\frac{-2}{5}$ **5.** $\frac{66}{7}$ **7.** 72 **9.** $\frac{1}{45}$

**Exercise Set 5.5**

**1.** 240 **3.** 105 **5.** 195 **7.** 72 **9.** 225

**Exercise Set 5.6**

**1.** $\frac{10}{3}$ **3.** $\frac{-2}{5}$ **5.** $\frac{11}{9}$ **7.** $\frac{9}{5}$ **9.** 0 **11.** $\frac{2}{3}$

**Exercise Set 5.7**

**1.** $\frac{1}{8}$ **3.** $\frac{49}{25}$ **5.** $\frac{2401}{1000}$ **7.** $\frac{8}{1331}$ **9.** $\frac{243}{1024}$ **11.** $\frac{20736}{625}$

**Exercise Set 5.8**

**1.** $\frac{64}{11}$ **3.** 2 **5.** $\frac{6}{5}$ **7.** $-8$ **9.** $\frac{3}{2}$ **11.** $\frac{24}{121}$ **13.** 512 **15.** $\frac{-83}{8}$

**Chapter Five Exercises**

**1.** $3 \cdot 5$ **3.** $11 \cdot 1$ **5.** $2 \cdot 2$ **7.** $2 \cdot 2 \cdot 3$ **9.** $5^2 \cdot 2^2$ **11.** $7 \cdot 2^3$
**13.** $2 \cdot 3 \cdot 13$ **15.** $7 \cdot 5$ **17.** $3 \cdot 13 \cdot 2^2$ **19.** $5^3 \cdot 2^2$ **21.** 3 **23.** 3 **25.** 16
**27.** 9 **29.** 14 **31.** none **33.** $\frac{3}{4}$ **35.** $\frac{13}{25}$ **37.** $\frac{7}{11}$ **39.** $\frac{1}{5}$ **41.** $\frac{7}{11}$ **43.** 6
**45.** $\frac{11}{4}$ **47.** $.7\overline{5}$ **49.** $.6\overline{6}$ **51.** .83 **53.** .03125 **55.** .375 **57.** $\frac{10}{21}$ **59.** $\frac{-1}{6}$
**61.** $\frac{9}{2}$ **63.** $\frac{2}{3}$ **65.** $-1$ **67.** $\frac{-39}{16}$ **69.** $\frac{9}{8}$ **71.** $\frac{7}{8}$ **73.** $\frac{15}{66} = \frac{5}{22}$ **75.** $\frac{-45}{14}$
**77.** $-28$ **79.** 72 **81.** 300 **83.** 60 **85.** 224 **87.** 12 **89.** 156 **91.** $\frac{14}{13}$
**93.** $\frac{1}{4}$ **95.** $\frac{-8}{15}$ **97.** $\frac{-12}{17}$ **99.** $\frac{13}{4}$ **101.** $\frac{1}{26}$ **103.** $\frac{8}{9}$ **105.** $\frac{43}{12}$ **107.** $\frac{29}{15}$
**109.** $\frac{73}{105}$ **111.** $\frac{29}{90}$ **113.** $\frac{55}{54}$ **115.** $\frac{31}{45}$ **117.** $\frac{-1}{10}$ **119.** $\frac{4}{9}$ **121.** $\frac{49}{25}$
**123.** $\frac{625}{81}$ **125.** $\frac{28561}{256}$ **127.** $\frac{262144}{729}$ **129.** $\frac{243}{32768}$

### Chapter Test *Form 5A*

**1.** 13, 17, 19, 23, 29 **3.** $2^3$, 3, 5 **5.** $2 \cdot 7$ or 14 **7.** $2^2 \cdot 3^2$ or 36 **9.** $\frac{1}{8}$ **11.** $\frac{2}{15}$
**13.** $\frac{35}{8}$ **15.** $\frac{2}{5}$ **17.** 2 **19.** $\frac{44}{60}$ or $\frac{11}{15}$ **21.** $\frac{61}{14}$

### CUMULATIVE EXERCISES Chapters 1–5 *Set A*

**1.** 0 **3.** \$30.65 **5.** Distributive **7.** 2 **9.** $-24.8$ **11.** $-26.4$ **13.** $2^3 \cdot 5 \cdot 31$
**15.** $2^4 \cdot 103$ **17.** $\frac{8}{11}$ **19.** $\frac{-55}{78}$ **21.** 1323 **23.** $\frac{567}{34}$ **25.** $-2$ **27.** $\frac{-31}{24}$
**29.** $\frac{-119}{16}$ **31.** 1009 **33.** 176 **35.** $-287$

### CUMULATIVE TEST Chapters 1–5 *Form A*

**1.** 109285 **3.** $\frac{4}{165}$ **5.** 135 **7.** $-32.048$ **9.** 0 **11.** $2^3 \cdot 5 \cdot 43$ **13.** $\frac{3}{16}$
**15.** $\frac{-5}{4}$ **17.** $\frac{71}{105}$ **19.** $\frac{49}{64}$ **21.** $-225.1$ **23.** $\frac{25}{49}$ **25.** 472

## CHAPTER SIX

### Exercise Set 6.1

**1.** 10000 **3.** 100000 **5.** 1000 **7.** 1000000000 **9.** $\frac{1}{10000000}$

### Exercise Set 6.2 A

**1.** 27 **3.** 8 **5.** 625 **7.** 1 **9.** $\frac{1}{16}$ **11.** 216 **13.** $\frac{1}{10}$ **15.** 7

### Exercise Set 6.2 B

**1.** 6 **3.** 5 **5.** 3 **7.** 3 **9.** $-1$ **11.** $-2$ **13.** 1 **15.** $-7$

### Exercise Set 6.3 A

**1.** 64 **3.** $\frac{1}{7}$ **5.** 64 **7.** 343 **9.** 256 **11.** $\frac{1}{216}$ **13.** $\frac{1}{256}$ **15.** 7

### Exercise Set 6.3 B

**1.** $\frac{1}{32}$ **3.** $\frac{16}{81}$ **5.** $\frac{36}{25}$ **7.** $\frac{25}{9}$ **9.** 5 **11.** $\frac{9}{4}$ **13.** 9 **15.** 36

### Exercise Set 6.4

**1.** $4^3 + 9(4^5)$ **3.** $7(4^4) + 3(4^2)$ **5.** $13(2^5) + 4(3^3) - 2(5^2)$ **7.** $17k^3 - 10g^5$ **9.** $8x^2$
**11.** $4y^4 + 8w^3$ **13.** $2q^2$ **15.** $4d^3 - 4g^5$ **17.** $10(3^{-2}) - 5(3^2)$ **19.** $3y - 5$

### Exercise Set 6.5

**1.** $3^8$ **3.** $2^5$ **5.** $(.7)^8$ **7.** $4^9$ **9.** $16^9$ **11.** $4^0$ **13.** $12^0$

**Exercise Set 6.6**

**1.** $10^6$ **3.** $6^0$ **5.** $3^{-2}$ **7.** $12^0$ **9.** $4^7$ **11.** $7^1$ **13.** $10^{-4}$ **15.** $5^{-5}$

**Exercise Set 6.7**

**1.** $2^{12}$ **3.** $7^0$ **5.** $10^{-14}$ **7.** $.3^{20}$ **9.** $6^4$ **11.** $8^{15}$ **13.** $2^{-8}$ **15.** $11^{-9}$

**Chapter Six Exercises**

**1.** 1000 **3.** $\frac{1}{1000} = .001$ **5.** 100000 **7.** 125 **9.** $\frac{1}{16}$

**11.** $\frac{1}{6}$ **13.** 4 **15.** 3 **17.** $-2$ **19.** 7 **21.** 4 **23.** 0

**25.** 2 **27.** 3 **29.** $19(4^2) - 6(4^3)$ **31.** $5(5^2) = 5^3$

**33.** $2(3^5)$ **35.** $9(9^2) + 3(7^2)$ **37.** cannot simplify

**39** $10(5^3) - 2(5^2)$ **41.** $3(8^2) + 2(8^3)$ **43.** $8x^3$

**45.** $-c^0 + 6d^3$ **47.** $2s^5 + 5t^5$ **49.** $17x^7 + 8x^5 - 6x^2$

**51.** $-3x^3 + 7y^2$ **53.** $6x - 4y + 8xy$ **55.** $6s^2 + 8s - 2st$ **57.** $(-2)^8$

**59.** $7^{10}$ **61** $3^{15}$ **63.** $16^4$ **65.** $\left(\frac{1}{3}\right)^9$ **67.** $5^4$

**69.** 10 **71.** $(.7)^{23}$ **73.** $a^{10}$ **75.** $y^2$ **77.** $s^0 = 1$

**79.** $a^8b^3$ **81.** $(xy)^4$ **83.** $x^{-1}z^5$ **85.** $(abc)^{-3}$

**87.** $7^2$ **89.** .5 **91.** $x^4$ **93.** $t^4$ **95.** $y^{-2}$ **97.** $a^{-2}$

**99.** $(.3)^3$ **101.** $x^{-3}$ **103.** $t^{-7}$ **105.** $(abc)^{-3}$

**107.** $5^4$ **109.** $(.2)^8$ **111.** $(.27)^0 = 1$ **113.** $y^{15}$

**115.** $t^{12}$ **117.** $w^{12}z^{20}$ **119.** $t^0w^0 = 1$ **121.** $a^9$

**123.** $3^{10}y^{15}$ **125.** $6^6a^{18}$ **127.** $5^4x^8y^4$

**CHAPTER TEST** *Form 6A*

**1.** 100000 **3.** 144 **5.** $\frac{1}{27}$ **7.** $\frac{1}{728}$ **9.** $-1$ **11.** 1 **13.** 8 **15.** $\left(\frac{7}{3}\right)^4$ **17.** $\frac{1}{25}$

**19.** 1 **21.** $\frac{49}{16}$ **23.** $23(6)^2$ **25.** $17x^2 - 3$ **27.** $17^8$ **29.** $x^{17}$ **31.** $5^6$

**33.** $(121)^{15}$ **35.** $(9)^{-12}$

**CUMULATIVE EXERCISES** **Chapters 1–6** *Set A*

**1.** 0 **3.** \$130.50 **5.** $7a + 7b$ **7.** 12 **9.** $-13$ **11.** $-3.7$ **13.** $5^3 \cdot 2^3$ **15.** $5^3$

**17.** $\frac{17}{9}$ **19.** $\frac{25}{6}$ **21.** 78 **23.** $\frac{-34}{567}$ **25.** $\frac{-99}{14}$ **27.** $\frac{-9}{20}$ **29.** $\frac{71}{15}$ **31.** 996

**33.** $\frac{1}{100}$ **35.** $-3 + 4y^5 + 6xy^5 - 8xy$ **37.** $6a^3 - 10a + a^2$ **39.** $\frac{1}{s^3}$ **41.** 1

**CUMULATIVE TEST** **Chapters 1–6** *Form A*

**1.** 125 **3.** $\frac{13}{2}$ **5.** 100 **7.** $-14.00$ **9.** 0 **11.** $5^2 \cdot 41$ **13.** $\frac{4}{9}$ **15.** $\frac{-1}{3}$

**17.** $\frac{9}{16}$ **19.** 30.5 **21.** $\frac{1}{16}$ **23.** $3(5^6)$ **25.** $\frac{b^3}{a^3}$ **27.** $\frac{1}{c}$

# CHAPTER SEVEN

**Exercise Set 7.2**

**1.** $43v + 19$ **3.** $7x + 2y$ **5.** $2s - r + 2t$ **7.** $9cd^2 - 4c^2d^2$ **9.** $8t^2 - 1 - 9t^3$
**11.** $0^3 - 40^5$ **13.** $8rs^5 - 6rs^3 + 9r^3s$ **15.** $8uv^3 - 4u^3v^3$ **17.** $4x^2 - 3x + 5$
**19.** $20a^2b - 8ab^2$

**Exercise Set 7.3**

**1.** $10$ **3.** $11c^3 + 7d^2 - 15a$ **5.** $11a^2 - 6b$ **7.** $-x + 3$ **9.** $-6a^2b^2$
**11.** $a^2b + 10a + 2$ **13.** $11c^5 + 14b^4 + 6bc - 16b + 4$ **15.** $20x + 3y + 4z$

**Exercise Set 7.4**

**1.** $a - 4b + 9$ **3.** $3c^2 - 1$ **5.** $-7c^3 - 12c$ **7.** $3y^2 + y - 5$
**9.** $6mn - n^2 - m^2 - 1$ **11.** $3x^3 - 6x^2 - 4x - 8$ **13.** $3y^2 - 5y - 7$
**15.** $-y - 7x^4y^2 + 3x^2 - 6x^4 + 3x^5y$

**Exercise Set 7.5**

**1.** $140x^8y^3z^2$ **3.** $12t^3s$ **5.** $.16c^3$ **7.** $-6a^2b^3$ **9.** $-10m^4n$ **11.** $p^3qr$ **13.** $4a^2b$
**15.** $26a^2b$ **17.** $-2a^2b^2c^2$ **19.** $a^5b^2c^2d^2$

**Exercise Set 7.6**

**1.** $5a + 5b$ **3.** $c^2d - 2cd^2$ **5.** $-40y^4z$ **7.** $28y^4 - 42y^2 + 14y$
**9.** $6x^3 + 18x^2 + 12x$ **11.** $6r + 14r^2$ **13.** $-2a^2b^3 - acb^2$ **15.** $6a^5b - 18a^4$

**Exercise Set 7.7**

**1.** $3x^2 + 2x - 8$ **3.** $x^2 + 8x + 16$ **5.** $6a^2 - 20a + 6$ **7.** $a^2 + 2ab + b^2$
**9.** $a^2 + 6a + 8$ **11.** $4r^2 + 4rs + s^2$ **13.** $b^2 + bc - 2c^2$ **15.** $4x^2 - 18x - 10$
**17.** $3x^2 - 2x - 8$ **19.** $x^2 - 16$ **21.** $s^2 - 8s + 15$ **23.** $a^2 - 2ab + b^2$
**25.** $a^2 - 2a - 8$ **27.** $9y^2 + 6y + 1$ **29.** $b^2 - bc - 2c^2$ **31.** $4x^2 - 22x + 10$

**Exercise Set 7.8 A**

**1.** $3x + 12y - 18z$ **3.** $2ab + 10ac - 14ad$ **5.** $ab - 4a + 3b - 12$
**7.** $3a^2 - 7ab - 6b^2$ **9.** $7y^2 - 62y - 9$ **11.** $m^2 + mq - 2q^2$

**Exercise Set 7.8 B**

**1.** $a^2 + 2ab + b^2$ **3.** $4a^2 + 4ac + c^2$ **5.** $d^2 + 6de + 9e^2$ **7.** $25s^2 + 10st + t^2$
**9.** $9q^2 + 24qt + 16t^2$ **11.** $49p^2 + 28pr + 4r^2$

**Exercise Set 7.8 C**

**1.** $a^2 - b^2$ **3.** $4s^2 - 9t^2$ **5.** $t^2 - 16r^2$ **7.** $u^2 - 49v^2$

**Exercise Set 7.8 D**

**1.** $a^2 + ac + bc - b^2$ **3.** $12de - 15df + 3dg + 4e^2 - 5ef + eg$
**5.** $5s^2 + 6st - 5su + t^2 - tu$

**Exercise Set 7.9**

**1.** $5x^2 - 6x$ **3.** $8x + 6$ **5.** $x^2 - 3$ **7.** $-7a^2 + 6a$ **9.** $ax + b$
**11.** $-17 - 16c + 1.5c^2$ **13.** $8w^2 - 3w - \frac{2}{w}$ **15.** $\frac{-3}{2}c - 3d$

## Chapter Seven Exercises

**1.** $10x$ **3.** $2y^2 - 7$ **5.** $19 - 5z$ **7.** $-2b^2 + 4$ **9.** $21 - 12x$ **11.** $10xy - 1$
**13.** $2x - 10b + a$ **15.** $12qt + 9q + 2t$ **17.** $106x^2 - 4y^2 - 7$ **19.** $7.8t - 6.3s$
**21.** $6t^2 - 4t - 6$ **23.** $7g^2h - 11.1$ **25.** $10x + 8$ **27.** $-2xy - 1$
**29.** $13y^2 - 10y - 1$ **31.** $7st - 8 - 6t$ **33.** $12x^2 - 24x + 11$ **35.** $13a^2 - 5ab$
**37.** $5.7 + 3.0x - x^2$ **39.** $-12z + 2y - 3$ **41.** $16b^2 - 9b + 4$ **43.** $42a + 3a^2 - 5.9$
**45.** $21y^2 + 11$ **47.** $s^2 + 5s$ **49.** $-2b^2 - 16b$ **51.** $-4.1s^2 - 7$
**53.** $5.6r^2 + 6r - 1.8$ **55.** $-3 + t + 10t^2$ **57.** $-11a - 5$ **59.** $-5k - 151 + 2$
**61.** $-25st + 6s^2 - 7t^2$ **63.** $16y$ **65.** $-14y^2$ **67.** $-42b^2$ **69.** $-60y^4$ **71.** $72x^3$
**73.** $-12x^2y^2$ **75.** $30x^4y$ **77.** $9x^2$ **79.** $9x^2$ **81.** $25z^4$ **83.** $35x^3$ **85.** $16x^8y^8$
**87.** $-27x^{12}$ **89.** $28x^2y$ **91.** $2x^4$ **93.** $6y^3$ **95.** $48a^3$ **97.** $30x - 40z$
**99.** $-42w + 14w^2$ **101.** $8x^2 + 6x$ **103.** $-60x - 45z + 15xz$ **105.** $-21a^3b^2 + 18ab$
**107.** $.08x^3y^2 - .6xy^3$ **109.** $30w^3 - 9w^3 + 15w^4$ **111.** $-28c + 42d$
**113.** $20r^2 + 6r - 8$ **115.** $-42a^3 + 18a^2 - 66a$ **117.** $15y^3 - 9y^2 - 3$
**119.** $6w^2x^3y^3 - 2wx^2y$ **121.** $35x^2y^3 - 10x^5y^6$ **123.** $42s^3y^2 - 126s^3y^3$
**125.** $-42x^2y^2 - 28x^4y + 70x^3y^3$ **127.** $y^2 - 7y + 12$ **129.** $w^2 + 8w - 48$
**131.** $10d^2 + d - 2$ **133.** $a^2 - 64$ **135.** $3x^2 + 28x - 20$ **137.** $2c^2 - cd - d^2$
**139.** $y^2 - 14y + 49$ **141.** $x^2 - 2xy + y^2$ **143.** $w^2 - z^2$ **145.** $12x^2 + 41x - 28$
**147.** $x^2 - 1$ **149.** $-2b^2 - 17b - 21$ **151.** $-15k - 2k^2 - 22$
**153.** $3xy + 15x - 4y - 20$ **155.** $4w^2 + 12w + 9$ **157.** $2x^4 - 5x^2 - 3$
**159.** $x^2 + 2xy^2 + y^4$ **161.** $4d^4 - 36$ **163.** $6x^2 - 0.2x - .20$ **165.** $.64 + 1.6x^2 + x^4$
**167.** $3a^3 + 11a^2 + 10a + 12$ **169.** $5x^3 + 22x^2 - 6x + 45$ **171.** $2r^3 + 3r^2t - t^3$
**173.** $a^3 + ba^2 - 6ab - 6b^2$ **175.** $5w^4 - 17w^3 - w^2 + 33w - 36$
**177.** $-21b^3 - 29b^2 + 22b + 20$ **179.** $-6y^3 + 29y^2 - 23y + 12$
**181.** $8t^4 - 6t^3 + 12t^2 - 12t - 8$ **183.** $3a^3 + a^2b + 2ab^2 - 6b^3$
**185.** $10a^2 - 27ay + 6ac + 18y^2 - 3cy - 28c^2$ **187.** $3abc^2 + 6bc - 1$
**189.** $3d + 2d^2 - \frac{4}{d}$

## Chapter Test *Form 7A*

**1.** $-6w + 7$, binomial **3.** $-7a^2 + b^2$, binomial **5.** $6a^2 - 7$ **7.** $7w^2 + 4w + 1$
**9.** $12x^3y^2$ **11.** $-8x^2y^4$ **13.** $-11y^3 + 33y^2 + 22y$ **15.** $x^2 - 4x - 21$
**17.** $x^2 - 3x - 10$ **19.** $6x^2 + 7x - 3$ **21.** $6x^3 + 7x^2 + 5x + 12$
**23.** $2a^3 - 3a^2b + b^3$ **25.** $a^3 + 9a^2 - a^2b - 8ab - b^2$ **27.** $2y + 1 - \frac{3}{y}$ **29.** $z + 2x$

## CUMULATIVE EXERCISES Chapters 1–7 *Set 7A*

**1.** \$95 **3.** $\frac{1}{75}$ **5.** 22.2 **7.** 6.25 **9.** 6.25 **11.** $3^2 \cdot 2^2$ **13.** $\frac{110}{9}$ **15.** 100
**17.** $\frac{1}{10}$ **19.** $-3.5$ **21.** 16 **23.** $\frac{4}{25}$ **25.** 1 **27.** $\frac{1}{X^3}$
**29.** $11T^3 - 6T + 7$ **31.** $7a^2bc^2 + 8a^2b^2$ **33.** 0
**35.** $-ab + 8ad^2 - 6c - 30d^2 + 3abd^2$ **37.** $2xy^2z + yz^2 - 3$

## CUMULATIVE TEST Chapters 1–7 *Form A*

**1.** associative **3.** commutative **5.** identity **7.** $>$ **9.** $>$ **11.** 5,37
**13.** $4x^2 - 3x + 3$ **15.** $x^2 + y^2 - 12$ **17.** $4x$ **19.** $-52x^2yz + 45xy$
**21.** $4xy^2 - 5y^3 - 7$ **23.** $y + xy - 1$

# CHAPTER EIGHT

## Exercise Set 8.1

**1.** $2n - 4$ **3.** $2n + 16$ **5.** $2(n + 16) = 34$ **7.** $\frac{1}{2}n$ or $\frac{n}{2}$ **9.** $\left(\frac{n}{2}\right) + 4$

**11.** $n - 3n$ or $-2n$ **13.** $2n + 7n$ **15.** $\frac{2}{3}n + n = 15$

## Exercise Set 8.2

**1.** 18 **3.** 20 **5.** 2 **7.** 6 **9.** 34 **11.** 15 **13.** 7 **15.** 6 **17.** 30 **19.** 3
**21.** 20 **23.** 3 **25.** 11

## Exercise Set 8.3

**1.** 18 **3.** 72 **5.** 13 **7.** 6 **9.** 50 **11.** 10 **13.** 480 **15.** 120 **17.** 36
**19.** 36 **21.** 80 **23.** 100 **25.** 11

## Exercise Set 8.4

**1.** 16 **3.** 18 **5.** 3 **7.** 9 **9.** 8 **11.** 10 **13.** 10 **15.** 4 **17.** 8 **19.** 5

## Exercise Set 8.5

**1.** 2 **3.** 6 **5.** 20 **7.** 32 **9.** 5 **11.** 3 **13.** $-3$ **15.** 15

## Exercise Set 8.6

**1.** $x = 5$ **3.** $y = -14$ **5.** $z = -11$ **7.** $c = 2.2$ **9.** $x = -1$ **11.** $x = -30$
**13.** $x = 10$ **15.** $x = 110$ **17.** $x = 20$ **19.** $w = -1.1$

## Exercise Set 8.7

**1.** $A = 2800$ **3.** $y = \sqrt{32}$ **5.** $b = 5$ **7.** $V = 301.44$ **9.** $L = 25$

## Exercise Set 8.8

**1.** $h = \frac{A}{b}$ **3.** $c = P - (a + b)$ **5.** $s = \frac{d}{t}$ **7.** $h = \frac{2A}{b}$ **9.** $R = \frac{I}{PT}$ **11.** $b = \frac{A}{h}$
**13.** $a = P - (b + c)$ **15.** $a = \sqrt{c^2 - b^2}$ **17.** $s = \sqrt{A}$ **19.** $P = C - T$

## Chapter Eight Exercises

**1.** $n + 5$ **3.** $2n + 2$ **5.** $3n - 5$ **7.** $\frac{1}{2}n + 10$ **9.** $2n = 24$ **11.** $n + 4 = 44$
**13.** $2n - 7 = 5$ **15.** $n - 5 = 25, n = 30$ **17.** $n - 4 = 28, n = 32$
**19.** $n + 15 = 27, n = 12$ **21.** $35 + n = 50, n = 15$ **23.** $25 + n = 32, n = 7$
**25.** $n - 5 = 2, n = 7$ **27.** $n + 8 = 0, n = -8$ **29.** $n - 16 = 20, n = 36$
**31.** $n - 1.5 = 3.5, n = 5$ **33.** $n - 3 = 17, n = 20$ **35.** $26 = n - 7, n = 33$
**37.** $11 = n + 5, n = 6$ **39.** $-9 = n + 3, n = -12$ **41.** $n + (-13) = 0, n = 13$
**43.** $n + (-2.5) = 0, n = 2.5$ **45.** $\frac{1}{3}n = 18, n = 54$ **47.** $\frac{n}{3} = 21, n = 63$
**49.** $\frac{2}{3}n = 50, n = 75$ **51.** $60 = \frac{2}{3}n, n = 90$ **53.** $\frac{n}{16} = 96, n = 1536$
**55.** $\frac{5}{4}n = 28, n = 22.4$ **57.** $\frac{2n}{3} = 45, n = 67.5$ **59.** $2n + 4 = 50, n = 23$
**61.** $\frac{1}{4}n = 32, n = 128$ **63.** $4(n + 7) = 44, n = 4$ **65.** $(n + 6) + 12 = 55, n = 37$

**67.** $7n - 22 = 64, n = 12.29$ **69.** $3(n - 8) = 42, n = 22$ **71.** $6n - 5 = 43, n = 8$
**73.** $9n - 10 = -64, n = -6$ **75.** $3(2n + 2.7) = 4.5, n = -0.6$
**77.** $\frac{2n}{5} + 7 = 43, n = 90$ **79.** $17 - 2n = 35, n = -9$ **81.** $12 - 4n = 48, n = -9$
**83.** $7.5 - 2n = 3.7, n = 1.9$ **85.** $10 - (n + 3) = -16, n = 23$
**87.** $5 - 5n = 35, n = -6$ **89.** $174 - 6n = 51, n = 20.5$ **91.** $n = 19$ **93.** $n = 22$
**95.** $y = 2$ **97.** $a = -10.6$ **99.** $y = -4$ **101.** $a = 14$ **103.** $x = -29$
**105.** $x = 2$ **107.** $x = .5$ **109.** $x = -11$ **111.** $x = -3.625$ **113.** $x = 48$
**115.** $x = 8$ **117.** $x = \frac{8}{12} = \frac{2}{3}$ **119.** $x = 4$ **121.** $V = 84$ **123.** $r = \frac{p - 2q}{2}$
**125.** $r = \sqrt{s}$ **127.** $t = \frac{s}{3r}$ **129.** $b = \frac{2A}{h} - c$

**Chapter Test** *Form 8A*

**1.** 27 **3.** −36 **5.** 18 **7.** −54 **9.** −4.5 **11.** −20 **13.** 3.12 **15.** −13
**17.** $x = 0$ **19.** $z = 9$ **21.** $y = 3$ **23.** $z = -3$ **25.** $x = \frac{-1}{7.4}$ **27.** $y = 4$
**29.** $d = 2r$

**CUMULATIVE EXERCISES Chapters 1–8** *Set A*

**1.** $\frac{1}{36}$ **3.** 1.44 **5.** −23.75 **7.** $-x^2 + 3x + 9$ **9.** $8y - 7$ **11.** $\frac{1}{100}$
**13.** $x^5$ **15.** $\frac{9}{121}$ **17.** $\frac{3}{5}$ **19.** $x^2 + 3x - 10$ **21.** $14a^3b^2 + 35a^2b^2 + 7abc$
**23.** $x^4 - 64$ **25.** $w - 4 + \frac{3}{w}$ **27.** $b = -8$ **29.** $x = 6$ **31.** $x = -100$
**33.** $n = -3$ **35.** 18,20 **37.** $x = 7$ **39.** $x = 1$ **41.** $x = \frac{-1}{2}$ **43.** 80
**45.** $a = 2c - b$

**CUMULATIVE TEST Chapters 1–8** *Form A*

**1.** $\frac{-6}{5}$ **3.** 56 **5.** 4.87 **7.** $2x^3$ **9.** $\frac{1}{15}$ **11.** $9x^2y - 6x^2z + 10x^2$
**13.** $3x^2 - 13x - 10$ **15.** $xy + 5y - 6$ **17.** $x = 16$ **19.** $n = 5$ **21.** $k = 22$

## CHAPTER NINE

**Exercise Set 9.2 B**

**1.** (3,0) **3.** (−2,−7) **5.** (0,−12) **7.** (−11,6) **9.** (1.5,−1.5)

**Exercise Set 9.2 C**

**1.** II **3.** x axis **5.** y axis **7.** IV **9.** II **11.** I **13.** I **15.** II

**Exercise Set 9.3**

(Answers will vary)

**Exercise Set 9.4**

**1.** $y = x + 4$ should go through (0,4) and (−4,0)
**3.** $y = 5x - 10$ should go through (0,−10) and (2,0)
**5.** $y = -3x + 2$ should go through (0,2) and $\left(\frac{2}{3},0\right)$
**7.** $y = \frac{1}{3}x + 2$ should go through (0,2) and (−6,0)
**9.** $y = \frac{-3}{4}x + 5$ should go through (0,5) and $\left(6\frac{2}{3},0\right)$
**11.** $y = 2x$ should go through (0,0) and (1,2)
**13.** $y = -4x$ should go through (0,0) and (1,−4)
**15.** $y = -1$ is a horizontal line 1 unit below the x axis.

**Exercise Set 9.5 A**

**1.** yes **3.** yes **5.** yes **7.** no **9.** yes **11.** no **13.** no

**Exercise Set 9.5 B**

**1.** 4 **3.** 5 **5.** 2 **7.** −7 **9.** −10 **11.** −4

**Exercise Set 9.6**

**1.** $m = 1, b = 4$ **3.** $m = 2, b = 0$ **5.** $m = 2, b = 4$ **7.** $m = -3, b = 2$
**9.** $m = 2, b = 8$ **11.** $m = \frac{2}{3}, b = -4$ **13.** $m = \frac{3}{-4}, b = 5$ **15.** $m = 0, b = 3$
**17.** no slope, no y intercept **19.** $m = 1, b = -3$ **21.** $m = 2, b = 0$

**Chapter Nine Exercises**

**11.** y axis **13.** III **15.** III **17.** IV **19.** x axis **21.** (3,8),(2,7),(1,6)
**23.** (1,7),(0,3),(−1,−1) **25.** (2,1),(1,3),(0,5) **27.** (1,−7),(0,−2),(−1,3)
**29.** (3,5),(2,0),(0,−10) **31.** (6,3),(3,1),(0,−1) **33.** (6,2),(0,−3),(−6,−8)
**35.** (4,−17),(0,−10),$\left(-1,\frac{-33}{4}\right)$ **37.** (3,5),(0,0),(−3,−5) **39.** (1,9),(2,6),(3,3)
**41.** yes **43.** yes **45.** no **47.** no **49.** yes **51.** no **53.** no **55.** no **57.** yes
**59.** no **61.** yes **63.** yes **65.** no **67.** yes **69.** no **71.** (2,9) **73.** (10,−3)
**75.** (−2,15) **77.** (−3,16) **79.** (2,10) **81.** (3,18) **83.** (21,14) **85.** (−3,18)
**87.** (3,7) **89.** (2,−4) **91.** $m = 1, b = 5$ **93.** $m = 4, b = 3$ **95.** $m = -2, b = 5$
**97.** $m = -5, b = -2$ **99.** $m = 5, b = -10$ **101.** $m = \frac{2}{3}, b = -1$
**103.** $m = \frac{5}{6}, b = -3$ **105.** $m = \frac{-7}{4}, b = -10$ **107.** $m = \frac{5}{3}, b = 0$
**109.** $m = -3, b = 12$ **111.** $m = \frac{-3}{2}, b = 4$ **113.** $m = 0, b = 7$

**Chapter Test Form 9A**

1.

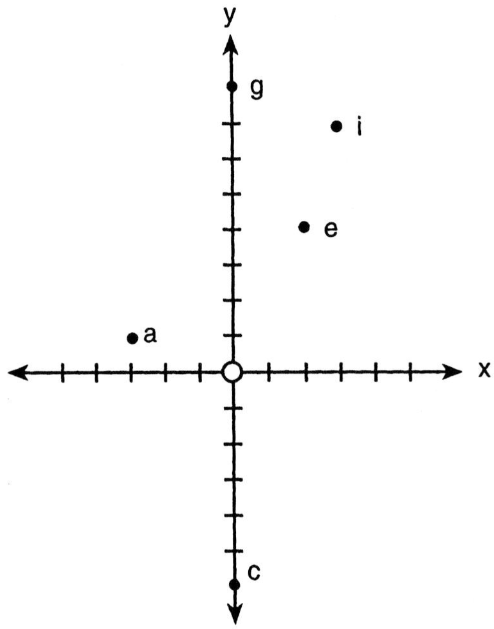


2. a. x axis  c. II  e. x axis  g. I  i. I

3. a. $y = -2$ is a horizontal line 2 units below the x axis.

c. $y = \frac{2}{3}x$ goes through (0,0) and (3,2)

5. $y = -14$  7. d  9. b  11. $m = -3, b = 2$

**CUMULATIVE EXERCISES Chapters 1–9** *Set A*

1. $-8$  3. 46656  5. $10y^2 - 9y + 2$  7. $2x^3 + 3x^2 - 6x + 6$  9. $a^5$  11. $\frac{113}{30}$

13. $\frac{7}{10}$  15. $\frac{1}{900}$  17. $a^2 + a - 12$  19. $16a^2 + 8a + 1$  21. $-2z^2 + 3z - 7$

23. $d = 3$  25. $y = -1$  27. $x = -30$  29. $4n - 3 = 25, n = 7$

31. $4n + 5 = 3n, n = -5$  33. $\frac{1}{3}n + 3 = 25, n = 66$  35. $x = 4$  37. $k = \frac{(z - y)}{2}$

39. $a = \frac{1}{2}$  41. IV  43. III  45. $y = -17$  47. $y = -7$  49. $m = 7, b = -2$

51. $m = \frac{2}{3}, b = 3$  53. the line goes through (0,5) and $\left(\frac{-5}{2},0\right)$

55. the line goes through (0,6) and (3,0)

**CUMULATIVE TEST Chapters 1–9** *Form A*

1. distributive  3. $18x^5$  5. $\frac{1}{7}$  7. $-5x^2 + 8x + 12$  9. $x^4 + x^2 - 12$  11. $x = 3$

13. 17  15. $w = \frac{(P - 2L)}{2}$  17. (0,2)  19. the line goes through (0,5) and $\left(\frac{5}{2},0\right)$

## CHAPTER TEN

### Exercise Set 10.1

**1.** $5x$ **3.** $13x$ **5.** $3xz$ **7.** $3rs$ **9.** $3x$

### Exercise Set 10.2

**1.** $3x(x + 4y + 3x^4)$ **3.** $5t^5(1 - 2t^2)$ **5.** $8r(r^3 + 2s + 3rs^3)$ **7.** $3t(2t + 1 + 8t^4)$
**9.** $3xy(1 + 2xy + 4x^3y^2)$ **11.** $6a(a + 4b + 3a^4)$ **13.** $3s(4t^3 - t^2 + 2s)$
**15.** Answers will vary.

### Exercise Set 10.3

**1.** $(m + n)(m - n)$ **3.** $2(c + d)(c - d)$ **5.** $7(a + 2b)(a - 2b)$
**7.** $9(2s^2 - t)(2s^2 + t)$ **9.** $(x + 11)(x - 11)$

### Exercise Set 10.4

**1.** $(y + 3)(y + 1)$ **3.** $(y - 3)(y + 1)$ **5.** $(z + 4)(z + 1)$
**7.** $(c + 1)(c + 1)$ or $(c + 1)^2$ **9.** $(c + 3)(c - 1)$ **11.** $(a + b)(a + b)$ or $(a + b)^2$

### Exercise Set 10.5

**1.** $4(x + 2y)(x - 2y)$ **3.** $5a(1 - 2a)$ **5.** $4(5 + x)(5 - x)$ **7.** $9x(x + 2)(x - 2)$
**9.** $2(w - 6)(w + 2)$

### Chapter Ten Exercises

**1.** $5a$ **3.** $3x$ **5.** $x$ **7.** $11c^2$ **9.** $3ab^2c$ **11.** $.3x^2$ **13.** $5yz$ **15.** $3b$ **17.** $\frac{1}{7}a$
**19.** $3r^2t^2$ **21.** $5a^2bc^2$ **23.** $x^2$ **25.** $3x(1 + 2x)$ **27.** $3x^2(2x^2 - 3)$ **29.** $3x^2(y - 2)$
**31.** $2a^2b(2 - 5b - 8b^2)$ **33.** $2.3a^2(1 + 2a)$ **35.** $\frac{1}{3}x^3(2x^2 - 5)$ **37.** $r(r - 4)$
**39.** $\frac{1}{2}pr(r - 3)$ **41.** $\frac{1}{2}h(b + x)$ **43.** $(a + b)(a - b)$ **45.** $(z + 4)(z - 4)$
**47.** $(6 - x)(6 + x)$ **49.** $9(x + 2)(x - 2)$ **51.** $(5k - 13)(5k + 13)$
**53.** $(x^2 - a)(x^2 + a)$ **55.** $(4a^4 - y^3)(4a^4 + y^3)$ **57.** $(ay - bz)(ay + bz)$
**59.** $(.2 - w^2)(.2 + w^2)$ **61.** $\left(\frac{11}{5} - y\right)\left(\frac{11}{5} + y\right)$ **63.** $(x - 3)(x - 2)$
**65.** $(y + 4)(y + 3)$ **67.** $(w - 6)(w + 2)$ **69.** $(a - 11)(a - 2)$ **71.** $(x + 8)(x + 4)$
**73.** $(w + 9)(w + 9)$ **75.** $(y - 9)(y + 8)$ **77.** $(w + 11)(W + 3)$
**79.** $(y + 11)(y - 2)$ **81.** $(b + 8)(b + 8)$ **83.** $(a - 15)(a + 4)$
**85.** $(d + 5)(d + 4)$ **87.** $(x - 36)(x - 2)$ **89.** $(y - 15)(y + 1)$

### Chapter Test *Form 10A*

**1.** $7$ **3.** $5x$ **5.** $y^2$ **7.** $3x(3x^2 - 5x + 1)$ **9.** $2z^2(25z^2 + 16 - 4z^4)$
**11.** $(9 - x)(9 + x)$ **13.** $(ay - cx)(ay + cx)$ **15.** $(x - 7)(x - 5)$
**17.** $(w + 12)(w + 3)$ **19.** $(y - 13)(y - 2)$

## CUMULATIVE EXERCISES Chapters 1–10 *Set A*

**1.** $\frac{-3}{7}$ **3.** 24.35 **5.** 6.1 **7.** 28 **9.** 64 **11.** $15a^2$ **13.** $-3c^2 - 10c + 1$

**15.** $20a^3 - 15a^2 + 25a$ **17.** $\frac{49}{4}$ **19.** $3x - 6$ **21.** $2x^2 + 3xy + y^2$ **23.** $\frac{29}{4}$

**25.** $c = \frac{(5 - b^2)}{4a}$ **27.** $a = 6$ **29.** $7n + 4 = 6n, n = -4$ **31.** $4a + 9 = 81, a = 18$

**33.** I **35.** II **37.** II **39.** $m = -7$ **41.** $m = \frac{5}{7}$ **43.** −8 **45.** 9

**47.** the line should go through (0,7) and $\left(\frac{7}{2},0\right)$ **49.** $9(x^2 + 3)$ **51.** $(x + 4)(x - 4)$

**53.** $(x - 2)(x - 1)$ **55.** $(a^2 + 4)(a - 2)(a + 2)$

## CUMULATIVE TEST Chapters 1–10 *Form A*

**1.** $\frac{-1}{41.5}$ **3.** 10 **5.** $y^3 + 2y^2 - 3y - 6$ **7.** $a - 4b + 2b^2$ **9.** $x = \frac{5}{3}$

**11.** $h = \frac{2A}{b}$ **13.** False, III **15.** False, $\frac{-5}{11}$ **17.** the line should go through (0,2) and (−6,0)

**19.** $(6y - 1)(6y + 1)$ **21.** $(y^2 + 1)(y + 1)(y - 1)$

# CHAPTER ELEVEN

### Exercise Set 11.1

**1.** $\frac{2x}{3y}$ **3.** $\frac{x}{2y^3}$ **5.** $\frac{3}{5x}$ **7.** $\frac{ab}{3c^4}$ **9.** $\frac{d}{10c}$ **11.** $\frac{s^2}{4r^3}$ **13.** $\frac{3x^3}{y^2}$ **15.** $ac^2$

### Exercise Set 11.2

**1.** $\frac{3}{25t}$ **3.** $\frac{8z}{5x}$ **5.** $\frac{9ad}{2}$ **7.** $2y^2x^2$ **9.** $\frac{y}{2}$

### Exercise Set 11.3

**1.** $\frac{5}{2x}$ **3.** $\frac{(10x^2y - 1)}{8}$ **5.** $\frac{(2 + x - 2y)}{10}$ **7.** $\frac{32y}{21}$ **9.** $\frac{-7t}{15}$ **11.** $\frac{9}{2y}$

**13.** $\frac{(5y + 2)}{6}$ **15.** $\frac{5m^2}{9}$ **17.** $\frac{(22x + 5)}{4}$ **19.** $\frac{-3}{a + 3}$

### Exercise Set 11.4

**1.** $x = -12$ **3.** $y = -5$ **5.** $x = 40$ **7.** $y = 17$ **9.** $z = 2$ **11.** $z = 7$

**13.** $z = -2$

**Chapter Eleven Exercises**

**1.** $\frac{6}{5}$ **3.** $\frac{6y^2}{7}$ **5.** $8x$ **7.** $\frac{2a}{5b}$ **9.** $\frac{-3bc}{5a}$ **11.** $6x^2$ **13.** $\frac{7x}{2}$ **15.** $12c^2$ **17.** 20
**19.** $\frac{3c^2}{7a}$ **21.** $\frac{-4tv}{5}$ **23.** $\frac{36y^3a}{x^3}$ **25.** $\frac{27xy}{100}$ **27.** $\frac{25}{4at^3x^3}$ **29.** $\frac{wk^4}{6}$ **31.** $2x$ **33.** $\frac{11}{y^2}$
**35.** $\frac{(2-7x)}{w}$ **37.** $\frac{(4+3x)}{5x}$ **39.** $\frac{(3y-3)}{x}$ **41.** $\frac{5}{2x}$ **43.** $\frac{(2+3x)}{x^2}$ **45.** $\frac{-1}{4x^2}$
**47.** $\frac{(8-y)}{y^2}$ **49.** $\frac{(5x^2-3x)}{10}$ **51.** $\frac{1}{2-y}$ **53.** $\frac{(7y-4)}{6t}$ **55.** $\frac{(y-1)}{15}$ **57.** 21
**59.** 35 **61.** −20 **63.** 23 **65.** 1 **67.** 1 **69.** 2 **71.** 30 **73.** 36.5 **75.** 18

**Chapter Test** *Form 11A*

**1.** 2 **3.** 4 **5.** $5y^2$ **7.** $\frac{a^2}{4}$ **9.** $\frac{9y}{4}$ **11.** $\frac{-7}{2x}$ **13.** $\frac{(x-2)}{(x+6)}$ **15.** 10 **17.** −18
**19.** $y = 43$

**CUMULATIVE EXERCISES Chapters 1–11** *Set A*

**1.** 16.044 **3.** 49 **5.** $\frac{9}{100}$ **7.** $-2a^2b^2 + a^2b$ **9.** $\frac{1}{42}$ **11.** $9x^2 - 12x + 4$
**13.** $-32x^4 + 28x^3 - 8x^2 + 28x$ **15.** $9z^2 + 2z - 4$ **17.** $n + n + 2 = 176$; 87, 89
**19.** $\frac{(312+x)}{5} = 80$; $x = 88$ **21.** $x = \frac{(-8-a)}{3}$ **23.** $m = 7, b = -5$
**25.** $m = \frac{1}{7}, b = \frac{-2}{3}$ **27.** $m = 8, b = 4$ **29.** the line goes through (−4,0) and (0,4)
**31.** the line goes through (0,1) and $\left(\frac{-3}{2},0\right)$ **33.** $8(x+3)$ **35.** $3(x+3)(x-3)$
**37.** $(4a-5)(4a+5)$ **39.** $(k-m)(k+m)$ **41.** $(x+7)(x-1)$ **43.** $\frac{4b}{3a^3}$
**45.** $\frac{7z}{3xy^4}$ **47.** $\frac{7}{12x}$ **49.** $\frac{4a}{3b}$ **51.** $\frac{40}{a^3}$ **53.** $\frac{(14+3a)}{2(a+2)}$
**55.** $x = 30$ **57.** $x = -4$

**CUMULATIVE TEST Chapters 1–11** *Form A*

**1.** 12.17 **3.** −49 **5.** $6x^2 - 3x - 3$ **7.** $3x^2(2x-1)$ **9.** $\frac{-49}{150}$
**11.** $49x^2 - 42x + 9$ **13.** $\frac{16a^2}{3b}$ **15.** $\frac{12xz}{y}$ **17.** $(a-1)(a+1)$ **19.** $115x$

## CHAPTER TWELVE

**Exercise Set 12.1**

**1.** 5,−5 **3.** 0,3 **5.** 1,−1 **7.** $\frac{3}{2}$,4

**Exercise Set 12.2**

**1.** 11,−11 **3.** 5,4 **5.** 3,2 **7.** 3,1 **9.** 5,−1 **11.** 3,−1 **13.** −4,−1 **15.** 6,−2
**17.** 6,−6 **19.** 15,−4

**Exercise Set 12.3**

**1.** $a = 1, b = 0, c = -121$ **3.** $a = 1, b = -9, c = 20$ **5.** $a = 5, b = -25, c = 30$ **7.** $a = 1, b = -4, c = 3$ **9.** $a = 1, b = -4, c = -5$ **11.** $a = 1, b = -2, c = -3$ **13.** $a = 1, b = 5, c = 4$ **15.** $a = 2, b = -8, c = -24$ **17.** $a = 1, b = 0, c = -36$ **19.** $a = 1, b = -11, c = -60$

Exercise Sets 12.4 A and 12.4 B have the same answers as Exercise Set 12.2

**Chapter Twelve Exercises**

**1.** 3,−1 **3.** 6,−2 **5.** −3,2 **7.** 4,−4 **9,** −6,3 **11.** −10,9 **13.** 12,1 **15.** $3,\frac{1}{2}$ **17.** 3,2 **19.** 5,−5 **21.** 5,2 **23.** −7,−3 **25.** −7,1 **27.** 6,−4 **29.** 6,1 **31.** −4,−3 **33.** 6,5 **35.** 15,−6 **37.** 0,7 **39.** 0,15 **41.** 0,5 **43.** 0,2 **45.** 5,−5 **47.** 4,−4 **49.** 6,7 **51.** 13,3

**Section 12.3**

**17.** $a = 1, b = -5, c = 6$ **19.** $a = 1, c = -25$ **21.** $a = 1, b = -7, c = 10$ **23.** 1, 10, 21 **25.** 1, 6, −7 **27.** 1, −2, −24 **29.** $w^2 - 7w + 6$, 1, −7, 6 **31.** $x^2 + 7x + 12$, 1, 7, 12 **33.** 1, −11, 30 **35.** 1, −9, −90 **37.** 1, −7, 0 **39.** $-x^2 + 15x$, −1, 15, 0 **41.** 3, −15, 0 **43.** $6y^2 - 12y$, 6, −12, 0 **45.** $x^2 - 25$, 1, 0, −25 **47.** $z^2 - 16$, 1, 0, −16 **49.** $y^2 - 13y + 42$, 1, −13, 42 **51.** $c^2 - 16c + 39$, 1, −16, 39

**Section 12.4**

**53.** −4,−1 **55.** 5,2 **57.** −3,1 **59.** −7,−3 **61.** 10,−1 **63.** 0,−5 **65.** 2,−2 **67.** −6,3 **69.** −7,4

**Chapter Test**

**1.** 5,6 **3.** 4,−4 **5.** −5,3 **7.** 5,2 **9.** 0,−12 **11.** 5,5 **13.** 1, 2, 0 **15.** $x^2 + 3x - 16$, 1, 3, −16

**CUMULATIVE EXERCISES Chapters 1–12** *Set A*

**1.** $\frac{-9}{17}$ **3.** $\frac{1}{8}$ **5.** −8 **7.** $7x^2y - 3xy^2$ **9.** $\frac{13}{30}$ **11.** $56x^2y^3 + 32x^3y^2 + 24xy$ **13.** $114t + 66sr$ **15.** yes **17.** no **19.** no **21.** $m = 13, b = -4$ **23.** $m = \frac{2}{3}, b = 3$ **25.** $y = 8 - x$ passes through (0,8) and (8,0) **27.** $y = 2x + 3$ passes through (0,3) and $\left(\frac{-3}{2},0\right)$ **29.** $8m^2n(1 + 3m^3n - 4mn^3)$ **31.** $2c^2d(3cd - 7c + 1)$ **33.** $(a + .5)(a - .5)$ **35.** $(x + 5)(x + 3)$ **37.** $(a - 4)(a + 1)$ **39.** $3x + 4 = 32 - x; x = 7$ **41.** $n + (n + 1) = 113$; 56, 57 **43.** $-6ab^2$ **45.** $\frac{9}{2(x - 2)}$ **47.** $x = 23$ **49.** $x = 57$ **51.** $\frac{1}{3}, \frac{-5}{2}$ **53.** −8,5 **55.** $\frac{2}{5}, \frac{-2}{5}$ **57.** $0, \frac{5}{8}$ **59.** −7,−5 **61.** 6,3 **63.** 7,−1 **65.** 11,−10

**CUMULATIVE TEST Chapters 1–12** *Form A*

**1.** $\frac{-3}{27} = \frac{-1}{9}$ **3.** $\frac{-4}{3}$ **5.** $\frac{(2x^2 - 6x - 18)}{x^2 + 3x}$ **7.** $1 - 4x + 4x^2$ **9.** $x = 28$ **11.** $(ab - z)(ab + z)$ **13.** 12, 14, 16 **15.** $\frac{-7}{3}, \frac{8}{5}$ **17.** 0,−9 **19.** 4,−4

## CHAPTER THIRTEEN

### Exercise Set 13.3 A

**1.** B = 40° **3.** A = 13° **5.** A = 78° B = 12° **7.** A = 30° B = 60°
**9.** A = 50° B = 40° **11.** A = 15° B = 75° **13.** A = 72° B = 18°

### Exercise Set 13.3 B

**1.** $\sqrt{90}$ **3.** 8 **5.** 2 **7.** $\sqrt{108}$ **9.** $\sqrt{147}$

### Exercise Set 13.4 A

**1.** $\sin A = \frac{9}{\sqrt{90}}$ $\cos A = \frac{3}{\sqrt{90}}$ $\tan A = 3$

**3.** $\sin B = \frac{4}{5}$ $\cos B = \frac{3}{5}$ $\tan B = \frac{4}{3}$

**5.** $\sin A = \frac{1}{\sqrt{2}}$ $\cos A = \frac{1}{\sqrt{2}}$ $\tan A = 1$

**7.** $\sin A = \frac{1}{2}$ $\cos B = \frac{\sqrt{108}}{12}$ $\tan B = \frac{6}{\sqrt{108}}$

**9.** $\sin A = \frac{12}{\sqrt{147}}$ $\cos A = \frac{\sqrt{3}}{\sqrt{147}}$ $\tan \frac{12}{\sqrt{3}}$

### Exercise Set 13.4 B

**1.** .5 **3.** .5 **5.** .1763 **7.** 1 **9.** 0 **11.** .7071 **13.** .8660 **15.** .9397

### Exercise Set 13.4 C

**1.** a = 12, b = −20.8, B = 60° **3.** a = 9.8, c = 17.1, A = 35°
**5.** a = 15.7, b = 43.2, A = 20° **7.** a = 11, c = 15.6, B = 45°
**9.** a = 19.6, c = 39.3, B = 60°

### Chapter Thirteen Exercises

**1.** 16 **3.** 45 **5.** 42 **7.** B = 45°, A = 45° **9.** $\sqrt{202}$ **11.** $\sqrt{7.2}$ **13.** 1596.64
**15.** $\sqrt{356}$ **17.** $\frac{5}{15}$ **19.** $\frac{12}{15}$ **21.** $\frac{5}{12}$ **23.** .9848 **25.** .3640 **27.** .9962
**29.** 153.21

### Chapter Test *Form 13A*

**1.** 74° **3.** 25°, 50° **5.** $\frac{16}{9}$ **7.** .5 **9.** 30° **11.** 4 **13.** $\sqrt{136}$ **15.** $\sqrt{65}$

### CUMULATIVE EXERCISES Chapters 1–13 *Set A*

**1.** .3535 **3.** $\frac{1}{27}$ **5.** 8.8 **7.** $-4m^2n + 3m^2n^2$ **9.** $-20x - 2$
**11.** $15ab^2 - 30a^2b^2 + 25a^2b$ **13.** −26a **15.** 20, 40, 30 **17.** −3 **19.** 12 **21.** 18
**23.** 27 **25.** $\frac{7}{2}$ **27.** m = −4, b = 1 **29.** the line should go through (0,−4) and (4,0)
**31.** $5(5xy - 3x - 2y)$ **33.** $(3y - 1)(3y + 1)$ **35.** $(y^2 + 8)(y^2 - 8)$
**37.** $(x - 14)(x + 2)$ **39.** $x(x - 2)(x - 2)$ **41.** 2,−2 **43.** −9,−3 **45.** −4,−1
**47.** $\frac{7cd^2}{2}$ **49.** $\frac{11b}{8}$ **51.** 10 **53.** $\sqrt{319}$ **55.** $\sqrt{18}$ **57.** 1 **59.** .8660 **61.** 60°

## CUMULATIVE TEST Chapters 1–13 *Form A*

**1.** associative **3.** $44y^2 - 3y - 2$ **5.** 236 **7.** $(a - 11)(a + 11)$ **9.** $m = 4, b = -2$
**11.** $\frac{57b}{2a^3c^8}$ **13.** $\frac{6q^3m^2}{p^2}$ **15.** 0,2 **17.** $-4, -3, -2, -1$ **19.** 53° **21.** .9063
**23.** .9848 **25.** .1736

# CHAPTER FOURTEEN

### Exercise Set 14.1

**1.** 54 **3.** 51 **5.** 60 **7.** 11 in. **9.** 25 ft. **11.** 7 yds **13.** 25 ft.
**15.** 1 ft. by 4 ft. **17.** 15.5 **19.** 10, 11, 12 **21.** 62.8 **23.** 47.1 ft. **25.** 219.8

### Exercise Set 14.2

**1.** 160 ft.$^2$ **3.** 25 m$^2$ **5.** 540 yds.$^2$ **7.** 5538.96 ft.$^2$ **9.** 35 yds.$^2$ **11.** 3 in.
**13.** 864 sq. in.

### Exercise Set 14.3

**1.** 4000 cu. in. **3.** 339.12 cu. in. **5.** 11,448.21 cu. yds. **7.** 3200 cu. ft.
**9.** 1695.6 cu. ft. **11.** 5572.45 cu. in **13.** 3846.5 cu. ft. **15.** 4000 cu. ft.

### Chapter Fourteen Exercises

**1.** 4 ft. by 3 ft. **3.** 8, 12, 10 **5.** 81.64 in. **7.** 20 yds. **9.** 14 ft.
**11.** build the rectangle **13.** 219.8 ft. **15.** 132.5 yds. by 132.5 yds. **17.** 154 ft.$^2$
**19.** 120 sq. ft. **21.** 80000 sq. ft. **23.** 556516 sq. ft. **25.** 13.5 sq. ft.
**27.** 1619.25 sq. rds. **29.** 1256 sq. in. **31.** 8′ **33.** 6 m **35.** 16 in. **37.** 21 yds.
**39.** 24 cu. ft. **41.** 75.36 cu. in. **43.** 1000 cu. in. **45.** 8.4 cu. yds. **47.** 12560 cu. ft.
**49.** 1766.25 cu. in. **51.** 1.96 cu. in. **53.** 6 in. **55.** 8 in. **57.** 5′ **59.** 8′
**61.** 301.44

### Chapter Test *Form 14A*

**1.** 16 ft. **3.** 25 ft. **5.** 36.8 ft. **7.** 12 ft. **9.** 9 in. **11.** 15 yds. **13.** 20.93 cu. in.
**15.** 24 cu. ft.

### CUMULATIVE EXERCISES Chapters 1–14 *Set A*

**1.** $-28$ **3.** 150 **5.** $16x^2 - 8x + 1$ **7.** $8x^3y^3 - 12x^2y^3 + 4x^2y^2$ **9.** $-x - 13$
**11.** 3 **13.** $-2$ **15.** $-4,-2$ **17.** $\frac{1}{3}, -\frac{1}{3}$ **19.** $3,-1$ **21.** 0,5 **23.** $8,-2$
**25.** $\frac{-18xy^2}{5z^2}$ **27.** $\frac{(11x - 8)}{(5x + 10)}$ **29.** $-40$ **31.** .9397 **33.** .7660 **35.** $\sqrt{96}$ **37.** 75°
**39.** 78.5 **41.** 210 **43.** 1436.03 **45.** 84.78

### CUMULATIVE TEST Chapters 1–14 *Form 14A*

**1.** 424 **3.** $-274.3$ **5.** $-3$ **7.** $a = \frac{(3c - 10)}{2}$ **9.** $(x + 7)(x - 3)$

**11.** $(a - b)(a - b)$ **13.** $2, \frac{-1}{2}$ **15.** 57° **17.** 7,12 **19.** .9135

## CHAPTER FIFTEEN

### Exercise Set 15.2

**1.** $(0,-2)$ **3.** $(-1,4)$ **5.** no solution, inconsistent **7.** $(-4,1)$ **9.** $(.56,-.08)$
**11.** $\left(\frac{7}{2},\frac{7}{2}\right)$

### Exercise Set 15.3

**1.** $(1,3)$ **3.** inconsistent **5.** $\left(\frac{-1}{3},\frac{-10}{3}\right)$ **7.** $(-3,-6)$ **9.** inconsistent
**11.** $(-2.7,-2.8)$

### Exercise Set 15.4

**1.** $\left(\frac{11}{3},\frac{-2}{3}\right)$ **3.** $(8,-21)$ **5.** inconsistent **7.** $(4,-1)$ **9.** $(2,-2)$ **11.** $(2,-2)$

### Chapter Fifteen Exercises

**1.** $(1,6)$ **3.** $(-6,-18)$ **5.** $(1,2)$ **7.** $(1,3)$ **9.** $\left(\frac{3}{7},\frac{17}{7}\right)$ **11.** $(3,4)$
**13.** $(-5,-4)$ **15.** $(5,10)$ **17.** $(2,4)$ **19.** $(-1,4)$ **21.** $(0,-2)$ **23.** inconsistent
**25.** $\left(\frac{5}{7},\frac{-2}{7}\right)$ **27.** $(2,4)$

### Chapter Test *Form 15A*

**1.** $(-1,3)$ **3.** $(-10,-36)$ **5.** $(1,-4)$ **7.** $\left(\frac{-19}{7},\frac{-1}{7}\right)$

### CUMULATIVE EXERCISES Chapters 1–15 *Set A*

**1.** $-438$ **3.** $55.812$ **5.** $18x^2 + 9x - 5$ **7.** $6a^3b - 8ab^2$ **9.** $-5x^2 + 43x - 48$
**11.** $x = \frac{-8}{3}$ **13.** $z = \frac{1}{4}$ **15.** $x = -8, x = \frac{-3}{2}$ **17.** $x = -2, x = -7$
**19.** $x = 0, x = 11$ **21.** $x = \frac{\pm 6}{5}$ **23.** $x = \frac{3 \pm \sqrt{33}}{2}$ **25.** $\frac{-21y}{2xz}$ **27.** $\frac{47y - 14}{16 - 8y}$
**29.** $x = -32$ **31.** $0.1736$ **33.** $1.1918$ **35.** $b = 20$ **37.** $A = 15°, C = 120°$
**39.** $140\ cm^2$ **41.** $8\ in^3$ **43.** $(2,2)$ **45.** $(7,11)$ **47.** $\left(-2,\frac{4}{3}\right)$

### CUMULATIVE TEST Chapters 1–15 *Form A*

**1.** $753$ **3.** $x = \frac{-16}{11}$ **5.** $x = \frac{112}{9}$ **7.** $(a - 4)(a - 12)$ **9.** $\frac{-2y^3}{3xz^2}$ **11.** $\sqrt{336}$
**13.** $(-2,-7)$ **15.** $(-34,57)$

# CHAPTER SIXTEEN

**Exercise Set 16.1**

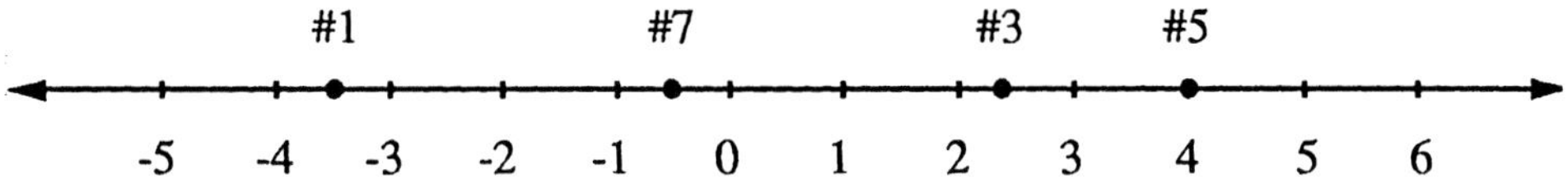


**Exercise Set 16.2**

**1.**

**3.** $\{x | x \varepsilon I, -3 \leq x \leq 3\}$

**5.**

**7.** $\{x | x \varepsilon R, -5 < x < 3\}$

**9.**

**Exercise Set 16.3**

**1.**

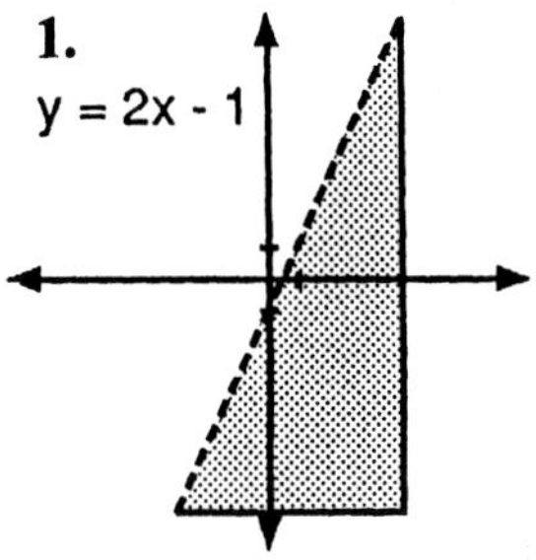


**3.**

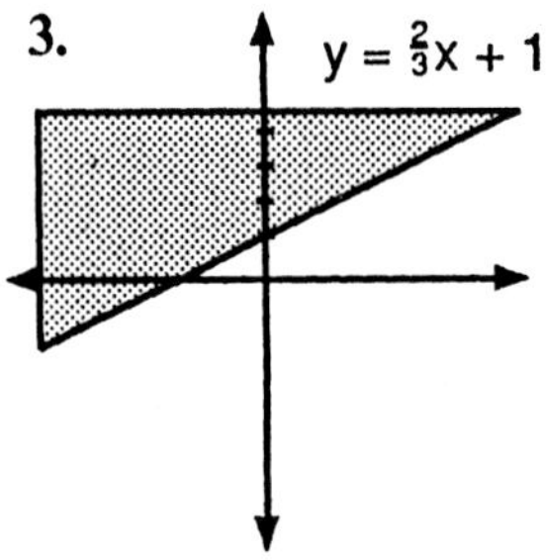


**5.**

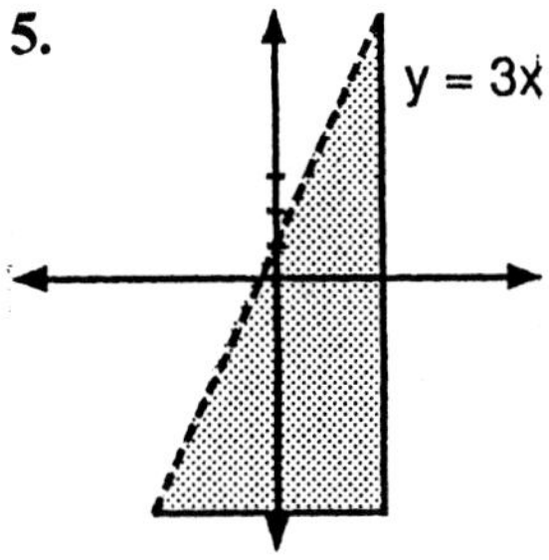


**7.**

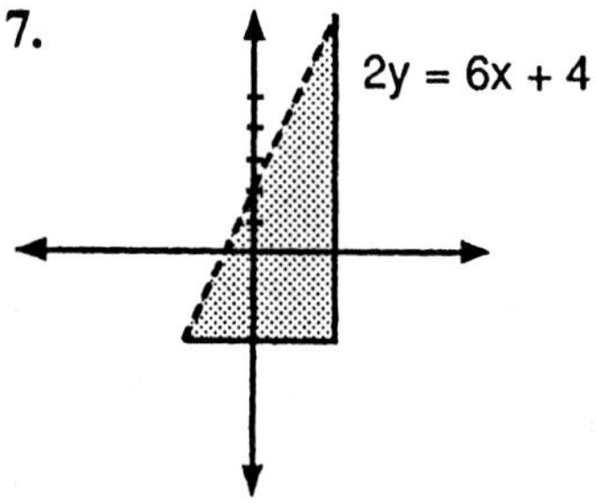


**9.**

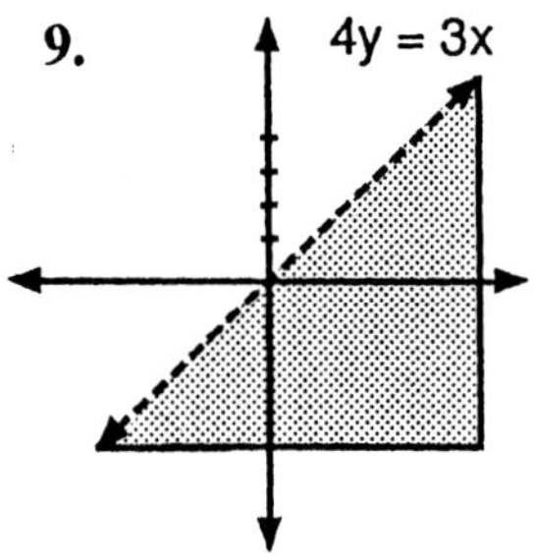


**11.**

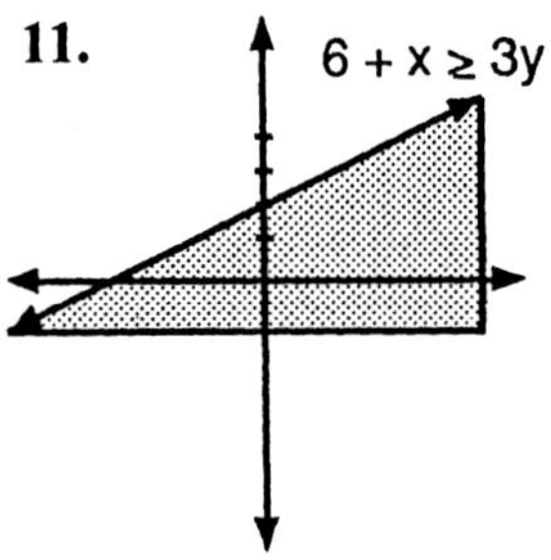


**13.**

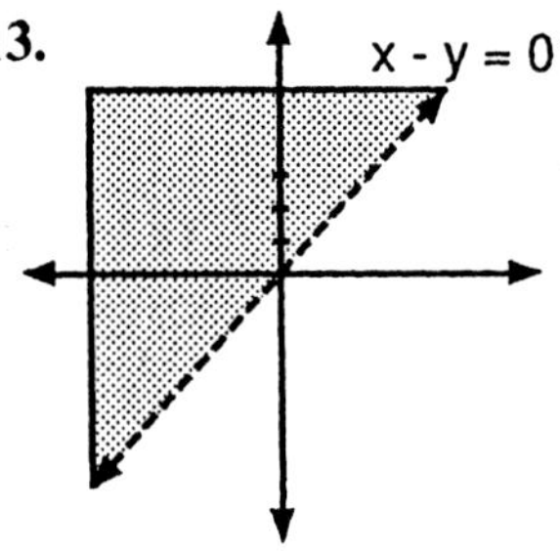


**Exercise Set 16.4**

**1.**

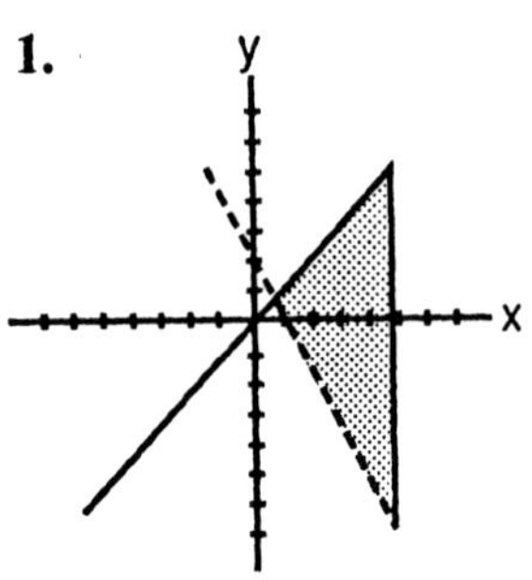


**3.**

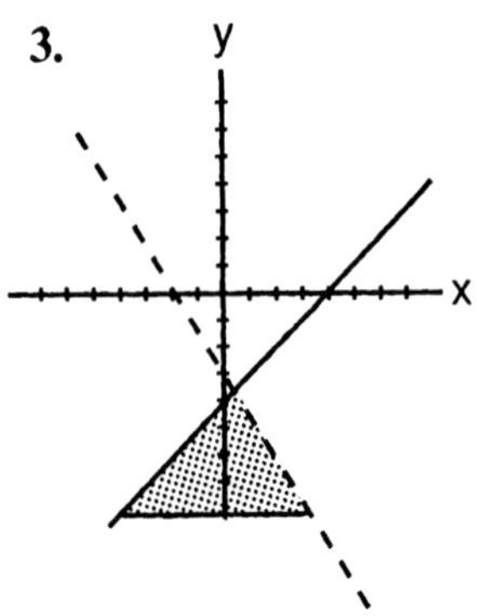


**5.**

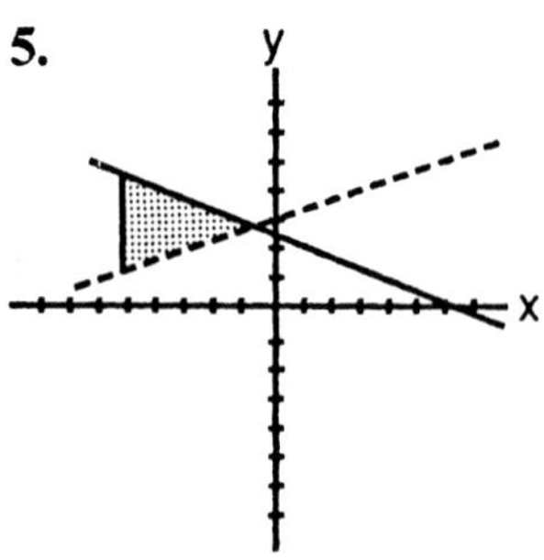

**7.**

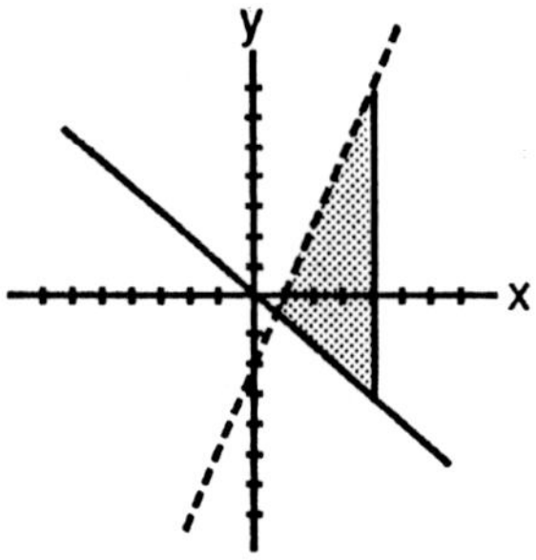


**9.**

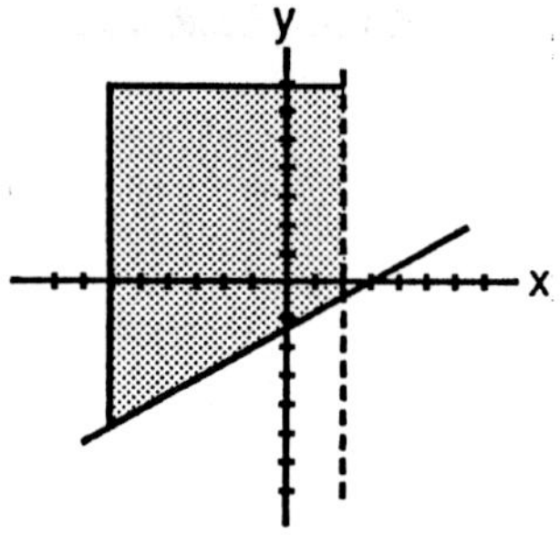


**11.**

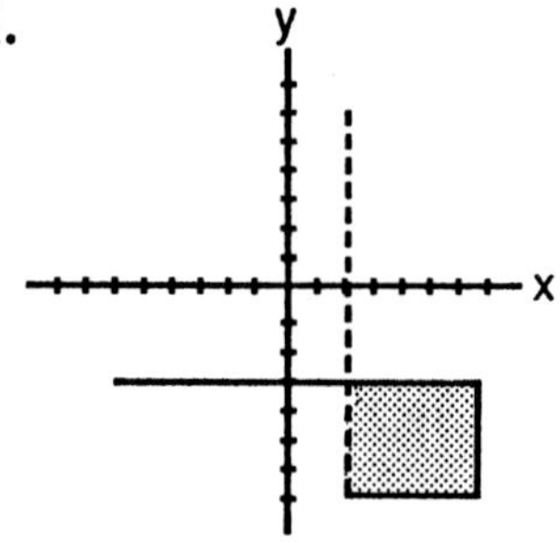


**13.**

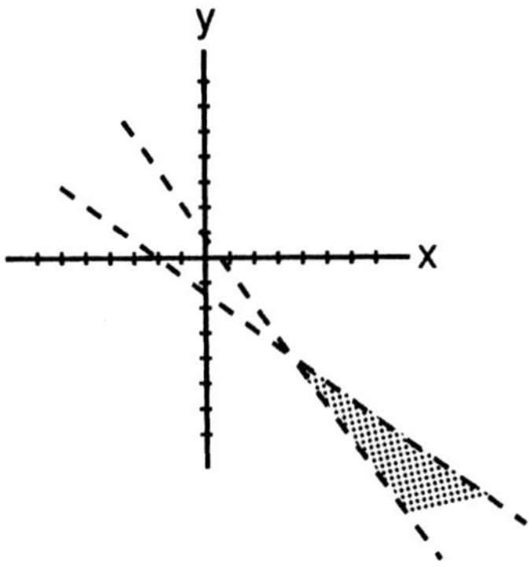


**Chapter Sixteen Exercises**

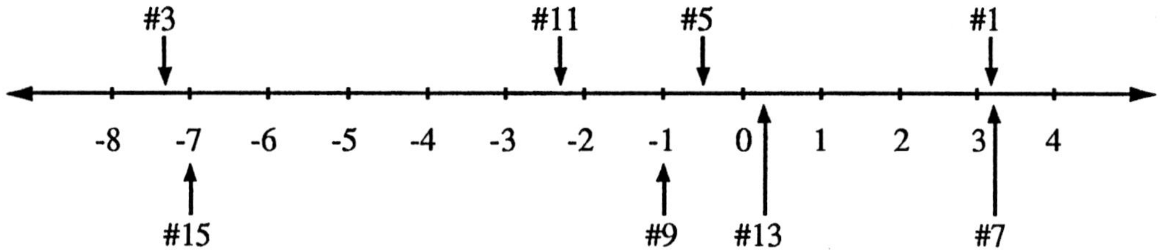


**17.**

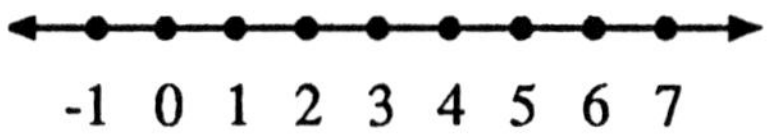


**19.**

**21.**

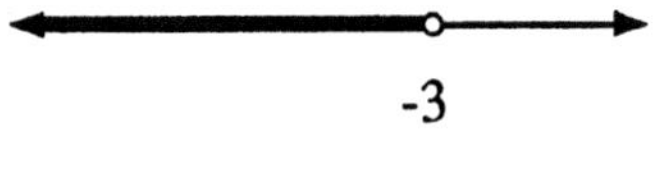


**23.**

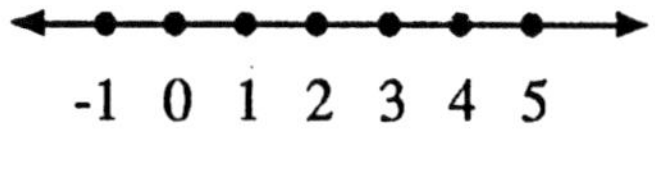


etc.

**25.**

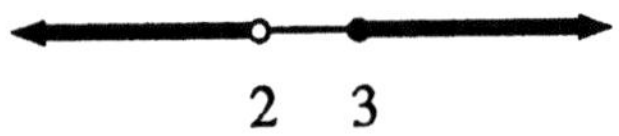


**27.** $\{x | x \in R, -2 \leq x \leq -3\}$ **29.** $\{x | x \in I, -5 \leq x \leq -2\}$

**31.** $\{x|x \varepsilon R, -7 < x \leq 5\}$ **33.** $\{x|x \varepsilon R, x < 7\}$

**35.**

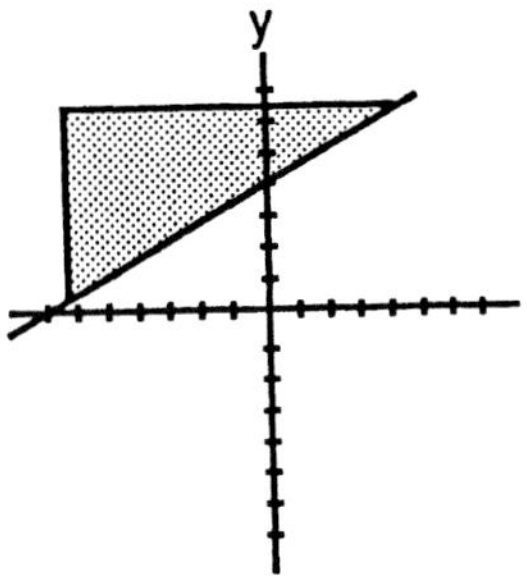


**37.**

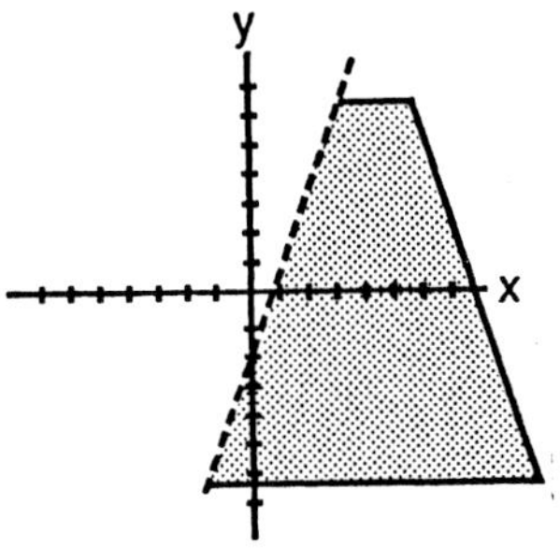


**39.**

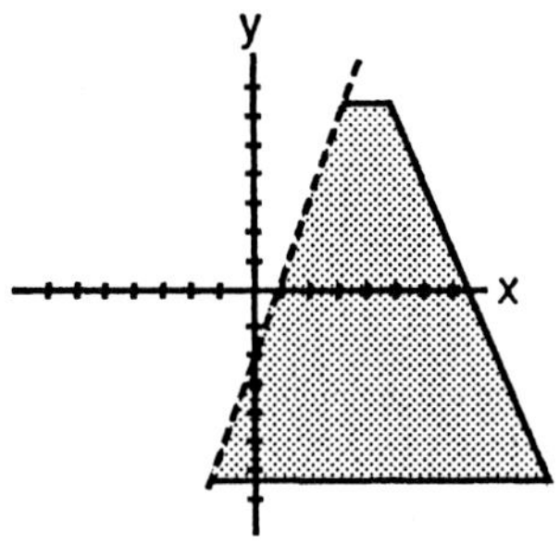


**41.**

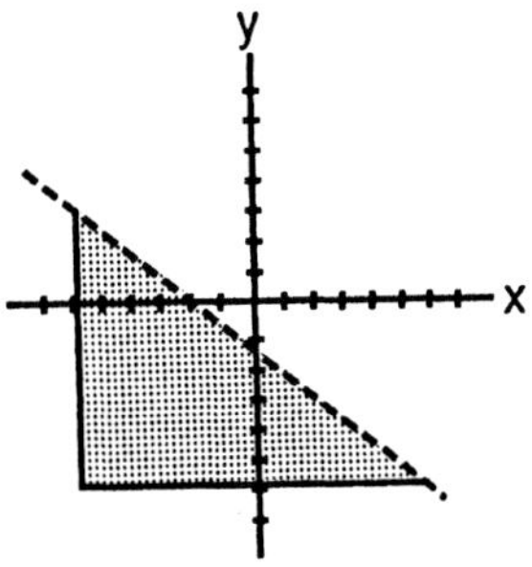


**43.**

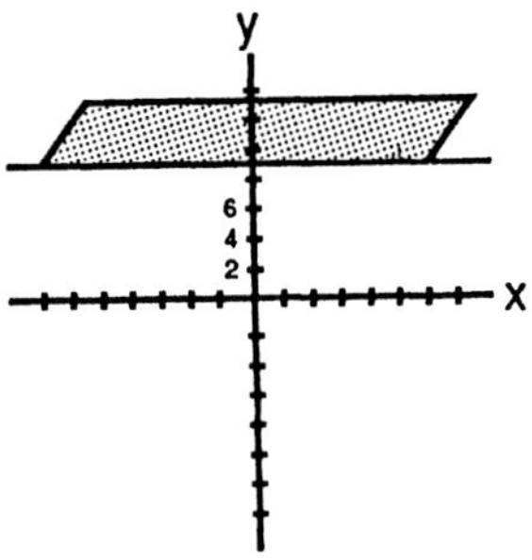


**45.**

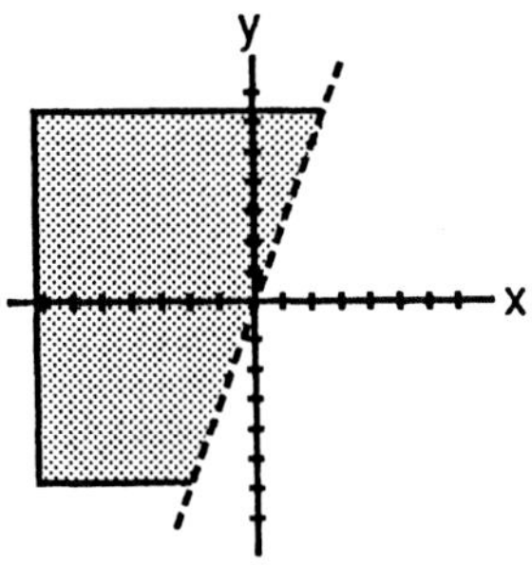


**47.**

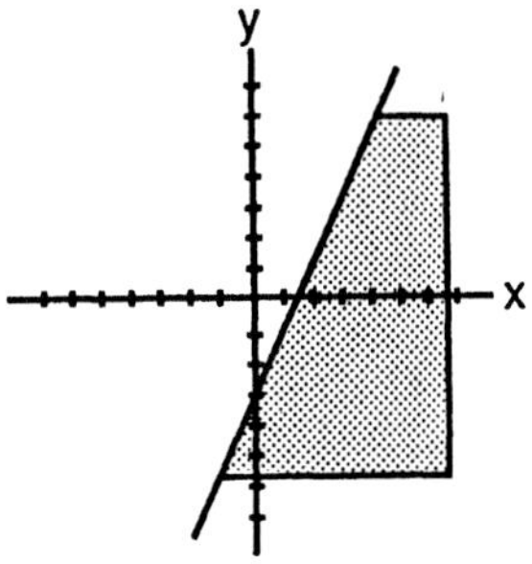


**49.**

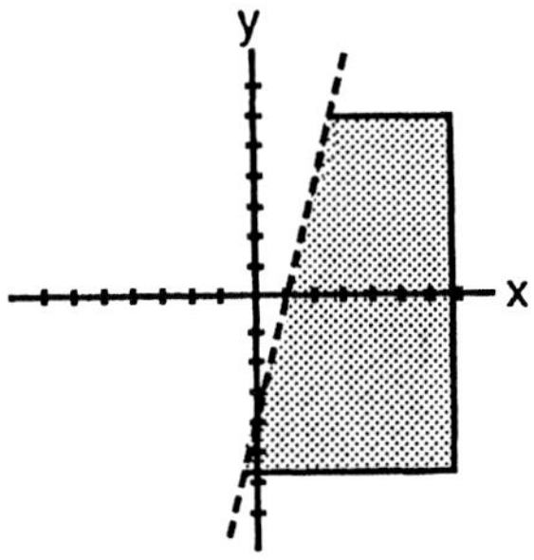


**51.**

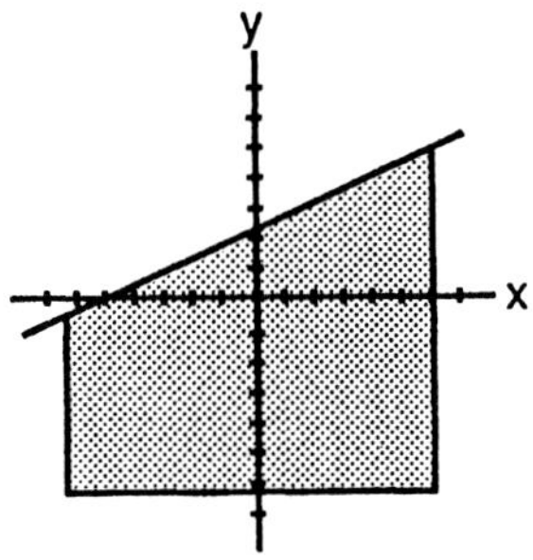


**53.**

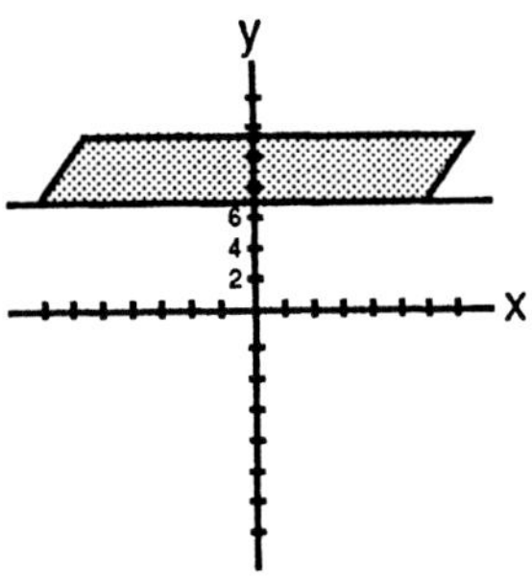


**55.**

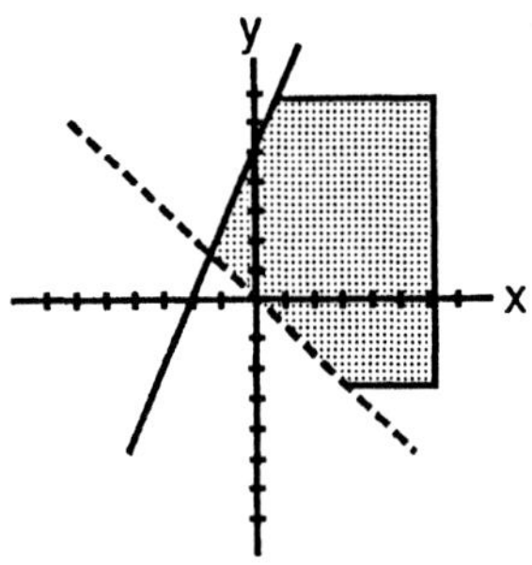


**57.**

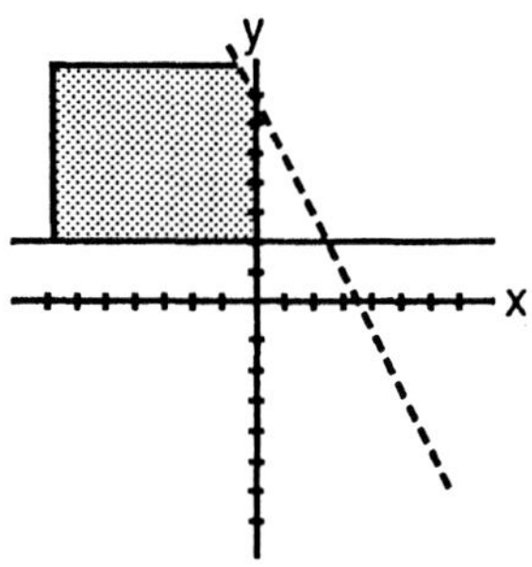

59.

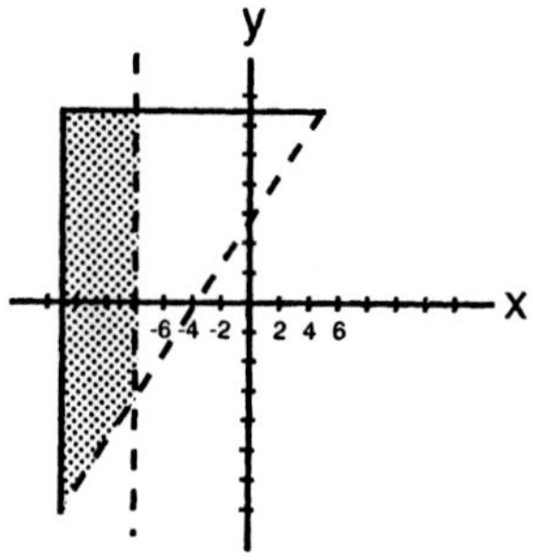


61.

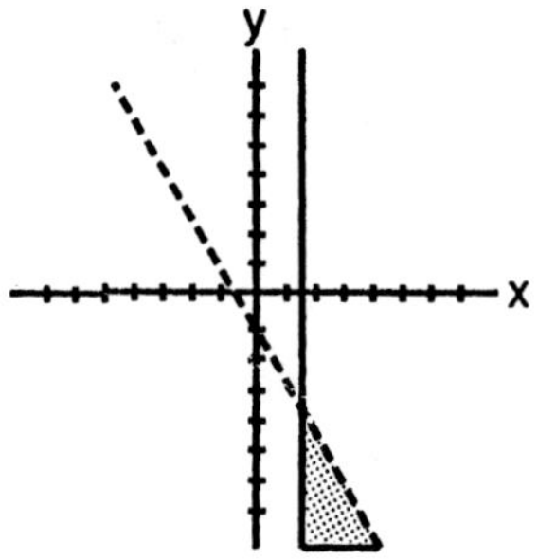


63.

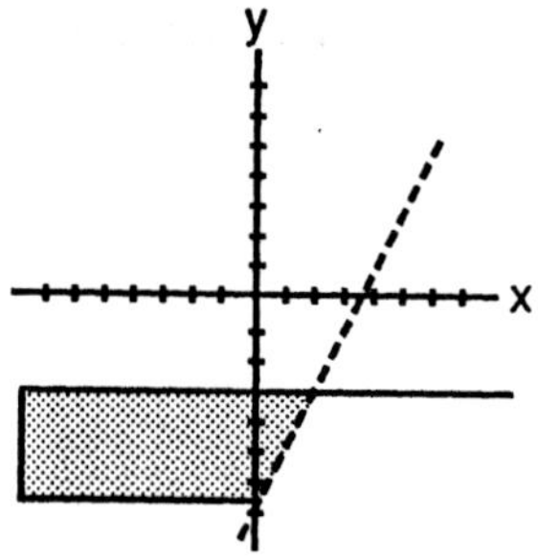


65.

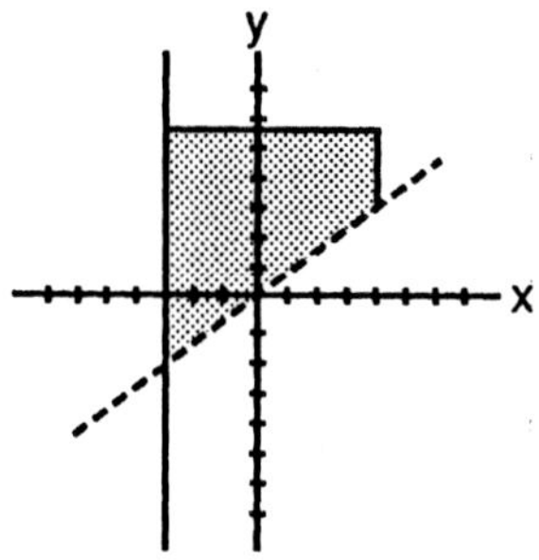


67.

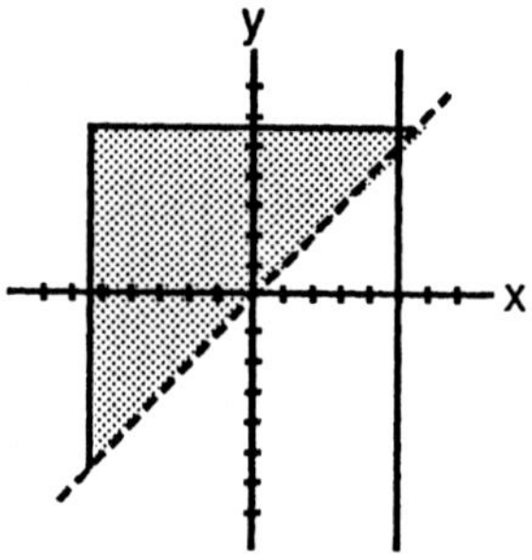


69.

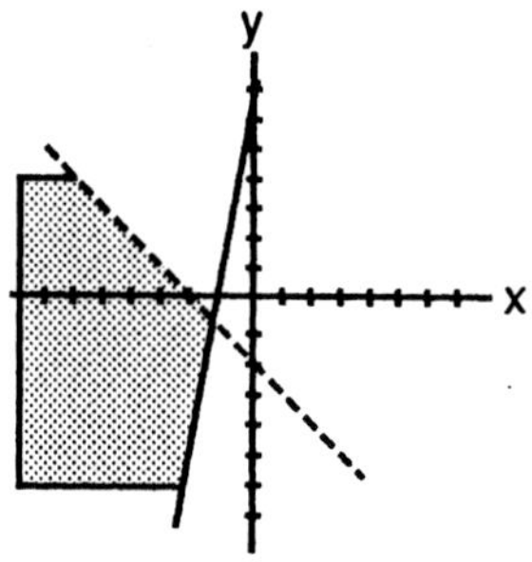


## Chapter Sixteen Test

1.

3.

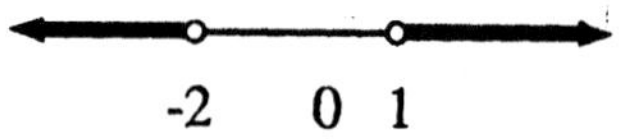


5.

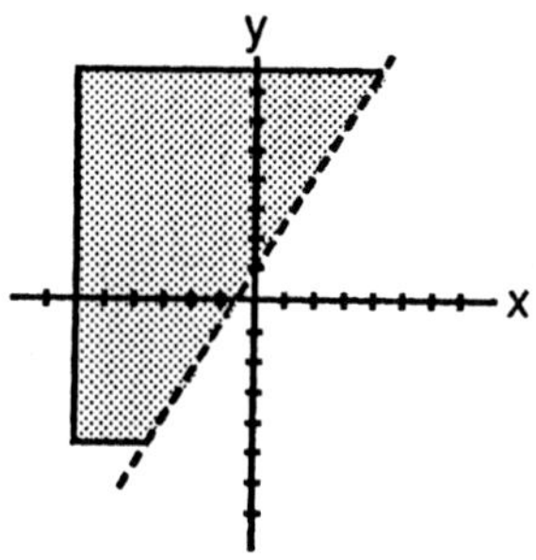


7.

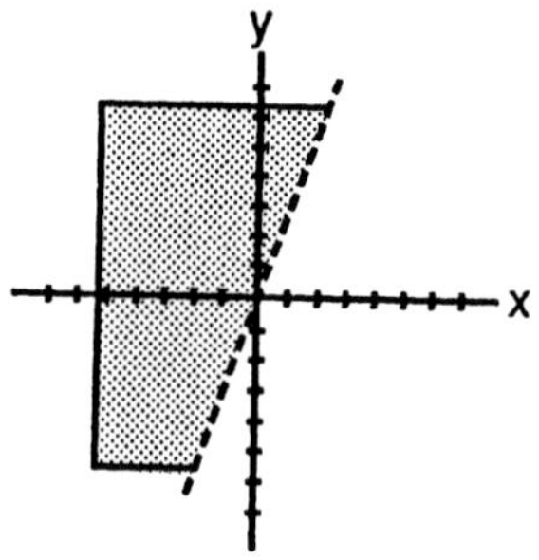


9.

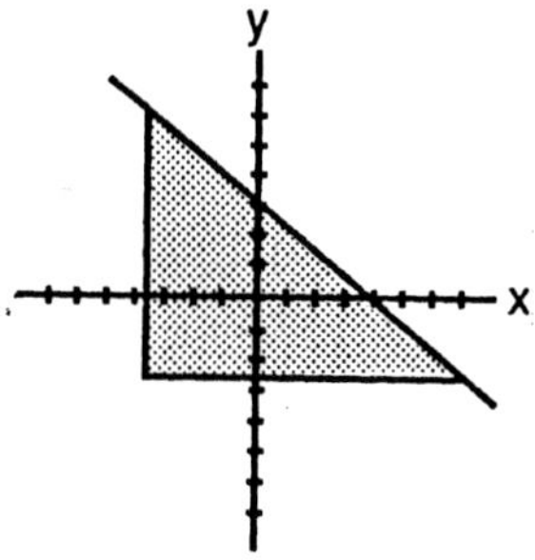


11.

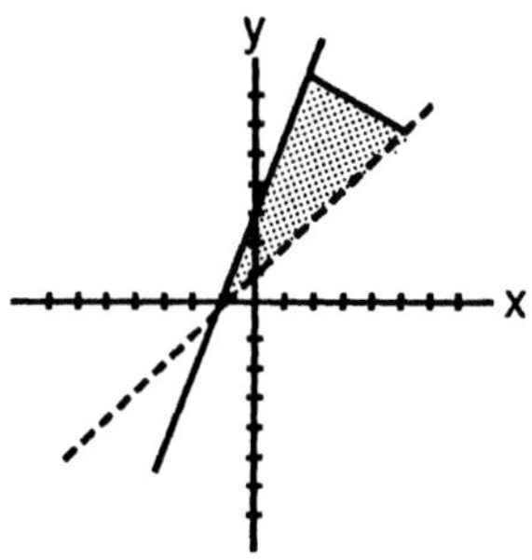


13.

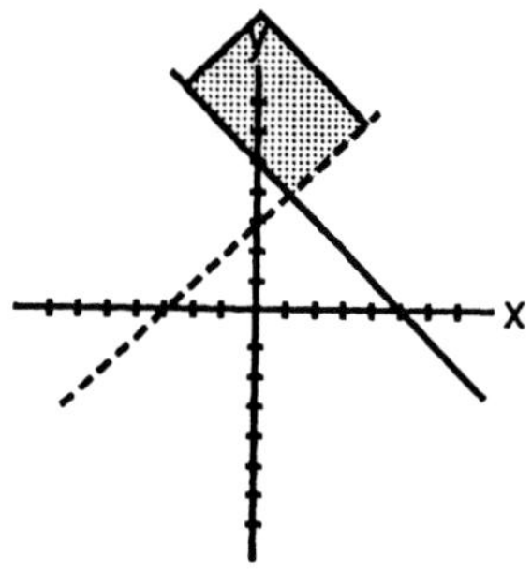

## CUMULATIVE EXERCISES Chapters 1–16 *Set A*

**1.** $-4x - 29$ **3.** $21x^2 + 11x - 2$ **5.** $a^2 - 5a - 36$ **7.** $3a^3 - 4a^2 + 8a + 8$

**9.** $\frac{xy^2}{(8z^3)}$ **11.** $\frac{(5a - 4b)}{12}$ **13.** $\frac{4xy}{5}$ **15.** $-4y$ **17.** $3x^2 - 5y$ **19.** $49a^6b^2$

**21.** 8 **23.** 3,−2 **25.** 2.2 **27.** 4,−2 **29.** 6,−3 **31.** $\frac{34}{15}$ **33.** 5,−1

**35.** 11,−7 **37.** −.7, .7 **39.** $0,\frac{2}{3}$ **41.** 9 **43.** $r = \frac{\sqrt{3V}}{\pi h}$ **45.** 0.052336

**47.** (2,−1) **49.** (2,5) **51.** (−2,−3)

**53.**

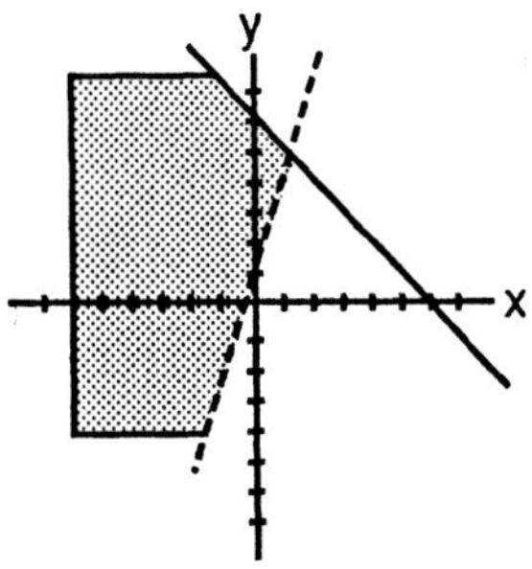


**55.**

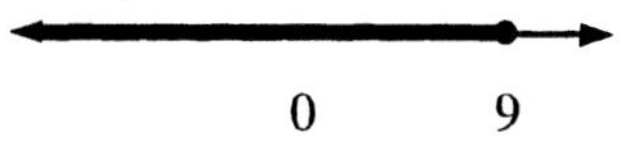


**57.** 26 **59.** 7, 12, 21 **61.** 7, 12

## CUMULATIVE TEST Chapters 1–16 *Form A*

**1.** $\frac{1}{3a}$ **3.** $x(x + 3)(x - 3)$ **5.** $(x + 8)(x - 8)$ **7.** $m = \frac{2}{3}, b = -5$ **9.** −3

**11.** $\frac{3}{(4xy^2)}$ **13.** 9, 11 **15.** .6018 **17.** 124°, 41° **19.** 175 sq. ft. **21.** 13

**23.**

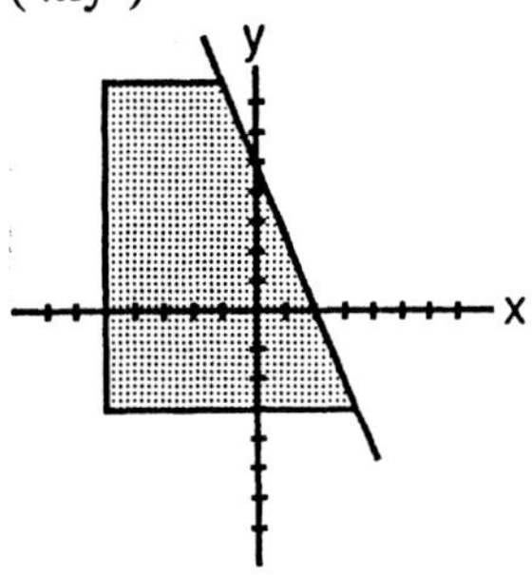


**25.**

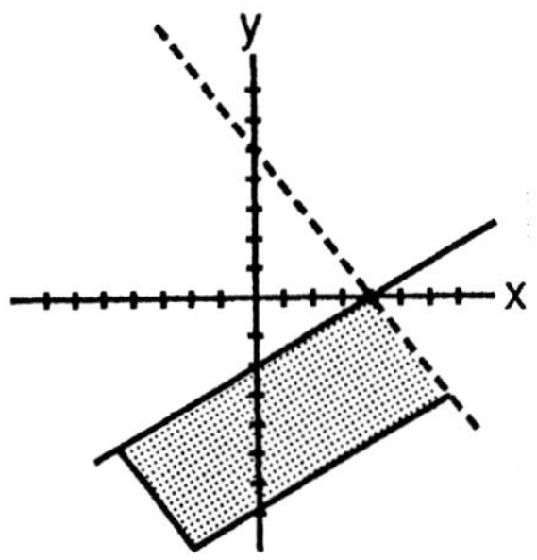

## CHAPTER SEVENTEEN

**Exercise Set 17.1**

**1.** $2t(6st + 2s - 3t^2)$ **3.** $(s + 4t)(s - 4t)$ **5.** $2t^2(1 + 2s)(1 - 2s)$
**7.** $(s + 3)(s + 1)$ **9.** $r(r + 6)(r + 1)$ **11.** $(x^2 + y)(x^2 - y)$
**13.** $x^2y^2(x + y)(x - y)$ **15.** $3(x + 1)^2$ **17.** $(2c - d)(c + d)$
**19.** $(7x + 4)(7x - 4)$ **21.** $4(4a + 3b)(4a - 3b)$ **23.** $(a + b)(a - 3b)$

**Exercise Set 17.2**

**1.** $(2x - 1)(x - 3)$ **3.** $(6a - b)(a - b)$ **5.** $6(x - 4y)(x + y)$
**7.** $(2y - 3)(y - 1)$ **9.** $(2x + 7)(x - 2)$ **11.** $7(a + 3b)(a - b)$
**13.** $(3x + 5)(x + 1)$ **15.** $(5m - 3)(m - 2)$ **17.** $(6s + 1)(s + 2)$
**19.** $(2q + 3)(2q + 1)$

**Exercise Set 17.3**

**1.** $\frac{4}{(x - 2)}$ **3.** $\frac{2}{xy}$ **5.** $\frac{2}{3}$ **7.** $\frac{1}{x + 1}$ **9.** $\frac{(x + 1)}{(x - 1)}$ **11.** $\frac{(x + 2)}{(x - 2y)}$ **13.** $\frac{x - y}{2}$
**15.** $\frac{(x + 1)}{(x - 4)}$ **17.** $2b$ **19.** $\frac{2b(a^2 + 2b)}{(a^2 + 2)}$ **21.** $x + 7$ **23.** $\frac{(4a - 1)}{(a - 2)}$

**Exercise Set 17.4**

**1.** $\frac{3}{8}$ **3.** $-3b^2$ **5.** $\frac{2a^2}{(a - 1)}$ **7.** $(x + 2y)(x - 1)$ **9.** $\frac{1}{2(x - y)(x - 2)}$
**11.** $(x + 1)(x + 3)$

**Exercise Set 17.5**

**1.** $\frac{-s^3}{r}$ **3.** 1 **5.** $(x + 8)(x - 3)$ **7.** $\frac{(a + 2)}{3a}$ **9.** $\frac{(a + b)}{2ab}$ **11.** $\frac{(x - 1)(x - 1)}{(3x + 2)}$

**Exercise Set 17.6**

**1.** $\frac{(7x + 20)}{8x}$ **3.** $\frac{(-3x + 21)}{(x^2 - 9)}$ **5.** $\frac{(b - 2ax)}{abx^2}$ **7.** $\frac{50}{(3x + 6)}$ **9.** $\frac{(8x + 9)}{(x^2 + 2x + 1)}$
**11.** $\frac{(15a^2 - 11a + 4ab + 9b)}{(6a^2 + 11a - 10)}$ **13.** $\frac{-2}{3(t + 6)}$

**Exercise Set 17.7**

**1.** $x = 3$ **3.** $a = 5$ **5.** $x = \frac{-5}{2}$ **7.** $d = -6$ **9.** $y = 5$ **11.** $t = \frac{5}{3}$
**13.** $d = 9$

**Chapter Seventeen Exercises**

**1.** $3(x + 2)$ **3.** $(x + 4)(x + 1)$ **5.** $(x + 4)(x - 1)$ **7.** $(x^2 + 4y)(x + 2y)(x - 2y)$
**9.** $3(x + 2)(x - 2)$ **11.** $9(3a + b)(3a - b)$ **13.** $(a + .5)(a - .5)$
**15.** $(3x + 1)(x - 1)$ **17.** $2(x + 3)(x + 1)$ **19.** $(2x + 1)(x - 5)$
**21.** $4(y + 3)(y + 2)$ **23.** $(3x + 1)(x - 2)$ **25.** $(4 + c)(1 - 3c)$ **27.** $2(z - 5)(z + 1)$
**29.** $(12s + 1)(s + 3)$ **31.** $(y - 11)(y + 7)$ **33.** $(2r - 1)(r + 3)$
**35.** $x(2x + y)(x - 2y)$ **37.** $\frac{4}{(a + 4b)}$ **39.** $\frac{(3x + 1)}{3}$ **41.** $\frac{1}{2}$

**43.** cannot be reduced **45.** $\frac{2}{x}$ **47.** $2z^2$ **49.** $(2c + 7)(2c - 7)$ **51.** $\frac{1}{4c}$ **53.** $x^2y$

**55.** $r^2 - 25$ **57.** 5 **59.** $18t^2r$ **61.** $(s - 1)^2$ **63.** $\frac{(2x^2 + 6x + 8)}{x(x^2 + 5x + 4)}$

**65.** $\frac{13}{(4a + 2b)}$ **67.** $\frac{(-x^2 - 6x + 3)}{(x^2 - 1)}$ **69.** $\frac{(-4x^2 + 6x + 4)}{(3x + 1)(x - 1)(x + 2)}$

**71.** $\frac{(5c^2 - 8d^2 + 6a^2)}{adc}$ **73.** $\frac{(6x + 28y)}{(x + 2y)(x - 2y)(x + 3y)}$ **75.** $\frac{(2b - c + ac + a^2)}{ac}$

**77.** $\frac{(-x^2 - x + 3)}{(x^2 - 2x)}$ **79.** $k = -4$ **81.** $z = 14$ **83.** no solution **85.** no solution

**87.** $c = -1$ **89.** $x = 10$ **91.** $x = 5$ **93.** $x = \frac{-8}{5}$ **95.** $x = \frac{-4}{3}$

**97.** $c = -20$

**Chapter Test** *Form 17A*

**1.** $\frac{x + y}{2}$ **3.** $\frac{(x + 1)}{(x - 3)}$ **5.** $\frac{3}{(2 + x)(x - 2)}$ **7.** $\frac{5}{(ab^2)(a - 2b)}$ **9.** $\frac{(7x^2 + 18x)}{(x^2 + 5x + 6)}$

**11.** $\frac{(2x^2 + 2x + 3)}{(x^2 + 5x + 4)}$ **13.** $x = -25$ **15.** $x = -10$

**CUMULATIVE EXERCISES Chapters 1–17** *Set A*

**1.** $10x - 18$ **3.** $x^2 - 2x - 3$ **5.** $x^2 + 2x + 1$ **7.** $a^3 - 2a^2 - 4a - 1$ **9.** $\frac{y^5}{288xz^4}$

**11.** $\frac{(2x - y + 2z)}{6}$ **13.** $\frac{ab}{6}$ **15.** $y^2 + 3y$ **17.** $4a^2 + b^3 - 7$ **19.** $81a^4b^{12}c^8$

**21.** $-8$ **23.** $-1,8$ **25.** 3.1 **27.** $-4,2$ **29.** 4 **31.** 2 **33.** $3,-1$ **35.** $-5,1$

**37.** $\pm 0.6$ **39.** $0,-2$ **41.** $-1$ **43.** $\sqrt{\frac{A}{\pi}}$ **45.** 0.8660 **47.** $x = 2, y = -1$

**49.** $x = 2, y = 5$ **51.**

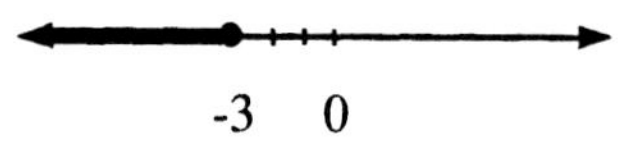


**53.**

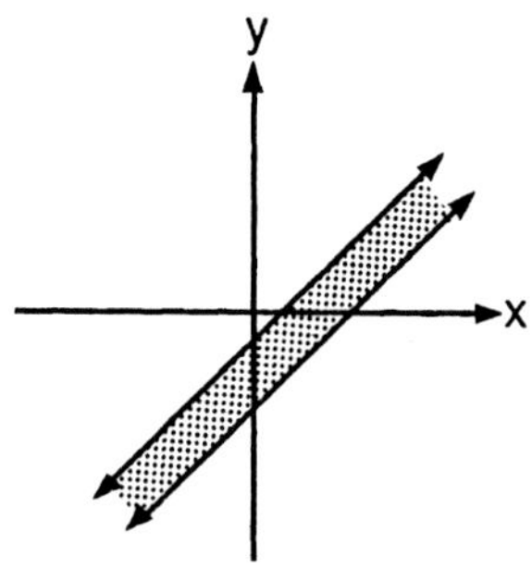


**55.** $x + 2$ **57.** $\frac{x^2}{(y)(x - 3)}$ **59.** 1 **61.** $\frac{(7x + 4)}{10}$ **63.** $\frac{(ab + b^2 - 12a)}{4b}$

**65.** $\frac{11}{2}$ **67.** 25

**CUMULATIVE TEST** *Form 17A*

**1.** $-2x - 4$ **3.** $y^2 - 10y + 25$ **5.** $\frac{(x - 7)}{4}$ **7.** $31x^2 - 7x - 5$

**9.** $x^3 + 3x^2a - 2xa - 6a^2$ **11.** 5 **13.** $7,-3$ **15.** 12 **17.** $x = 3, y = 5$

**19.** (1,3) **21.** $\frac{1}{2}$ **23.** 0.6018 **25.** 70,650

## CHAPTER EIGHTEEN

### Exercise Set 18.1

**1.** 5 **3.** 2 **5.** 3 **7.** 8 **9.** .2 **11.** no solution **13.** $-6$ **15.** 50 **17.** 0.2 **19.** 2 **21.** 3

### Exercise Set 18.2

**1.** $(125)^{1/3} = 5$ **3.** $(64)^{1/6} = 2$ **5.** $(243)^{1/5} = 3$ **7.** $(512)^{1/3} = 8$ **9.** $(0.04)^{1/2} = .2$ **11.** $(-81)^{1/4}$ has no solution **13.** $-(216)^{1/3} = -6$ **15.** $(2500)^{1/2} = 50$ **17.** $(.0016)^{1/4} = 0.2$ **19.** $(16)^{1/4} = 2$ **21.** $-(-27)^{1/3} = 3$

### Exercise Set 18.3

**1.** $\sqrt{6}$ **3.** 3 **5.** 7 **7.** 4 **9.** $3\sqrt{3}$ **11.** $\sqrt[3]{9}$ **13.** $\sqrt[4]{24}$ **15.** 12 **17.** 8

### Exercise Set 18.4

**1.** $4\sqrt{2}$ **3.** $5\sqrt{3}$ **5.** $b\sqrt{a}$ **7.** $2y\sqrt[4]{4y}$ **9.** $3b$ **11.** $11b^3$ **13.** $6a^2\sqrt{a}$ **15.** $3\sqrt[3]{4}$ **17.** $y^2$ **19.** $2ab^2\sqrt{3a}$ **21.** $xy\sqrt[5]{x^2y}$

### Exercise Set 18.5 A

**1.** $5\sqrt{5}$ **3.** $26\sqrt{y}$ **5.** cannot be simplified **7.** $17\sqrt{5} - 7\sqrt{3}$ **9.** $9\sqrt{a} + 3\sqrt{b}$ **11.** $5\sqrt{cb} + 3\sqrt{c}$ **13.** $-5\sqrt{ac} + 4\sqrt{a}$

### Exercise Set 18.5 B

**1.** $17\sqrt{2} + 5\sqrt{3}$ **3.** $18a\sqrt{a}$ **5.** $36\sqrt{2}$ **7.** $17\sqrt{7}$ **9.** $2\sqrt{6} + 4\sqrt{2}$ **11.** $10\sqrt{6}$ **13.** $-4\sqrt{5}$ **15.** $3\sqrt[5]{2}$

### Exercise Set 18.6

**1.** $10 + 6\sqrt{30}$ **3.** $20\sqrt{6} - 4\sqrt{2}$ **5.** $5x + x^2\sqrt{5}$ **7.** $2b\sqrt{b} + 10\sqrt{3b} - b\sqrt{3} - 15$ **9.** $3s - 7\sqrt{3s} + 6$ **11.** $10\sqrt{5} + 30$ **13.** $a - 2\sqrt{ab} + b$ **15.** $36x - 4$

### Exercise Set 18.7

**1.** 3 **3.** $\frac{1}{10}$ **5.** $7\sqrt{a}$ **7.** $\frac{a\sqrt{7}}{7}$ **9.** $3\sqrt{x}$ **11.** $\frac{7\sqrt{xyz}}{z}$ **13.** $2x\sqrt{6x}$ **15.** $\frac{2}{3x}$ **17.** $\frac{(\sqrt{y} - y\sqrt{x})}{y}$ **19.** $\frac{(3\sqrt{3} - x)}{(3 - x)}$ **21.** $\frac{(x\sqrt{x} - x)}{(x - 1)}$ **23.** $\frac{(5 - \sqrt{5x})}{5 - x}$ **25.** $\frac{\sqrt{y^2 + 3y}}{y + 3}$ **27.** $\frac{(\sqrt{9 + 3y})}{y + 3}$ **29.** $\frac{(3xy\sqrt{x} - 3xy\sqrt{y})}{(x - y)}$

### Chapter Eighteen Exercises

**1.** 12 **3.** no solution **5.** $-4$ **7.** 3 **9.** $-3$ **11.** .2 **13.** 5 **15.** $-6$ **17.** 6 **19.** 23 **21.** 8 **23.** 7 **25.** $\sqrt{21}$ **27.** 6 **29.** $\sqrt{70}$ **31.** 3 **33.** $\sqrt[12]{a^7}$ **35.** 3 **37.** $\sqrt{13}$ **39.** $\sqrt[4]{4}$ **41.** $\sqrt{z}$ **43.** $6\sqrt{2}$ **45.** $6\sqrt{3}$ **47.** $3xy^3\sqrt{2xy}$ **49.** $3ay^2\sqrt[3]{a}$ **51.** $12c^2d\sqrt{cd}$ **53.** $4a^4b^4\sqrt{2b}$ **55.** $16\sqrt{2}$ **57.** $-2\sqrt{x}$ **59.** 0 **61.** $xy^2 + xy + x\sqrt[3]{y}$ **63.** $6m$ **65.** 4 **67.** $3\sqrt{x} - 15$ **69.** $10 + 2\sqrt{5}$ **71.** $3b\sqrt{a} + 9a$ **73.** $\sqrt{7y}$ **75.** $\frac{3x\sqrt{y}}{y}$ **77.** $\frac{4x(\sqrt{x} - 2)}{(x - 4)}$ **79.** $\frac{(x + 5)(\sqrt{x} - 5)}{(x - 25)}$ **81.** $\frac{(4x + 2)(\sqrt{2} - 4x)}{(2 - 16x^2)}$

**Chapter Test** *Form 18A*

**1.** 26 **3.** no solution **5.** no solution **7.** 5 **9.** 12 **11.** 3 **13.** $-5\sqrt{5} + 4\sqrt{7}$
**15.** 7 **17.** 2xy **19.** 15 **21.** $a + 2b\sqrt{a} + b^2$ **23.** $3\sqrt{xy}$ **25.** $\frac{(3 - x)(\sqrt{3} + x)}{3 - x^2}$

**CUMULATIVE EXERCISES Chapters 1–18** *Set A*

**1.** $x + 18$ **3.** $y^2 - 3y - 10$ **5.** $s^2 + 2st + t^2$ **7.** $b^3 + b^2 + b - 3$
**9.** $\frac{v^3}{16t}$ **11.** $\frac{(15s - 2t)}{12}$ **13.** $-5ab$ **15.** $2y^2 - y + 1$
**17.** $10b^2 - 7c^3d$ **19.** $\frac{(4x^2 - 2x)}{(x - 1)(x + 1)}$ **21.** $x = 0$ **23.** $c = -2, c = 3$
**25.** $x = -1.25$ **27.** $b = 4, b = -2$ **29.** $y = -1$ **31.** $y = -1$
**33.** $x = -3, x = 1$ **35.** $x = 5, x = -1$ **37.** $c = \pm.9$ **39.** $x = 0, x = -2$
**41.** $r = \sqrt{\frac{V}{\pi}}$ **43.** .5774 **45.** $x = -7, y = -19$ **47.**

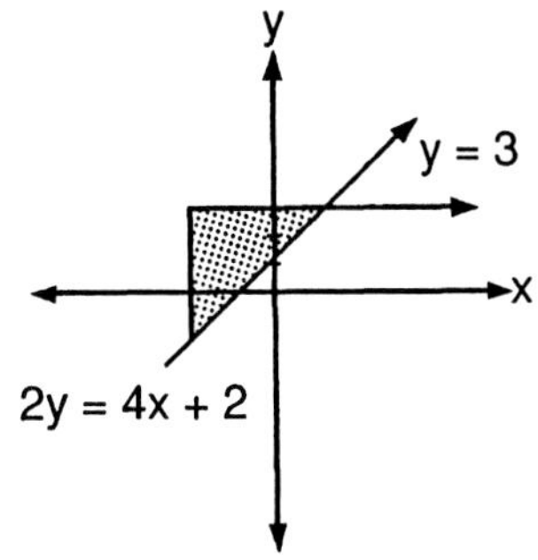


**49.** $a^3(a - 1)$ **51.** $\frac{-(a^2 + b^2)}{(a^2 - b^2)}$ **53.** $x = -11$ **55.** b = 80 units **57.** $7\sqrt{ac} - 7c$
**59.** $\frac{2a^2}{b^2}\sqrt{a}$ **61.** $\frac{4\sqrt{xy}}{y}$ **63.** $a - 2\sqrt{ab} + b$

**CUMULATIVE TEST Chapters 1–18** *Form A*

**1.** $14x + 14y + 56$ **3.** $6xy$ **5.** $\frac{(2x + xy)}{14}$ **7.** $3x^2y^3 + xy^2 - 7xy + 3$ **9.** $x = 7$
**11.** $x = -5, x = 1$ **13.** $x = 0$ **15.** $h = \frac{(P - 2b)}{2}$ **17.** 14.39 **19.** $x = 3$
**21.**

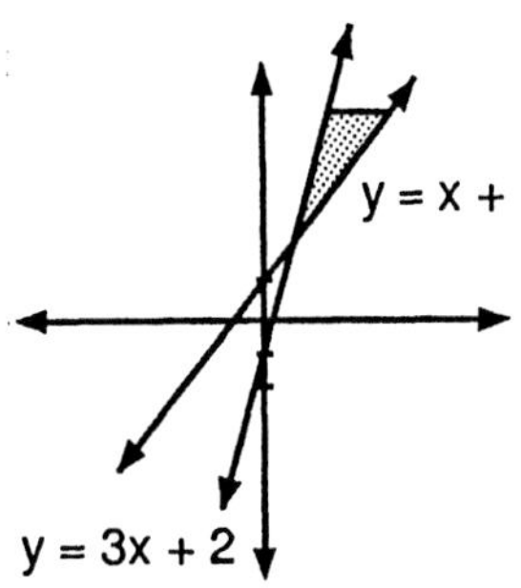


**23.** $\frac{(-8x - 1)}{3x}$ **25.** $10a\sqrt{a} - 8\sqrt{a}$ **27.** $\frac{\sqrt{3x}(\sqrt{x} + 2)}{(x + 2)}$

## Squares and Square Roots

| Number | Square | Square Root | Number | Square | Square Root |
|---|---|---|---|---|---|
| 1 | 1 | 1.000 | 51 | 2,601 | 7.141 |
| 2 | 4 | 1.414 | 52 | 2,704 | 7.211 |
| 3 | 9 | 1.732 | 53 | 2,809 | 7.280 |
| 4 | 16 | 2.000 | 54 | 2,916 | 7.348 |
| 5 | 25 | 2.236 | 55 | 3,025 | 7.416 |
| 6 | 36 | 2.449 | 56 | 3,136 | 7.483 |
| 7 | 49 | 2.646 | 57 | 3,249 | 7.550 |
| 8 | 64 | 2.828 | 58 | 3,364 | 7.616 |
| 9 | 81 | 3.000 | 59 | 3,481 | 7.681 |
| 10 | 100 | 3.162 | 60 | 3,600 | 7.746 |
| 11 | 121 | 3.317 | 61 | 3,721 | 7.810 |
| 12 | 144 | 3.464 | 62 | 3,844 | 7.874 |
| 13 | 169 | 3.606 | 63 | 3,969 | 7.937 |
| 14 | 196 | 3.742 | 64 | 4,096 | 8.000 |
| 15 | 225 | 3.873 | 65 | 4,225 | 8.062 |
| 16 | 256 | 4.000 | 66 | 4,356 | 8.124 |
| 17 | 289 | 4.123 | 67 | 4,489 | 8.185 |
| 18 | 324 | 4.243 | 68 | 4,624 | 8.246 |
| 19 | 361 | 4.359 | 69 | 4,761 | 8.307 |
| 20 | 400 | 4.472 | 70 | 4,900 | 8.367 |
| 21 | 441 | 4.583 | 71 | 5,041 | 8.426 |
| 22 | 484 | 4.690 | 72 | 5,184 | 8.485 |
| 23 | 529 | 4.796 | 73 | 5,329 | 8.544 |
| 24 | 576 | 4.899 | 74 | 5,476 | 8.602 |
| 25 | 625 | 5.000 | 75 | 5,625 | 8.660 |
| 26 | 676 | 5.099 | 76 | 5,776 | 8.718 |
| 27 | 729 | 5.196 | 77 | 5,929 | 8.775 |
| 28 | 784 | 5.292 | 78 | 6,084 | 8.832 |
| 29 | 841 | 5.383 | 79 | 6,241 | 8.888 |
| 30 | 900 | 5.477 | 80 | 6,400 | 8.944 |
| 31 | 961 | 5.568 | 81 | 6,561 | 9.000 |
| 32 | 1,024 | 5.657 | 82 | 6,724 | 9.055 |
| 33 | 1,089 | 5.745 | 83 | 6,889 | 9.110 |
| 34 | 1,156 | 5.831 | 84 | 7,056 | 9.165 |
| 35 | 1,225 | 5.916 | 85 | 7,225 | 9.220 |
| 36 | 1,296 | 6.000 | 86 | 7,396 | 9.274 |
| 37 | 1,369 | 6.083 | 87 | 7,569 | 9.327 |
| 38 | 1,444 | 6.164 | 88 | 7,744 | 9.381 |
| 39 | 1,521 | 6.245 | 89 | 7,921 | 9.434 |
| 40 | 1,600 | 6.325 | 90 | 8,100 | 9.487 |
| 41 | 1,681 | 6.403 | 91 | 8,281 | 9.539 |
| 42 | 1,764 | 6.481 | 92 | 8,464 | 9.592 |
| 43 | 1,849 | 6.557 | 93 | 8,649 | 9.644 |
| 44 | 1,936 | 6.633 | 94 | 8,836 | 9.695 |
| 45 | 2,025 | 6.708 | 95 | 9,025 | 9.747 |
| 46 | 2,116 | 6.782 | 96 | 9,216 | 9.798 |
| 47 | 2,209 | 6.856 | 97 | 9,409 | 9.849 |
| 48 | 2,304 | 6.928 | 98 | 9,604 | 9.899 |
| 49 | 2,401 | 7.000 | 99 | 9,801 | 9.950 |
| 50 | 2,500 | 7.071 | 100 | 10,000 | 10.000 |

## Trigonometric Functions of Angles

| Angle | Sin(θ) | Cos(θ) | Tan(θ) |
|---|---|---|---|
| 1° | 0.0175 | 0.9998 | 0.0175 |
| 2° | 0.0349 | 0.9994 | 0.0349 |
| 3° | 0.0523 | 0.9986 | 0.0524 |
| 4° | 0.0698 | 0.9976 | 0.0699 |
| 5° | 0.0872 | 0.9962 | 0.0875 |
| 6° | 0.1045 | 0.9945 | 0.1051 |
| 7° | 0.1219 | 0.9925 | 0.1228 |
| 8° | 0.1392 | 0.9903 | 0.1405 |
| 9° | 0.1564 | 0.9877 | 0.1584 |
| 10° | 0.1736 | 0.9848 | 0.1763 |
| 11° | 0.1908 | 0.9816 | 0.1944 |
| 12° | 0.2079 | 0.9781 | 0.2126 |
| 13° | 0.2250 | 0.9744 | 0.2309 |
| 14° | 0.2419 | 0.9703 | 0.2493 |
| 15° | 0.2588 | 0.9659 | 0.2679 |
| 16° | 0.2756 | 0.9613 | 0.2867 |
| 17° | 0.2924 | 0.9563 | 0.3057 |
| 18° | 0.3090 | 0.9511 | 0.3249 |
| 19° | 0.3256 | 0.9455 | 0.3443 |
| 20° | 0.3420 | 0.9397 | 0.3640 |
| 21° | 0.3584 | 0.9336 | 0.3839 |
| 22° | 0.3746 | 0.9272 | 0.4040 |
| 23° | 0.3907 | 0.9205 | 0.4245 |
| 24° | 0.4067 | 0.9135 | 0.4452 |
| 25° | 0.4226 | 0.9063 | 0.4663 |
| 26° | 0.4384 | 0.8988 | 0.4877 |
| 27° | 0.4540 | 0.8910 | 0.5095 |
| 28° | 0.4695 | 0.8829 | 0.5317 |
| 29° | 0.4848 | 0.8746 | 0.5543 |
| 30° | 0.5000 | 0.8660 | 0.5774 |
| 31° | 0.5150 | 0.8572 | 0.6009 |
| 32° | 0.5299 | 0.8480 | 0.6249 |
| 33° | 0.5446 | 0.8387 | 0.6494 |
| 34° | 0.5592 | 0.8290 | 0.6745 |
| 35° | 0.5736 | 0.8192 | 0.7002 |
| 36° | 0.5878 | 0.8090 | 0.7265 |
| 37° | 0.6018 | 0.7986 | 0.7536 |
| 38° | 0.6157 | 0.7880 | 0.7813 |
| 39° | 0.6293 | 0.7771 | 0.8098 |
| 40° | 0.6428 | 0.7660 | 0.8391 |
| 41° | 0.6561 | 0.7547 | 0.8693 |
| 42° | 0.6691 | 0.7431 | 0.9004 |
| 43° | 0.6820 | 0.7314 | 0.9325 |
| 44° | 0.6947 | 0.7193 | 0.9657 |
| 45° | 0.7071 | 0.7071 | 1.0000 |

## Trigonometric Functions of Angles

| Angle | Sin(θ) | Cos(θ) | Tan(θ) |
|---|---|---|---|
| 46° | 0.7193 | 0.6947 | 1.0355 |
| 47° | 0.7314 | 0.6820 | 1.0724 |
| 48° | 0.7431 | 0.6691 | 1.1106 |
| 49° | 0.7547 | 0.6561 | 1.1504 |
| 50° | 0.7660 | 0.6428 | 1.1918 |
| 51° | 0.7771 | 0.6293 | 1.2349 |
| 52° | 0.7880 | 0.6157 | 1.2799 |
| 53° | 0.7986 | 0.6018 | 1.3270 |
| 54° | 0.8090 | 0.5878 | 1.3764 |
| 55° | 0.8192 | 0.5736 | 1.4281 |
| 56° | 0.8290 | 0.5592 | 1.4826 |
| 57° | 0.8387 | 0.5446 | 1.5399 |
| 58° | 0.8480 | 0.5299 | 1.6003 |
| 59° | 0.8572 | 0.5150 | 1.6643 |
| 60° | 0.8660 | 0.5000 | 1.7321 |
| 61° | 0.8746 | 0.4848 | 1.8040 |
| 62° | 0.8829 | 0.4695 | 1.8807 |
| 63° | 0.8910 | 0.4540 | 1.9626 |
| 64° | 0.8988 | 0.4384 | 2.0503 |
| 65° | 0.9063 | 0.4226 | 2.1445 |
| 66° | 0.9135 | 0.4067 | 2.2460 |
| 67° | 0.9205 | 0.3907 | 2.3559 |
| 68° | 0.9272 | 0.3746 | 2.4751 |
| 69° | 0.9336 | 0.3584 | 2.6051 |
| 70° | 0.9397 | 0.3420 | 2.7475 |
| 71° | 0.9455 | 0.3256 | 2.9042 |
| 72° | 0.9511 | 0.3090 | 3.0777 |
| 73° | 0.9563 | 0.2924 | 3.2709 |
| 74° | 0.9613 | 0.2756 | 3.4874 |
| 75° | 0.9659 | 0.2588 | 3.7321 |
| 76° | 0.9703 | 0.2419 | 4.0108 |
| 77° | 0.9744 | 0.2250 | 4.3315 |
| 78° | 0.9781 | 0.2079 | 4.7046 |
| 79° | 0.9816 | 0.1908 | 5.1446 |
| 80° | 0.9848 | 0.1736 | 5.6713 |
| 81° | 0.9877 | 0.1564 | 6.3138 |
| 82° | 0.9903 | 0.1392 | 7.1154 |
| 83° | 0.9925 | 0.1219 | 8.1443 |
| 84° | 0.9945 | 0.1045 | 9.5144 |
| 85° | 0.9962 | 0.0872 | 11.4300 |
| 86° | 0.9976 | 0.0698 | 14.3007 |
| 87° | 0.9986 | 0.0523 | 19.0811 |
| 88° | 0.9994 | 0.0349 | 28.6363 |
| 89° | 0.9998 | 0.0175 | 57.2900 |
| 90° | 1.0000 | 0.0000 | — |

# Index